Power Electronics

Second Edition

Cyril W. Lander

School of Electronic and Electrical Engineering
Leicester Polytechnic

McGRAW-HILL BOOK COMPANY

London · New York · St Louis · San Francisco · Auckland · Bogotá · Guatemala
Hamburg · Lisbon · Madrid · Mexico · Montreal · New Delhi · Panama · Paris
San Juan · São Paulo · Singapore · Sydney · Tokyo · Toronto

Published by

McGRAW-HILL Book Company (UK) Limited

MAIDENHEAD · BERKSHIRE · ENGLAND

British Library Cataloguing in Publication Data

Lander, Cyril W.
Power electronics. — 2nd ed.
1. Power semiconductors
I. Title
621.3815'2   TK871.85

ISBN 0-07-084162-4

Library of Congress Cataloguing-in-Publication Data

Lander, Cyril W.
Power electronics. — 2nd ed.
Bibliography: p.
Includes index.
1. Power electronics. I. Title.
TK7881.15.L36 1987   621.3815'2   86-27304

ISBN 0-07-084162-4

2345 AP 8987

Printed in Great Britain at The Alden Press, Oxford

CONTENTS

Preface   ix

Preface to the Second Edition   xi

Chapter 1. Rectifying Devices   1

1-1 The diode   1
1-2 The thyristor   3
1-3 The triac   5
1-4 The power transistor   6
1-5 Thyristor gating requirements   10
     1-5-1 Requirements of firing circuits   13
     1-5-2 Typical firing circuits   13
     1-5-3 Control features of firing circuits   16
1-6 Ratings   17
1-7 Cooling   19
1-8 Gate turn-off thyristor   22
1-9 Power MOSFET   25
1-10 Device comparisons   27
1-11 Worked examples   28

Chapter 2. Rectifying Circuits   37

2-1 Circuit nomenclature   37
2-2 Commutating diode   38
2-3 Single-phase half-wave (or single-way)   38
2-4 Bi-phase half-wave (or single-way)   42
2-5 Single-phase bridge (or double-way)   44
     2-5-1 Uncontrolled   44
     2-5-2 Fully-controlled   46
     2-5-3 Half-controlled   48
2-6 Three-phase half-wave (or single-way)   49
2-7 Six-phase half-wave (or single-way)   54
2-8 Three-phase bridge (or double-way)   60
2-9 Twelve-pulse circuits   65
2-10 Transformer rating   66
2-11 Summary   68
2-12 Worked examples   68

Chapter 3. Converter Operation 92

3-1 Overlap 92
3-2 Power factor 99
3-3 Inversion 99
3-4 Regulation 102
3-5 Equations for p-pulse converter 105
3-6 Worked examples 107

Chapter 4. D.C. Line Commutation 125

4-1 Parallel capacitance 125
4-2 Resonant turn-off 130
4-3 Coupled pulse 135
4-4 Commutation by another load-carrying thyristor ... 138
4-5 Summary with formulae 145
4-6 Worked examples 147

Chapter 5. Frequency Conversion 164

5-1 Cycloconverter 164
    5-1-1 Principle 164
    5-1-2 Blocked group operation 167
    5-1-3 Circulating current mode 174
    5-1-4 Control 177
5-2 Envelope cycloconverter 178
5-3 Single-phase centre-tapped inverter 180
5-4 Single-phase bridge inverter 183
5-5 Three-phase bridge inverter 189
5-6 Constant-current source inverter 195
5-7 Inverter devices 197
5-8 Inverter reverse power flow 199
5-9 Worked examples 199

Chapter 6. Some Applications 220

6-1 Contactor 220
6-2 Heating 221
6-3 Voltage regulation 230
6-4 Voltage multipliers 232
6-5 Standby inverters 233
6-6 Parallel connections 234
6-7 Series connections 236
6-8 Electrochemical 239
6-9 H.V.D.C. transmission 242
6-10 Switched mode power supplies 244
6-11 Worked examples 246

Chapter 7. Harmonics   259

7-1   Harmonic analysis   259
7-2   Load aspects   261
7-3   Supply aspects   263
7-4   Filters   266
    7-4-1   Rectifier output smoothing   267
    7-4-2   Inverter output filtering   268
    7-4-3   A.C. line filters   269
    7-4-4   Radio-interference suppression   270
7-5   Worked examples   270

Chapter 8. D.C. Machine Control   293

8-1   Basic machine equations   293
8-2   Variable-speed drives   296
8-3   Control feedback loops   301
8-4   Traction drives   306
8-5   Industrial application considerations   313
8-6   Worked examples   313

Chapter 9. A.C. Machine Control   330

9-1   Basic machine equations   330
    9-1-1   Synchronous machine   330
    9-1-2   Cage induction motor   334
    9-1-3   Slip-ring induction motor   341
9.2   Motor speed control by voltage regulation   342
9-3   Constant-voltage inverter drives   343
9-4   Constant-current inverter drives   352
9-5   Motor speed control via the cycloconverter   356
9-6   Transistorized inverter drives   357
9-7   Slip-ring induction motor control   359
9-8   Brushless synchronous machines   362
9-9   Inverter-fed synchronous motor drives   364
9-10 Reluctance and stepper motor drives   367
9-11 Drive considerations   370
9-12 Worked examples   372

Chapter 10. Protection   382

10-1 Current   382
10-2 Voltage   386
10-3 Gate turn-off thyristor and MOSFET   391

Tutorial Problems   393

Answers 397

Glossary of Terms 398

References 404

Bibliography 404

Index 410

PREFACE

The application of semiconductor devices in the electric power field has been steadily increasing, and a study of power electronics (as it is commonly called) is now a feature of most electrical and electronic engineering courses.

The power semiconductor devices, such as the diode, thyristor, triac, and power transistor, are used in power applications as switching devices. The development of theory and application relies heavily on waveforms and transient responses, which distinguishes the subject of power electronics from many other engineering studies. The aim of the author has been to produce a student text which explains the use of the power semiconductor devices in such applications as rectification, inversion, frequency conversion, d.c. and a.c. machine control, and the many non-motor applications.

In choosing material to include in this text, it was decided that material relating to the physics of semiconductor devices, electronic control circuitry involved in (for example) firing circuits, and control system theory is adequately covered elsewhere and would unnecessarily lengthen the book. Material relating to these areas has only been included where it is essential to the proper understanding of the applications of, and circuits used in, power electronic systems.

The treatment and level of the material in this book is intended for students following courses ranging from the higher technician level to final-year first degree. This book will also provide useful background material for post-graduate master's degree courses.

Engineers concerned with the supply and utilization of electricity need to be aware of the effects which power electronics equipment has on both the supply and load. It is hoped that this text will fulfil this need, particularly for those whose student days date back to the era before the development of the power semiconductor devices.

It is the author's view that a technical book requires worked examples to reinforce the understanding of the subject, and to this end 167 worked examples have been included. It is suggested that initially the reader attempts the worked examples without reference to the solution, as this will identify gaps in his/her knowledge. Much of the routine arithmetic has been omitted from the solutions, as this should be within the competence of the reader to solve. All the mathematical work in the solutions has been carried out with the aid of an electronic hand calculator. All the worked examples and problems have been devised by the author.

The order of the chapters is that in which the author has taught the subject to final-year BSc students, although it is recognized that protection may well be included at an earlier stage. A glossary of terms in use in power electronics has been included.

I am grateful to my wife Audrey and children Karen and Stephen for the patience they have shown during the long hours spent at home in preparing the manuscript for this book. My grateful thanks are also extended to all those teachers, colleagues, students and friends who over the years have inspired and helped me.

CYRIL LANDER

PREFACE TO THE SECOND EDITION

Since the original edition was published, new devices have become commercially available and are being used in power electronic equipment. Also, further developments have taken place in the applications of the power electronic devices. The opportunity has been taken in this new edition to include the new devices and applications as additional material, retaining all of the original material from the first edition.

I am grateful for the many encouraging and favourable comments received from lecturers and students who are using my book, and hope that the updating of the material will enhance its value in the study of power electronics. I would also like to acknowledge with thanks the help and assistance received from the staff of McGraw-Hill, the publishers of this book.

CYRIL LANDER

RECTIFYING DEVICES

A rectifying device is one which permits current flow in one direction only, being able to withstand a potential difference without current flow in the opposite direction. The four major devices in power rectification are the diode, thyristor, triac and the power transistor; the latter three have the additional ability to withstand a potential difference in either direction, and are thus controllable rectifying devices (Refs. 1, 2).

1-1 THE DIODE

The active material from which the semiconductor power diode is formed is silicon, a semiconducting material, that is, a material which is classified as being between the insulating and conducting materials, its resistance decreasing with temperature rise.

Silicon is an element in group IV of the periodic table, and has four electrons in the outer orbit of its atomic structure. If an element from group V is added, that is, an element having five outer orbit electrons, then a free electron is present in the crystal structure. The free electrons allow greatly increased conduction, and as the electron is negatively charged such a material is known as an N-type semiconductor.

If to silicon is added an impurity element from group III, that is, an element having three outer orbit electrons, then a gap or hole appears in the crystal structure which can accept an electron. This gap can be considered to provide a positively charged carrier known as a *hole*, which will allow greatly increased conduction, the material so doped being known as a P-type semiconductor.

The order of doping (addition of impurity) is in the order of 1 part in 10^7 atoms. In N-type semiconductors, the majority carriers of current are electrons, the minority carriers being holes. The reverse applies to the P-type semiconductor. Depending on the degree of doping, the conductivity of the N- or P-type semiconductor is very much increased compared to the pure silicon.

The diode shown in Fig. 1-1 is formed by the junction within a single crystal of P- and N-type materials. At the junction, the free electrons of N and the free holes of P combine, leaving the N side with a positive charge and the P side with a negative charge. Hence, a potential barrier exists across the junction having a value of the order of 0.6 V.

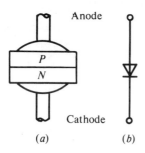

Figure 1-1 The diode. (*a*) Structure. (*b*) Symbol.

The diode characteristic is shown in Fig. 1-2, and with reference to Fig. 1-1*a*, a positive voltage applied to *P* (the anode) with respect to *N* (the cathode) will result in current flow once the potential barrier of 0.6 V is overcome, giving an overall forward volt-drop of the order of 0.7 V at its rated current. The application of a reverse voltage will move the mobile carriers of holes and electrons away from the junction in the *P* and *N* sides respectively, hence preventing current flow and allowing the junction to withstand the applied voltage without conduction. The junction experiences a high electric field gradient and hence can be considered as having capacitance. Thermal agitation does rupture some of the bonds in the crystal, resulting in minority carriers which permit a small reverse current flow shown as a leakage current. An increase in the reverse voltage will lead to an increase in the acceleration rate of the minority carriers across the junction. Eventually the minority carriers have sufficient energy to remove others by collision, when avalanche multiplication takes place and the junction is broken down, giving the reverse breakdown characteristic. Typically the leakage current is a few milliamperes.

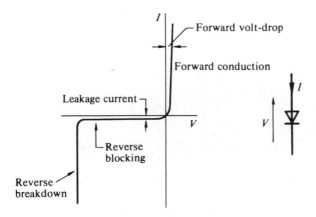

Figure 1-2 Diode characteristic.

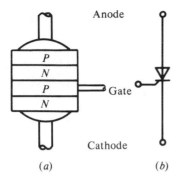

Anode

P
N
P
N

Gate

Cathode

(a)   (b)

Figure 1-3 The thyristor. (a) Structure. (b) Symbol.

1-2 THE THYRISTOR

The thyristor is a four-layer *P-N-P-N* device, with a third terminal, the gate, as shown in Fig. 1-3. A 2000 V, 300 A device would typically be a silicon wafer of diameter 30 mm and thickness 0.7 mm.

The characteristic of the *P-N-P-N* device without any external connection to the gate is shown in Fig. 1-4. The thyristor in this condition may be considered as three diodes in series, with directions such as to prevent conduction in either direction. The reverse characteristic, that is, with the cathode positive, exhibits similar features to the diode. The forward characteristic, that is, with the anode positive, exhibits no current flow other than leakage until the breakover voltage of the centre control junction is exceeded. The forward and reverse breakover voltages are similar in magnitude, due to, in the reverse blocking state, almost all the voltage appearing at the anode *P-N* junction, the cathode *P-N* junction breaking over at

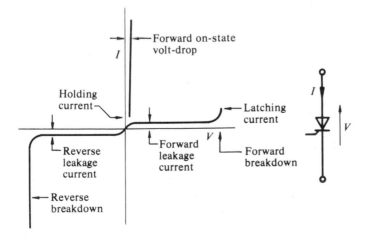

Figure 1-4 The thyristor characteristic with no gate current.

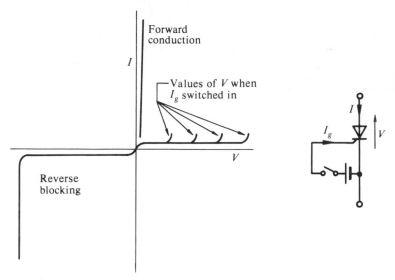

Figure 1-5 Thyristor characteristic with gate current.

about 10 V. Once breakover in the forward direction occurs, the centre P slice is neutralized by the electrons from the cathode, and the device acts as a conducting diode having two junctions giving a forward volt-drop approximately double that of the diode. In order for the thyristor to attain and retain the on-state, the anode current must reach its latching level, and not fall below its holding level as shown in Fig. 1-5. The latching current is typically double the holding current, but both are low, being much less that 1% of the full-load rated value.

The thyristor, when forward-biased (anode positive), can be switched into the on-state by injecting current into the gate terminal relative to the negative cathode, as illustrated in Fig. 1-5. The action of the gate current is to inject holes into the inner P slice, which, together with the electrons from the cathode N layer, breakover the centre control junction, switching the thyristor into the on-state. Once the anode current has exceeded the latching level, the gate current can cease, the thyristor remaining in the on-state, irrespective of conditions in the gate circuit.

To turn off the thyristor, the anode current must be reduced below the holding level, and a relatively long time allowed to elapse for the thyristor control junction to recover its blocking state, before a forward voltage can again be applied without conduction. More typically, to turn off the thyristor, the anode current is driven into reverse by the external circuitry, when for a very brief period a reverse current flows as shown in Fig. 1-6, permitting charge movement within the PN layers, allowing the two outer junctions to block any further reverse current after the storage charge has been recovered. The stored charge is due to the presence of the current carriers in the junction region. The central control junction will, however, not block the re-application of a forward voltage until a further time has elapsed, sufficient to allow recombination of the carriers at this junction. Typically 10 to

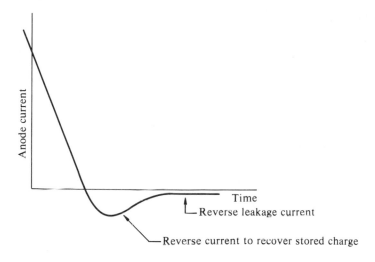

Figure 1-6 Typical current waveform during turn-off

$100\,\mu s$ must elapse before the forward voltage can again be applied without break-down. The storage charge could typically be $20\,\mu C$ for a 20 A thyristor.

The thyristor described in this section is that which was the first to be developed, and can be referred to as the conventional thyristor. More recent developments have seen the introduction of the gate turn-off thyristor which can be turned off by removal of current from the gate, unlike the conventional thyristor which can only be turned off by reduction of its anode current to near zero. The gate turn-off thyristor is described fully later in Sec. 1-8.

A further device which has been developed is the combination of a thyristor with a reverse conducting diode on one silicon wafer. This device will always conduct in the reverse direction, but is controllable (as with the normal thyristor) in the forward direction. Applications of this reverse conducting thyristor are in inverter circuits such as that shown in Fig. 5-20, where this single device can be used instead of the parallel connection of one thyristor and one diode.

1-3 THE TRIAC

The triac is a five-layer device, as shown in Fig. 1-7, having a *P-N-P-N* path in either direction between terminals T_1 and T_2, and can hence conduct in either direction as the symbol clearly indicates. Electrically, the triac performs in one device that which would require two thyristors in the inverse-parallel connection shown in Fig. 1-7c.

The triac can be switched into the on-state by either positive or negative gate current, but is most sensitive if positive current is injected when T_2 is positive, and

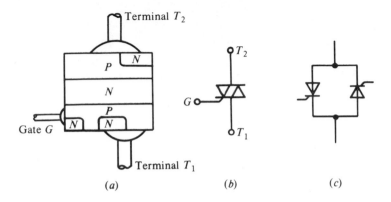

Figure 1-7 The triac. (*a*) Structure. (*b*) Symbol. (*c*) Thyristor equivalent.

negative current when T_1 is positive. However, in practice, negative gate current is always used as shown with the characteristic in Fig. 1-8.

Both the maximum steady-state and transient ratings described in Sec. 1-6 are inferior to those of the thyristor.

1-4 THE POWER TRANSISTOR

The transistor is a three-layer N-P-N or P-N-P device as shown in Figs. 1-9 and 1-10. Within the working range, the collector current I_C is a function of the base current I_B, a change in base current giving a corresponding amplified change in the collector current for a given collector-emitter voltage V_{CE}. The ratio of these two currents is in the order of 15 to 100. Related to the circuit symbol of Fig. 1-9*b*, the transistor

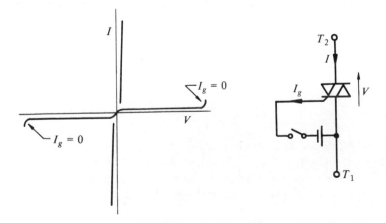

Figure 1-8 The triac characteristic.

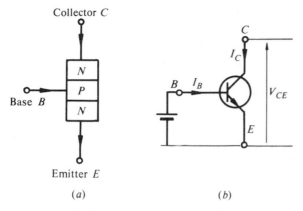

Figure 1-9 *N-P-N* transistor. (*a*) Structure. (*b*) Symbol with current directions.

characteristic is shown in Fig. 1-11. In a similar manner to the other devices, a breakdown level is reached with increasing voltage, when avalanche breakdown will occur. A reversal of the collector-emitter voltage will break down the base-emitter junction at a low voltage, say 10 V, hence the transistor is not operated in this reverse mode. A diode in series with the transistor will enable it to be used in circuits where reverse voltages are encountered.

The *P-N-P* transistor shown in Fig. 1-10 exhibits similar characteristics to the *N-P-N* transistor, the current and voltage directions being reversed.

The power loss in the transistor is a function of the product of the collector-emitter voltage with the collector current. With reference to Fig. 1-12, if the base current is varied to control the load current in the collector circuit, then large voltages can appear at the transistor. For example, if $V = 200$ V and (say) the base current I_B were adjusted to give 10 A into a load of 10 Ω, then the transistor would drop 100 V. This would give the transistor a power loss of 1 kW and an overall

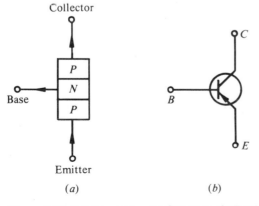

Figure 1-10 *P-N-P* transistor. (*a*) Structure. (*b*) Symbol.

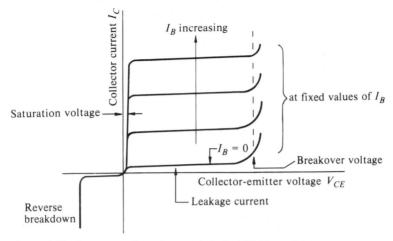

Figure 1-11 Common emitter characteristic for *N-P-N* transistor.

efficiency of 50%. This condition is unacceptable from both the viewpoint of device loss and rating, as well as overall efficiency.

In practice, for power applications, the transistor is operated as a switch. With zero base current, it is effectively an open-circuit condition as shown in Fig. 1-13*a*. With a base current which takes the device into saturation, it is effectively a closed switch. As the transistor is a controlled device, it is essential to profile the base current to the collector current. In order to retain control when in the saturated state and so avoid excessive base charge, the base current should be just sufficient to maintain saturation. At turn-on, initially, the base current should be high so as to give a fast turn-on. Any change in the collector current must be matched by a change in base current. At turn-off, the base current should be reduced at a rate which the collector current can follow, so as to avoid secondary breakdown (see Sec. 10-1). In the off-state, a small reverse current is maintained to avoid spurious collector current. As a switch, the transistor power losses are small, being due to the small leakage current in the open position and the saturation voltage (shown in

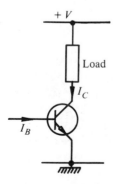

Figure 1-12 Transistor controlled load.

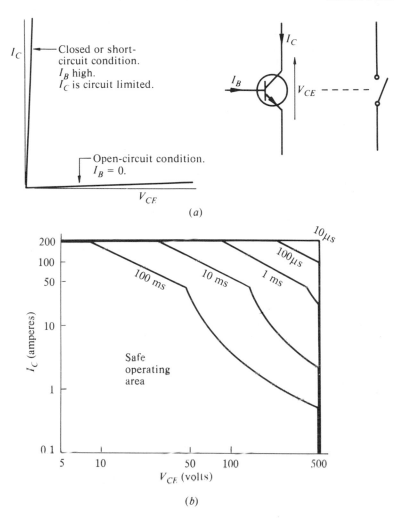

Figure 1-13 Transistor as a switch. (*a*) On- or off-states. (*b*) Typical safe operating area.

Fig. 1-11) with the collector current when in the closed position. Typically, the saturation voltage is 1.1 V for a silicon power transistor.

To exploit the transistor fully without overheating during switching, use can be made of the safe operating area characteristic shown in Fig. 1-13*b*. When switching between the two states shown in Fig. 1-13*a*, it is essential that the instantaneous voltage and current values must at all times during the switching period be within the rectangular area shown in Fig. 1-13*b*. Only the very shortest switching time is almost rectangular, the high instantaneous power loss which can be tolerated being progressively restricted for longer switching times as shown in Fig. 1-13*b*, by the corner being placed outside the safe operating area. Note that the scales of the safe operating area are logarithmic.

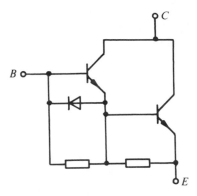

Figure 1-14 Darlington arrangement of power transistor.

The switching losses of the transistor can be high. During switching, both voltage across and current through the transistor can be high, the product of voltage and current with the switching time giving the energy loss for one switching operation. A high switching frequency can mean the predominant loss is that due to switching. The exact switching loss is a function of the load circuit parameters as well as the form of the base current change.

The transistor can switch considerably faster than the thyristor, typically a switching time of less than 1 to 2 μs being possible. The base drive requirements of the transistor are more onerous than the gate requirements of the thyristor; for example, a 30 A thyristor may require a 0.1 A pulse for turn-on, whereas a 30 A transistor may require 2 A base current continuously during the on-period. The overload rating of the power transistor is considerably less than that of the thyristor. The transistor by control of the base current is capable of turn-off while carrying load current, whereas in the thyristor the gate loses control after turn-on.

The current gain of the power transistor can be considerably improved if the base drive current is obtained from another transistor in what is known as a *Darlington arrangement* as shown in Fig. 1-14. The driver transistor can be incorporated on the same silicon chip, with overall current gains of 250 being possible, but with a longer switching time.

1-5 THYRISTOR GATING REQUIREMENTS

The gate-cathode characteristic of a thyristor is of a rather poor *P-N* junction. There will be a considerable range of characteristic within a given production batch, individual thyristors having characteristics as shown in Fig. 1-15*a*. All thyristors can be assumed to have a characteristic lying somewhere between the low and high resistance limits. The minimum level of current and voltage required to turn on the thyristor is a function of the junction temperature; an indication of these minimum levels is shown in Fig. 1-15*a*.

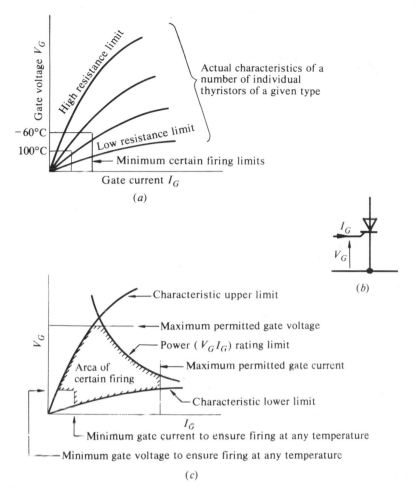

Figure 1-15 Gate characteristic of the thyristor. (*a*) Range of characteristic. (*b*) Circuit reference. (*c*) Limits.

The current into, and the voltage at, the gate are both subject to maximum values, but turn-on requirements demand they also exceed certain minimum levels. The product of gate voltage and current gives a power level to which a maximum is set. Fig. 1-15*c* shows these limits imposed on the gate-cathode characteristic, giving the area into which must be fitted the gate firing signal for certain firing into the on-state to take place.

The final stage of the gate firing (triggering) network shown in Fig. 1-16*a* will consist of a transformer for isolation, a resistance R_1 to limit the gate current, and a resistance R_2 to limit the gate voltage when the thyristor is in the off-state.

The Thévenin equivalent circuit of the firing network, shown in Fig. 1-16*b*, is taken as a voltage E in series with a resistance R_G.

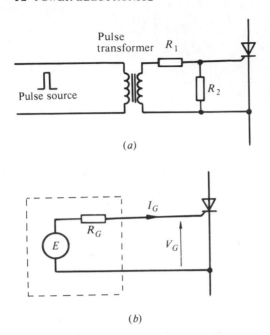

(a)

(b)

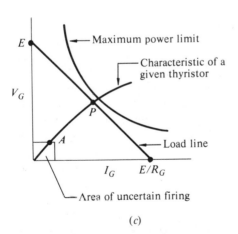

(c)

Figure 1-16 Firing (trigger) network. (*a*) Desirable network connection. (*b*) Thévenin's equivalent circuit for firing network. (*c*) Firing network load line.

The relationship in the steady state between the gate voltage V_G and the gate current I_G is defined by the values of E and R_G along the load line shown in Fig. 1.16*c*. When the firing signal is initiated, the gate current will grow along the line of the characteristic for that thyristor until in the steady state the load line at point P is reached. However, before point P is reached, the thyristor will have turned on, most likely in the region of point A. The parameters of the firing network must be so chosen that the load line is above A but within the maximum power limit. Typically the value of E will be 5 to 10 V, having an associated maximum current of 0.5 to 1 A.

1-5-1 Requirements of firing circuits

To turn on a thyristor positively in the shortest time, it is desirable to have a gate current with a fast rise time up to the maximum permitted value. This rise time is best achieved by pulse techniques, where the firing circuit generates a fast rise pulse of sufficient length to allow the anode current enough time to reach its latching value. The advantage of the pulse is that much less power is dissipated in the gate compared to a continuous current, and the instant of firing can be accurately timed.

Reference to the simple rectifying circuit of Fig. 2-4 shows that an essential requirement of the firing circuit in a.c. supply applications is that the thyristor shall be turned on at a time related to the a.c. supply voltage phase. Also, the phase of the firing pulse in relation to the a.c. supply voltage zero must be capable of variation.

A practical commercial firing circuit will typically have a characteristic pulse such as in Fig. 1-17a, a rise time of 1 μs from a voltage source of 10 V, capable of delivering 1 A, the peak pulse load line being as shown in Fig. 1-17b. A pulse length of 10 μs with a rise to 2 V in 1 μs may suffice for many applications, whereas other applications may require lengths up to 100 μs. The firing circuit will normally reset itself after the first pulse, to give a succession of pulses spaced (say) 400 μs up to the end of the half cycle as shown in Fig. 1-17c. In the rectifier circuit, conditions may not be right for conduction to take place at the first pulse, the second and successive pulses being available to turn on the thyristor.

Many rectifier configurations require the simultaneous firing of two thyristors, the cathodes of which are at different potentials. To overcome this problem, the final stage of the firing circuit will be a transformer with two or more isolated outputs.

Reverse gate current must be prevented to avoid excessive gate power dissipation. Also, injection of gate current when the thyristor is reverse-biased will increase the leakage current, and is best avoided.

1-5-2 Typical firing circuits

Figure 1-18a shows a crude firing arrangement, the object of which is to control the load voltage to the waveform shape of Fig. 1-18b. The gate current $i_g \simeq (v_{supply})/R$ and, as the sinusoidal voltage rises from zero, the gate current will eventually reach a level such as to turn on the thyristor, this occurring at (say) the angle α shown in Fig. 1-18. Variation of the firing delay angle α will vary the load power. This simple firing arrangement has so many shortcomings that its practical use is negligible. The turn-on angle will vary from cycle to cycle as temperature and other changes occur in the thyristor. Further, the turn-on will be slow and will not

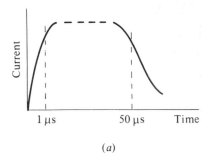

(a)

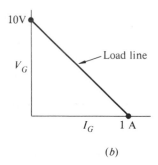

(b)

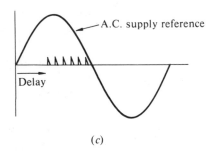

(c)

Figure 1-17 Typical firing circuit output pulse characteristic. (a) Pulse shape. (b) Load line. (c) Firing pulse reference to a.c. supply.

occur near the zero voltage, nor will extend beyond the supply voltage maximum ($\alpha = 90°$). This demonstrates a clear need for firing circuits giving a pulse output timed to the a.c. cycle as shown in Fig. 1-17c.

Simple but practical firing circuits are those using the switching action of a transistor as in Fig. 1-19. Both circuits are supplied from the a.c. source, which is clipped by the Zener diode Z to give a level voltage to the $R_1 C_1$ series circuit. Resistance R_2 drops the voltage difference between the supply and the Zener diode. In both circuits the voltage on the capacitor C_1 will rise exponentially at a rate determined by the value of R_1.

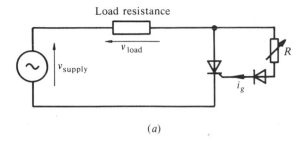

Load resistance

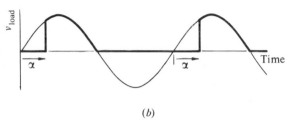

(a)

(b)

Figure 1-18 Simple rectifier circuit. (a) A crude firing arrangement. (b) Single phase control using one thyristor.

In the circuit of Fig. 1-19a, the reference voltage to the base of the transistor is determined by the resistor chain S. Initially with zero voltage on C_1 the transistor is held off but, when the voltage on C_1 at the emitter reaches a high enough level, the transistor starts to conduct. Feedback action via the transformer winding raises the base current so the transistor turns hard on, discharging C_1 rapidly into the thyristor gate via the transformer. The unijunction transistor of Fig. 1-19b exhibits a characteristic of no conduction until the voltage on C_1 reaches a particular value, then the unijunction transistor will change to its conducting state, allowing C_1 to discharge into the thyristor gate. The rapid discharge of C_1 gives a fast rise pulse into the thyristor gate.

Both of the transistor firing circuits will reset after C_1 is discharged, building up to give a second and further pulses. At the zero point in the a.c. supply cycle, the circuit will completely discharge, hence the initial growth of voltage at C_1 is timed from the supply zero. By adjustment of R_1, the time of the first output pulse can be controlled up to 180° delay in a waveform such as shown in Fig. 1-18b.

The simple circuits of Fig. 1-19 give a pulse length and rise time which is adequate for most passive resistive loads, where small variations in firing angle from cycle to cycle can be tolerated. Additional circuits can be added in series with R_1 to inhibit firing by preventing charging current flow to C_1, or to incorporate some desired remote automatic control.

The more sophisticated firing circuits contain many more stages in their electronic circuitry. Such circuits may, for example, rely on the interrelationship between a ramp voltage and the external control voltage to initiate accurately, at the same time in each cycle, the start of the pulse generator.

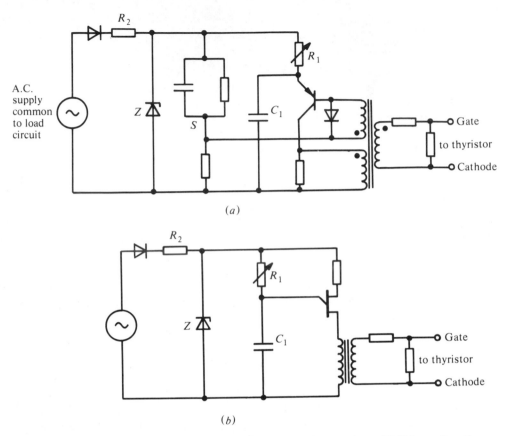

Figure 1-19 Simple firing circuits using transistors. (*a*) Using transistor. (*b*) Using unijunction transistor.

For applications other than those connected with a fixed frequency supply, the firing circuits include oscillators to initiate the pulse starts and finishes.

1-5-3 Control features of firing circuits

The more complex systems utilizing thyristors as the power control elements will include, for example, closed-loop links, polyphase supply, automatic control of current or motor torque level, inhibiting loops to ensure against malfunction of operation due to simultaneous firing of different groups. The control characteristic is designed to give a defined relationship between firing delay angle and input voltage, say, either linear or cosine. These aspects are shown diagrammatically in Fig. 1-20 which can include (as shown) inputs to limit the length of the pulse train, or to ensure a pulse will appear at the thyristor gate at a given angle, in particular, just before the time when control would be lost in the inverting mode of converter operation.

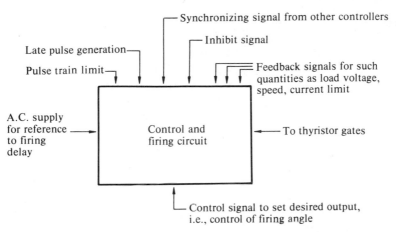

Figure 1-20 Typical features required in a more complex a.c. converter system.

The example given above relates to a.c. converter systems, but similar control features will be required in the many other systems involving d.c. supply, or variable-frequency supply with the many varied load features. The later chapters will describe many of these systems.

1-6 RATINGS

The earlier sections describe in broad outline the characteristics of the various rectifying devices. In practice, many other aspects of the device may need to be considered, and these all contribute to specifying the device rating.

The turn-off mechanism described in relation to Fig. 1-6 is rated by reference to Fig. 1-21. Given a forward diode or thyristor current I_F, which is turned off at a given di/dt rate, the current will go into reverse until the carrier storage charge Q_{rr} is recovered. At defined values of I_F and di/dt, a particular device will have rated values of Q_{rr}, the reverse recovery charge, associated with a reverse recovery time t_{rr} and a reverse recovery current I_{rr}.

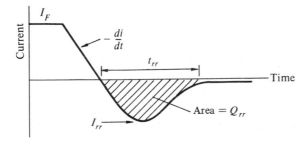

Figure 1-21 Typical turn-off condition.

The usual turn-on mechanism with gate current has been described, but it is possible to turn on a thyristor by an excessive rate of rise of forward voltage. Considering the junction to act as a capacitor in the off-state, a displacement current $i = C\,dv/dt$ (capacitor charging current) will exist. Given a high enough rate of rise of voltage, say $100\,\text{V}/\mu\text{s}$, the minority carriers will be accelerated across the junction at a rate high enough to fire (trigger) the thyristor into the on-state, even though there is an absence of gate current. A thyristor will have a rated value of dV/dt which must not be exceeded.

When turned on in the usual manner, the junction of the thyristor will breakover first in the region of the gate electrode. If the total anode current were established immediately, the current density in this region would be excessive, resulting in damage by overheating. The rate of rise of the anode current must be limited to the time taken for the junction breakdown to spread completely across the slice, a time of typically $10\,\mu\text{s}$. A thyristor will have a dI/dt rating which is not to be exceeded.

The junction temperature of the device is normally rated not to exceed $150\,^{\circ}\text{C}$ for the diode, $125\,^{\circ}\text{C}$ for the thyristor, and between 150 and $200\,^{\circ}\text{C}$ for the power transistor. The thermal resistance of the junction to base (stud) is quoted in the rating.

The device loss is approximately a function of the square of the current, hence r.m.s. values can be used to rate the device for most waveform shapes. The rated current is that value of current which under steady-state conditions will result in the rated temperature being achieved at the junction, when the base or stud (heat sink) on the device is at its specified temperature. When the device is carrying a cyclic current repetitive at 50 or 60 Hz, the temperature change over each cycle is small enough for the r.m.s. value of the current to be used. Given a particular current waveshape, such as a complete half sinewave, the mean value of this waveshape over 180° conduction is sometimes given as the device rating.

Under heavy short-time overloads, the heat generated within the device will be mostly stored in the thermal mass of the silicon, giving a rise in temperature, as very little of the heat will be dissipated. Assuming the device to have a power loss proportional to the square of the current i, then the sum of the i^2 values over a given time t, that is, $\int i^2 dt$, can be related to the temperature rise. The device will have an $\int i^2 dt$ rating related to the permitted temperature rise above the steady-state maximum temperature. The overload condition is assumed to occur after the device has been carrying its permitted rated current for a long period, the junction then being at its rated temperature.

The characteristics of the diode, thyristor, triac and power transistor shown in Figs. 1-2, 1-4, 1-8 and 1-11, all indicate a breakdown voltage above which the device will switch from the non-conducting state into the on-state, resulting frequently in device destruction. Each device will have rated continuous and repetitive values of voltage which it will withstand in the reverse direction without breakdown. The thyristor will have similar ratings given for the forward direction. These voltages are known as the repetitive peak reverse and peak forward (off-state) voltages. In practice, occasional once-only transient voltages will appear in a circuit. The thyristor (and other devices) will have ratings for these transient or non-repetitive voltages which it can withstand without breakdown. A particular device will have an on-state forward volt-drop quoted for a particular stated current value.

In addition to those aspects covered above, the power transistor ratings will include the collector-to-base current gain, frequency and switching time considerations.

Reference to Sec. 1-5 will indicate the gate ratings of the thyristor gate in respect to the maximum rated values of voltage and current. In addition, a particular thyristor will have average and peak gate power ratings which are not to be exceeded. The holding and latching current levels of a thyristor were explained in Sec. 1-2, these being maximum quoted values in a particular thyristor rating.

1-7 COOLING

The sources of loss in power semiconductor devices may be listed:

1. The loss during forward conduction, which is a function of the forward volt-drop and conduction current. This is the major source of loss at mains and lower-frequency operation.
2. The loss associated with the leakage current during the blocking state.
3. The loss occurring in the gate circuit as a result of the energy input from the gating signal. In practice, with pulse firing these losses are negligible.
4. The switching loss, that is, the energy dissipated in the device during turn-on and turn-off which can be significant when switching is occurring at a relatively high frequency, say, at 1 kHz.

The losses will lead to heat generation within the device, and consequently a rise in temperature until the rate of heat dissipation matches the loss. The heat generated in the junction area is transferred to the base, and then to a heat sink. The temperature level to which the base can be allowed to rise is not high enough for much heat dissipation to take place by radiation, hence most of the heat transfer takes place by convection, to air in the case of the metal-finned heat sink illustrated in Fig. 1-22a. Where the size of the fins is prohibitive, or the level of heat dissipation is very high, the heat sink can be cooled by a compact water-cooled assembly as illustrated in Fig. 1-22b.

Heat transfer will take place from a higher-temperature region to a lower-temperature region. The heat transfer P is proportional to the temperature difference, the ratio being known as the thermal resistance R.

$$P = \frac{T_1 - T_2}{R} \tag{1-1}$$

where T_1 and T_2 are the temperatures of the hot and cold bodies respectively. The

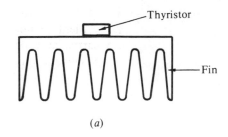

(a)

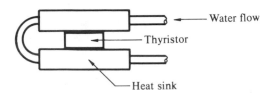

Figure 1-22 Typical cooling arrangements. (a) Air-cooled. (b) Water-cooled.

unit of power is the watt, temperature is in °C, and the thermal resistance in °C/W.

The heat flow is from the junction to the base, to the heat sink, to the surroundings, and the total thermal resistance will be the series addition of the individual thermal resistances of each section. The total thermal resistance from the junction to the surroundings (ambient) is

$$R_{ja} = R_{jb} + R_{bh} + R_{ha} \tag{1-2}$$

where R_{jb}, R_{bh} and R_{ha} are the thermal resistances from junction to base, base to heat sink, and heat sink to ambient.

The virtual junction temperature T_{vj} is

$$T_{vj} = T_a + PR_{ja} \tag{1-3}$$

where T_a is the ambient temperature. Figure 1-23 illustrates the circuit equivalent to the heat transfer, enabling the intermediate temperatures of base and heat sink to be determined. Note that *virtual junction* is used to avoid defining exactly what is the junction in a multi-layer device.

The above calculations all refer to steady-state continuous-current conditions, when the loss is constant, the temperature having settled to a steady value. The thermal storage capacity of the semiconductor devices is low, and temperature changes will occur over the period of a cyclic change. With steady-state conditions at 50 Hz, the thermal resistance values quoted at the rated mean power loss will be such as to ensure the maximum allowable temperature is not exceeded during the cyclic period.

During short-period transients, such as overload or fault conditions, the temperature rise of the junction must be calculated by taking into account the thermal storage capacity of the device. The thermal conditions are such that the heat

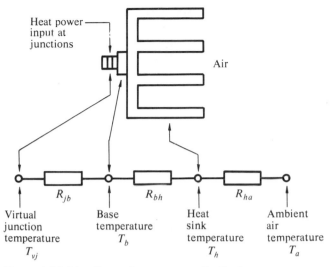

Figure 1-23 Heat flow and temperature distribution.

generated by the losses is partly stored in the thermal mass by an increase in temperature, the rest being dissipated by transfer to the cooling sink. At a junction temperature θ above ambient, taking a short period δt over which the temperature rises $\delta\theta$, the energy balance is

energy input in losses = increase in stored thermal energy + energy

dissipated to the cooling sink (ambient)

$$P\delta t = A\delta\theta + B\theta\delta t$$

where P = power loss in device

A = thermal storage capacity in joules of stored energy per °C rise

B = power dissipated per °C rise

In the limit $P = A(d\theta/dt) + B\theta$

The solution to this equation, assuming θ is zero at time $t = 0$, is the well-known exponential equation

$$\theta = \theta_{max}(1 - e^{-t/T}) \tag{1-4}$$

where $\theta_{max} = P/B$ = final steady temperature rise   (1-5)

and $T = A/B$ = thermal time constant   (1-6)

The simple exponential-rise curve of Eq. (1-4) is appropriate for a homogeneous material, such as a copper conductor where the spread of heat within the material is rapid, but care must be taken in applying it to the junction temperature rise. The power loss in the thyristor occurs nonuniformly over the junction area; the higher

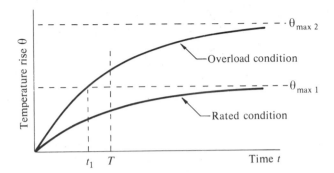

Figure 1-24 Junction temperature rise during overload conditions.

the transient overload, the smaller will be the mass experiencing a temperature rise, hence no single value can be given to its thermal storage capacity. Silicon is not a good conductor of heat, hence the heat generated will have little spread during a short transient compared to a longer transient of lesser magnitude.

Using Eq. (1-4) to illustrate overload conditions, the final maximum temperature rise θ_{max} will be higher because the value of P in Eq. (1-5) will be higher. Reference to Fig. 1-24 shows the temperature rise from the initial cold conditions giving θ_{max1} as the rated rise, and θ_{max2} for the overload condition. In practice, the junction must not exceed θ_{max1}, otherwise the device will be destroyed, hence the overload condition can be tolerated up to the time t_1 when the junction has reached its maximum rated value. θ_{max2} is only a mathematical value to define the overload curve, not a practical obtainable value.

As the conditions within the device are so complex during overload, the concept of transient thermal impedance is used for predicting conditions, rather than the simple exponential rise of Eq. (1-4). The transient thermal impedance Z_{th} associated with a given time is defined as

$$Z_{th} = \frac{\text{Temperature difference (rise)}}{\text{Power loss in device over the defined time}} \qquad (1\text{-}7)$$

Given a short-time overload, manufacturer's data for a given device will quote a transient thermal impedance for this time which can be used in Eq. (1-7). This makes overload calculations as simple as those for steady-state rated conditions using Eq. (1-1). The transient thermal impedance is used for those applications where repetitive high-power dissipation is encountered.

1-8 GATE TURN-OFF THYRISTOR

The conventional thyristor as described in Sec. 1-2 has over the years been developed such that two new devices of the thyristor family are now available, the asymmetrical thyristor and the gate turn-off thyristor.

The conventional thyristor has two *P-N* junctions which can block high voltages in one or other direction, this being an essential requirement for applications in the rectifier circuits described in Chapter 2. However, for the inverter circuits described in Chapter 5 the reverse blocking capability is not needed.

To reduce the time taken for the thyristor to recover its blocking state after turn-off the silicon can be made thinner at the expense of it losing its ability to block a reverse voltage, this device now being known as the asymmetrical thyristor. In the inverter circuits a diode is connected in parallel with the thyristor so the loss of the reverse blocking capability is of little consequence, but the switching time is reduced to a few microseconds compared to the tens of microseconds for the conventional thyristor.

The conventional thyristor can only be turned off by effectively reducing the anode current to zero, but the gate turn-off thyristor, as its name implies, has a structure such that it can be turned off by removing current from the gate. Turn-on is achieved by injecting current into the gate as in the conventional thyristor.

The more complex structure of the gate turn-off thyristor compared to the conventional thyristor is shown in Fig. 1-25. The circuit symbol for the gate turn-off thyristor is an extension from the conventional thyristor showing the dual role of the gate terminal. Referring to Fig. 1-25*d*, the gate turn-off thyristor has highly doped *N* spots in the *P* layer at the anode, the plus sign indicating high

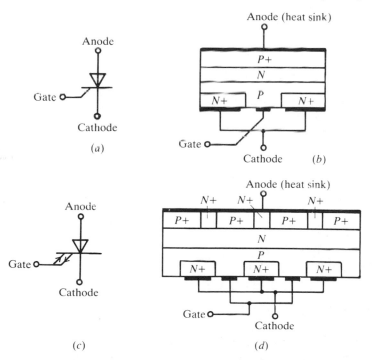

Figure 1-25 Thyristor structure. (*a*) Conventional thyristor symbol. (*b*) Conventional thyristor *P-N-P-N* structure. (*c*) Gate turn-off thyristor symbol. (*d*) Gate turn-off thyristor structure.

doping levels. The gate-cathode structure is interdigitated, that is, each electrode is composed of a large number of narrow channels closely located.

With the gate turn-off thyristor, in the absence of any gate current, a positive voltage at the anode with respect to the cathode is withstood at the centre N-P junction in a like manner to the conventional thyristor, but a reverse voltage with the cathode positive will break down the anode junction at a low level in a similar manner to the asymmetrical thyristor.

The turn-on conditions for the gate turn-off thyristor are similar to the conventional thyristor, but because of the differing structure the latching current is higher. The interdigitated nature of the gate results in a very rapid spread of conduction in the silicon, but it is necessary to maintain the gate current at a high level for a longer time to ensure that latching takes place. To minimize the anode-cathode voltage drop it is advantageous to maintain a low level of gate current throughout conduction otherwise the on-state voltage and hence conduction losses will be slightly higher than necessary.

The thyristor remains on after removal of the gate current because the internal mechanism of carrier multiplication is self-maintaining provided the anode current is above the latching level. With the gate turn-off thyristor it is possible to cause the carrier multiplication to cease by removing holes from the P region, which causes the conducting area to be squeezed towards the anode N spots into the area under the cathode electrode furthest from the gate electrode, until all the conducting paths are extinguished. Once the cathode current has ceased a gate-anode current persists for a short time until the device gains its blocking state. The magnitude of the gate current for turn-off is of the order of one-fifth to one-third of the anode current; hence it is considerably higher than the turn-on magnitude. The time of turn-off is shorter than with the other thyristors.

The gating requirements of the gate turn-off thyristor are summarized by the circuit shown in Fig. 1-26a. At turn-on a current is injected into the gate. At turn-off a negative voltage is placed across the gate-cathode of the order of 10 V, thus removing current from the gate. The turn-off voltage must be less than the gate-cathode reverse breakdown but high enough to extract the charge necessary to

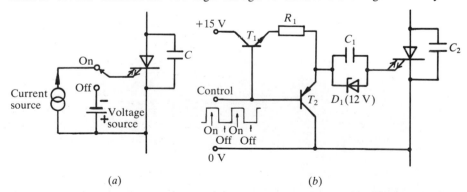

(a)   (b)

Figure 1-26 Gate turn-off thyristor gate circuit. (a) Basic requirement. (b) Simple gate control circuit.

bring about turn-off. The turn-off physics of the device is complex but basically the gate charge must be extracted rapidly with a peak gate current value near to the value of the anode current, this current being established in much less time than one microsecond. To limit the rate at which the anode voltage rises at turn-off a snubber capacitor is connected across the thyristor.

A simple gate control circuit is shown in Fig. 1-26b. Positive current to the base of transistor T_1 allows current flow into the gate via R_1 and C_1, with the initial value being set by R_1. The Zener diode D_1 conducts when its breakdown voltage is reached, thus holding the charge on C_1 to (say) 12 V, allowing a small continuous gate current to flow from the 15 V supply as ideally required. Reversal of the control current will turn transistor T_2 on, T_1 going off. With T_2 on, the capacitor C_1 now discharges via T_1, removing gate current and turning the thyristor off. The capacitor C_2 across the thyristor limits the dV/dt rise of the anode-cathode voltage.

1-9 POWER MOSFET

The power metal oxide semiconductor field-effect transistor (MOSFET) is a device derived from the field-effect transistor (FET) for use as a fast-acting switch at power levels. Unlike the bipolar transistor which is current controlled, the MOSFET is a voltage-controlled device. Referring to Fig. 1-27a, the main terminals are the drain and source, the current flow from drain to source being controlled by the gate to source voltage.

Figure 1-27b shows the cross-section of a part of a MOSFET. With zero gate to source voltage, a positive voltage at the drain relative to the source will result in current of up to possibly a few hundred volts being blocked. If a sufficiently positive voltage, approximately 3 V, is applied to the gate, a negative charge is induced on the silicon surface under the gate which causes the P layer to become an induced N layer, allowing electrons to flow. Hence, a positive gate voltage sets up a surface channel for current flow from drain to source. The gate voltage determines the depth of the induced channel and in this manner determines the current flow.

The characteristic of the MOSFET is shown in Fig. 1-27d to the circuit reference of Fig. 1-27c. At very low values of drain-source voltage the device has a constant resistance characteristic, but at the higher values of drain-source voltage the current is determined by the gate voltage. However, in power applications the drain-source voltage must be small in order to minimize the on-state conduction losses. The gate voltage is thus set at a high enough level to ensure that the drain current limit is above the load current value, that is, the device is operating in the constant resistance condition. The gate voltage must be limited to a maximum value of approximately 20 V.

The silicon dioxide which insulates the gate from the body of the transistor is an insulator with negligible leakage current. Once the gate charge is established there is no further gate current giving a very high gain between the output power

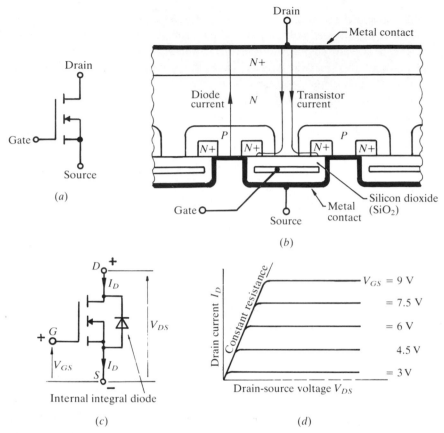

Figure 1-27 Power MOSFET. (*a*) Normal symbol, *N* channel. (*b*) Simplified cross-sectional structure. (*c*) Electrical circuit. (*d*) Output characteristic.

and control power. Reference to Fig. 1-27*b* shows that in the opposite direction, that is, from the source to drain, there is a *P-N* path, which means that there is a diode integral with the transistor from the source to drain as shown in Fig. 1-27*c*. For the inverter circuits described later in Chapter 5, it will be found that a reverse connected diode is essential; hence this integral diode is a bonus.

The absence of any stored charge makes very fast switching possible, with on and off times being much less than one microsecond. The on-resistance of the MOSFET is a function of the voltage breakdown rating, with typical values being 0.1 Ω for a 100 V device and 0.5 Ω for a 500 V device. The resistance is always higher for the higher-rated voltages but the actual resistance will vary according to the device structure. Figure 1-27*b* shows only a very small section of the inter-digitated structure.

The power MOSFET can be directly controlled from microelectronic circuits and is limited to much lower voltages than the thyristor, but is easily the fastest acting device. Above approximately 100 V conduction losses are higher than for the bipolar transistor and the thyristor, but the switching loss is much less. The MOSFET has a positive temperature coefficient for resistance; hence paralleling of devices is relatively simple. In terms of current and voltage capability the MOSFET is inferior to the current-controlled devices of the bipolar transistor and thyristor family of devices.

1-10 DEVICE COMPARISONS

The manner in which the devices are used in power electronic equipment is as a switch which is either open or closed. Ideally as a switch the device would have

unlimited voltage and current ratings,
instant turn-on and turn-off times,
zero leakage current,
zero conduction and switching losses,
zero gate firing power requirement,
ability to withstand current overloads and voltage transient,
easy to protect against spurious turn-on and fault conditions,
low cost and ease of assembly.

In practice the many devices have relative merits which make them more suitable for one application than another. In some areas there is overlap with the choice of device not being clear-cut. The important criteria in circuit applications depend very often on the parameters of ratings, conduction losses, switching losses, switching times, control strategy, and finally of cost.

The conventional thyristor has the highest ratings of all devices, is robust, has low conduction losses, is inexpensive, but is slow to turn on and cannot be turned off other than by cessation of its load current. For applications linked to the public electricity supply at 50 Hz or 60 Hz, such as the rectifiers described in Chapter 2, the conventional thyristor is the first choice, its capability of withstanding high forward and reverse voltages being essential to this application.

For those applications involving the production of an alternating voltage from a direct voltage source, the inverters as described in Chapter 5, all of the devices compete, the one being selected often being a function of the switching rate. Where the highest rates of switching are required, above 100 kHz, the MOSFET is the only device. In the 20 to 100 kHz range the bipolar transistor is competitive, having lower cost, lower conduction losses, but higher switching losses than the MOSFET. In the range up to 15 kHz the thyristor family, particularly the gate turn-off thyristor and the asymmetrical thyristor, is competitive due to their robustness, low conduction losses, and superior overload and transient capability.

The transistor family can operate at temperatures up to 200°C whereas the thyristor family is limited to 125°C. The cost of the losses and the cooling requirement are frequently important criteria in selection. The less demanding firing requirements of the voltage-controlled MOSFET can be a deciding factor when compared to the more demanding current-controlled bipolar transistor and thyristor devices.

Protection of the devices against fault conditions is easiest with the thyristor family. This has been one of the factors limiting the progress of the transistors into the very highest rated equipment.

Research and development work is constantly being undertaken to improve the present devices and develop new devices closer to the ideal electronic switch. A new device which links the high gate impedance and fast turn-on of the power MOSFET with the regenerative latching action of the thyristor and its low on-state loss is under active development.

1-11 WORKED EXAMPLES

Example 1-1

The thyristor in Fig. 1-28 has a latching current level of 50 mA and is fired by a pulse of length 50 μs. Show that without resistance R the thyristor will fail to remain on when the firing pulse ends, and then find the maximum value of R to ensure firing. Neglect the thyristor volt-drop.

SOLUTION Without R the thyristor current i will grow exponentially, given after a time t as

$$i = I(1 - e^{-t/T})$$

where $I = 100/20 = 5$ A, $T = 0.5/20 = 0.025$ s, the time constant. Substituting values, after 50 μs, $i = 10$ mA. Note that in this case as 50 μs $\ll T$, the initial value of $di/dt = 100/0.5$ could have been assumed constant over 50 μs, giving $i = 10$ mA. After 50 μs the thyristor has failed to reach its latching level, being $(50 - 10)$ mA below the required level of 50 mA. The addition of R will permit an immediate current of $100/R$ to be established,

$$\text{required value of } R \text{ is } \frac{100}{(50 - 10) \times 10^{-3}} = 2.5 \text{ k}\Omega.$$

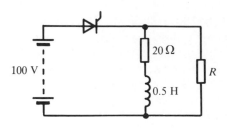

Figure 1-28

Example 1-2

During turn-on and turn-off of a power transistor the current-voltage and current-time relationships were as shown in Fig. 1-29. Calculate the energy loss during both turn-on and turn-off periods, and the mean power loss if the transistor is being switched at a frequency of 1 kHz.

SOLUTION Energy dissipation $= \int_0^t vi \, dt$

$$i_c = \frac{200}{80 \times 10^{-6}} t = 2.5 \times 10^6 t \text{ A, and also}$$

$$i_c = 100 - \frac{100}{40} v_{ce} \text{ from zero to 100 A, giving}$$

$$v_{ce} = 40 - 10^6 t, \text{ from } t = 0 \text{ to } t = 80 \times \frac{100}{200} = 40 \, \mu s;$$

for the time 40 to 80 μs, $i_c = 100$ mA and v_{ce} is zero.
Hence the loss is zero in this period.

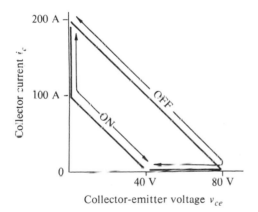

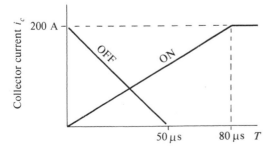

Figure 1-29

$$\text{Turn-on energy} = \int_{t=0}^{t=40\,\mu s} (40 - 10^6 t)(2.5 \times 10^6 t)\,dt = 27\,\text{mJ}$$

$$\text{Turn-off } i_c = 200 - \frac{200}{50 \times 10^{-6}} t = 200 - (4 \times 10^6) t \text{ A}$$

$$v_{ce} = 80 - \frac{80}{200} i_c = 1.6 \times 10^6 t \text{ V}$$

$$\text{Turn-off energy} = \int_{t=0}^{t=50\,\mu s} (1.6 \times 10^6 t)[200 - (4 \times 10^6 t)]\,dt = 133\,\text{mJ}$$

Mean power loss at 1 kHz $= (27 + 133) \times 10^{-3} \times 1000 = 160\,\text{W}$

Example 1-3

A thyristor has a forward characteristic which may be approximated over its normal working range to the straight line shown in Fig. 1-30. Estimate the mean power loss for
(i) a continuous on-state current of 23 A,
(ii) a half sinewave of mean value 18 A,
(iii) a level current of 39.6 A for one-half cycle,
(iv) a level current of 48.5 A for one-third cycle.

SOLUTION (i) At 23 A, the on-state voltage from Fig. 1-30 is

$$1 + \frac{23 \times 1.1}{60} = 1.42 \text{ V}$$

$$\text{Loss} = 23 \times 1.42 = 32.7 \text{ W}$$

(ii) The maximum value of the sinewave $= 18\pi$ A

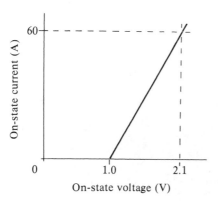

Figure 1-30

From Fig. 1-30 at any current i, $v = 1.0 + \dfrac{1.1}{60}i$

Over one cycle the total base length is 2π; from 0 to π, $i = 18\pi \sin x$; and from π to 2π, $i = 0$.

$$\text{Mean power} = \frac{1}{2\pi}\int_0^\pi vi\,dx = \frac{1}{2\pi}\int_0^\pi \left(1.0 + \frac{1.1}{60}18\pi \sin x\right)18\pi \sin x\,dx = 32.6\,\text{W}.$$

(iii) The mean power loss will be half the instantaneous power loss over the half cycle when the current is flowing.

$$\text{Mean power} = [39.6(1.0 + \frac{1.1}{60}39.6)]/2 = 34.2\,\text{W}.$$

(iv)

$$\text{Mean power} = [48.5(1.0 + \frac{1.1}{60}48.5)]/3 = 30.5\,\text{W}.$$

It is worth noting that the r.m.s. values of the currents quoted above are 23 A, 28 A, 28 A, and 28 A respectively. The closeness of the power loss in each case shows that r.m.s. values of a cyclic current can be used for rating purposes, but that the continuous current rating is somewhat lower.

Example 1-4

For a particular thyristor, manufacturer's data give a loss of 400 W for a rectangular current waveform of varying length as below:

Mean current A	138	170	196	218	250	305
Length of current in degrees	30	60	90	120	180	360

Determine the r.m.s. current value at each condition.

SOLUTION Figure 1-31 shows the current waveform with the current length shown as ϕ.

$$\text{R.M.S. value} = \left(\frac{1}{2\pi}\int_0^\phi I_{pk}^2\,d\phi\right)^{1/2} = I_{pk}\left(\frac{\phi}{2\pi}\right)^{1/2}$$

$$\text{Mean value} = I_{pk}\frac{\phi}{2\pi}$$

giving values:

ϕ	30	60	90	120	180	360
I_{mean}	138	170	196	218	250	305
I_{pk}	1656	1020	784	654	500	305
I_{rms}	478	416	392	377	354	305

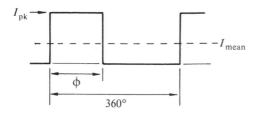

Figure 1-31

It can be seen here that the r.m.s. value of the current gives a better consistent guide to the rating than does the mean value, but note must be taken of the current waveform in selecting a thyristor for a particular application.

Example 1-5

Two different conditions during the turn-on of a thyristor are shown in Fig. 1-32, which gives the voltage and current changes. Plot the waveform of power loss for each case and the turn-on energy loss.

If the rise time t_r of the turn-on is defined as that for the voltage to fall from 90% to 10% of its initial value, determine the value of n in a formula $(V_{max} I_{max} t_r)/n$ joules which will give approximately the correct value of the loss.

If the thyristor is being turned on at a frequency of 3 kHz, determine the mean power loss due to turn-on losses.

SOLUTION The power curve is calculated as the product of the instantaneous values of voltage and current. The loss curves are plotted in Fig. 1-32 with the given voltage and current variations.

The energy loss is given by the area under the power curve, by (say) counting the number of squares and multiplying by the scaling factors.

For Fig. 1-32a, energy loss $\simeq 9.2$ mJ.

For Fig. 1-32b, energy loss $\simeq 6.5$ mJ.

For case (a), $t_r = 1.48 - 0.44 = 1.04\,\mu s$.

For case (b), $t_r = 2.15 - 0.94 = 1.21\,\mu s$.

For each case, $V_{max} = 600\,V$, $I_{max} = 78\,A$.

Hence, for (a) the formula gives $n = 5.29$ and for (b) gives $n = 8.7$. In general, for the fastest turn-on condition n will be of the order of 4 upwards, which will give the highest peak power loss.

Mean power loss at 3 kHz is $9.2 \times 10^{-3} \times 3 \times 10^3 = 27.6\,W$ for case (a),
$$6.5 \times 10^{-3} \times 3 \times 10^3 = 19.5\,W \text{ for case } (b).$$

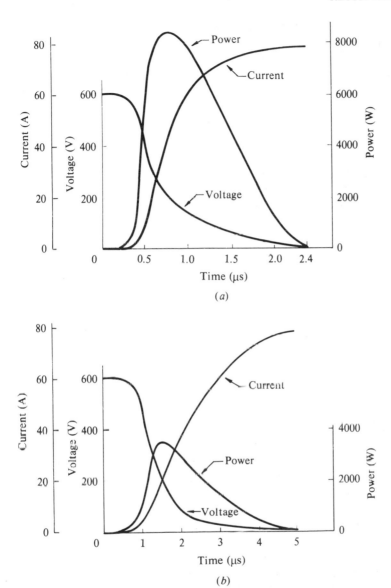

Figure 1-32

Example 1-6

During the turn-off of a thyristor the voltage and current waveforms are as shown in Fig. 1-33. Plot the power loss curve, calculate the energy loss and the reverse recovered charge.

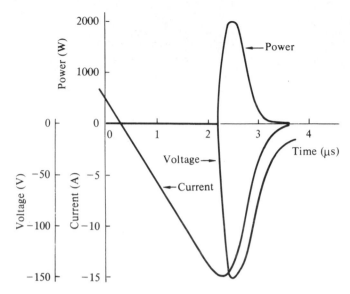

Figure 1-33

SOLUTION From the graph of power loss drawn in Fig. 1-33 from the given data, the energy loss $\simeq$ 1 mJ. Note that this loss is normally very much less than the turn-on loss.

From the graph of current, the recovered charge $= \int i \, dt =$ area under current curve $\simeq 23 \, \mu C$.

Example 1-7

A thyristor with a steady power loss of 30 W has a junction to heatsink thermal resistance of 0.7 °C/W. Determine the maximum value of the thermal resistance the heatsink can have if the ambient temperature is 40 °C and the junction temperature is limited to 125 °C. Give the base temperature at this condition.

SOLUTION Using Eq. (1-1),
the total thermal resistance $= (125 - 40)/30 = 2.83$ °C/W,
and the thermal resistance of the heatsink $= 2.83 - 0.7 = 2.13$ °C/W,
and the temperature at the base $= 40 + (30 \times 2.13) = 104$ °C.

Example 1-8

A thyristor of thermal resistance 1.8 °C/W is mounted on a heatsink of thermal resistance 2.0 °C/W. Calculate the maximum power loss of the thyristor if the junction temperature is not to exceed 125 °C in an ambient of 40 °C.

SOLUTION Using Eq. (1-1), the loss $= (125 - 40)/(2.0 + 1.8) = 22.4$ W.

Example 1-9

A thyristor has a thermal capacity of 0.1 J/°C and a thermal resistance of 0.9 °C/W. Calculate the permissible short-time power loss for the temperature rise not to exceed 40°C over (i) 0.01 s, (ii) 0.1 s, (iii) 1 s.
Suggest values of transient thermal impedances that might be used for these calculations.

SOLUTION Power dissipated for 1 °C rise = 1/(thermal resistance)
$$= 1/0.9$$
$$= 1.1 \text{ W/}^\circ\text{C},$$

and from Eq. (1-6), the thermal time constant $T = \dfrac{0.1}{1.11} = 0.09$ s.

(i) If after 0.01 s the temperature rise is 40 °C, then from Eq. (1-4),

$$40 = \theta_{max}(1 - e^{-0.01/0.09}), \text{ giving } \theta_{max} = 380\,^\circ\text{C}.$$

From Eq. (1-5), the power loss = $380 \times 1.11 = 422$ W.
Similar calculations give

(ii) for 0.1 s, $\theta_{max} = 59.6\,^\circ\text{C}$, with a power loss of 66.2 W,

(iii) for 1 s, $\theta_{max} = 40\,^\circ\text{C}$, with a power loss of 44.4 W.

Note that one second is very much longer than the time constant of 0.09 s, hence a load of one second duration must be treated as a steady-state condition, not an overload condition. The 40°C rise quoted would be relative to the cooling fin, which has a very long time constant and may be considered as being at a fixed temperature during overloads.
Taking these overload power values and using Eq. (1-7), then for

0.01 s value, transient thermal impedance = 40/422 = 0.09 °C/W,

0.1 s value, transient thermal impedance = 40/66.2 = 0.60 °C/W,

1 s value, transient thermal impedance = 40/44.4 = 0.9 °C/W.

These calculations for the transient thermal impedance and overload conditions put quantitative values to the theory developed in Sec. 1-7 but, as explained, conditions are too complex to be based on a simple exponential rise curve.

Example 1-10

In a particular application, during switch-on from cold conditions of 40°C, a thyristor experiences a surge giving a power loss of 2000 W for 10 ms. Calculate the junction temperature if the transient thermal impedance for this time is 0.03 °C/W.

SOLUTION Using Eq. (1-7), the temperature rise after 10 ms is

$$2000 \times 0.03 = 60\,^\circ\text{C}.$$

Junction temperature = ambient + rise = 40 + 60 = 100 °C, which is a safe value.

Example 1-11

A thyristor has with its heatsink a thermal resistance of 0.2 °C/W steady state and a 100 ms value of 0.05 °C/W. What power loss can the thyristor tolerate for 100 ms if the junction temperature is not to exceed 125 °C following a steady power loss of 300 W, the ambient temperature being 30 °C?

SOLUTION Using Eq. (1-3), the steady junction temperature at 300 W loss = 30 + (300 × 0.2) = 90 °C. Hence, during 100 ms overload the junction temperature can rise another 125 − 90 = 35 °C.

Additional power loss $= \dfrac{35}{0.05} = 700\,\text{W}.$

Total overload power = 300 + 700 = 1000 W.

TWO

RECTIFYING CIRCUITS

A rectifier circuit is one which links an a.c. supply to a d.c. load, that is, it converts an alternating voltage supply to a direct voltage. The direct voltage so obtained is not normally level, as from a battery, but contains an alternating ripple component superimposed on the mean (d.c.) level.

The various circuit connections described, although all giving a d.c. output, differ in regard to the a.c. ripple in the output, the mean voltage level, efficiency, and their loading effects on the a.c. supply system.

2-1 CIRCUIT NOMENCLATURE

Rectifying circuits divide broadly into two groups, namely, the half-wave and full-wave connections.

The half-wave circuits are those having a rectifying device in each line of the a.c. supply, all cathodes of the varying devices being connected to a common connection to feed the d.c. load, the return from the load being to the a.c. supply neutral. The expression *half-wave* describes the fact that the current in each a.c. supply line is unidirectional. An alternative to the description *half-wave* is to use the expression *single-way* in describing these circuits.

The full-wave circuits are those which are in effect two half-wave circuits in series, one feeding into the load, the other returning load current directly to the a.c. lines, eliminating the need to employ the a.c. supply neutral. The expression *full-wave* is used because the current in each a.c. supply line, although not necessarily symmetrical, is in fact alternating. The full-wave circuits are more commonly called *bridge circuits,* but alternatively are also known as *double-way circuits.*

The control characteristics of the various circuits may be placed broadly into one of three categories: namely, uncontrolled, fully-controlled, and half-controlled.

The uncontrolled rectifier circuits contain only diodes, giving a d.c. load voltage fixed in magnitude relative to the a.c. supply voltage magnitude.

In the fully-controlled circuits all the rectifying elements are thyristors (or power transistors). In these circuits, by suitable control of the phase angle at which the thyristors are turned on, it is possible to control the mean (d.c.) value of, and to reverse, the d.c. load voltage. The fully-controlled circuit is often described as a *bidirectional converter,* as it permits power flow in either direction between supply and load.

The half-controlled rectifier circuits contain a mixture of thyristors and diodes which prevent a reversal of the load voltage, but do allow adjustment of the direct (mean) voltage level. The half-controlled and uncontrolled (diode only) circuits are often described as *unidirectional converters,* as they permit power flow only from the a.c. supply into the d.c. load.

Pulse-number is a manner of describing the output characteristic of a given circuit, and defines the repetition rate in the direct voltage waveform over one cycle of the a.c. supply. For example, a six-pulse circuit has in its output a ripple of repetition rate six times the input frequency, that is, the fundamental ripple frequency is 300 Hz given a 50 Hz supply.

2-2 COMMUTATING DIODE

Many circuits, particularly those which are half- or uncontrolled, include a diode across the load as shown in Fig. 2-1. This diode is variously described as a free-wheeling, flywheel, or by-pass diode, but is best described as a commutating diode, as its function is to commutate or transfer load current away from the rectifier whenever the load voltage goes into a reverse state.

The commutating diode serves one or both of two functions; one is to prevent reversal of load voltage (except for the small diode volt-drop) and the other to transfer the load current away from the main rectifier, thereby allowing all of its thyristors to regain their blocking state.

2-3 SINGLE-PHASE HALF-WAVE (OR SINGLE-WAY)

Although the uncontrolled single-phase half-wave connection shown in Fig. 2-2*a* is very simple, the waveforms of Figs. 2-2*b* and *c* illustrate fundamentals which will constantly recur in the more complex circuits.

The assumption is made that the magnitude of the supply voltage is such as to make the diode volt-drop negligible when conducting. The waveforms are developed on the assumption that the diode will conduct like a closed switch when its anode voltage is positive with respect to its cathode, and cease to conduct when its current falls to zero, at which time it acts like an open switch. The turn-on and turn-off times of the diode, being only a few microseconds, may be taken as instantaneous times in relation to the half cycle time for a 50 Hz supply.

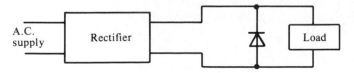

Figure 2-1 Position of commutating diode.

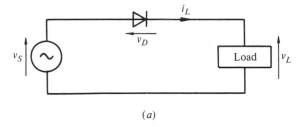

(a)

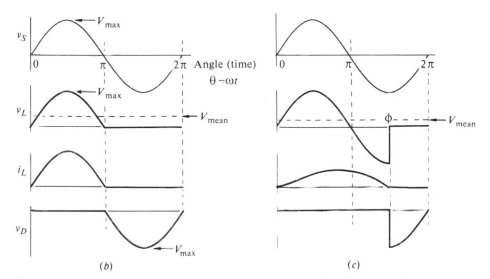

Figure 2-2 Single-phase half-wave circuit. (a) Connection. (b) Waveforms when the load is pure resistance. (c) Waveforms when the load contains some inductance.

The waveforms shown in Fig. 2-2b are for a load of pure resistance, the supply voltage v_S being sinusoidal of peak value V_{max}. Immediately prior to v_S going positive, there is no load current, hence no load voltage, the negative supply voltage appearing across the diode. As v_S goes positive, the diode voltage changes to anode positive relative to the cathode, current flow then being possible. Neglecting the small diode volt-drop, the load current $i_L = v_S/R$ until the current falls to zero at the end of the positive half cycle. As the diode prevents reverse current, the entire supply negative voltage appears across the diode.

The load voltage waveform in Fig. 2-2b has the mean value of a half sinewave, namely,

$$V_{max}/\pi \tag{2-1}$$

which could be calculated from

$$V_{mean} = (1/2\pi)\int_{\theta=0}^{\theta=\pi} V_{max}\sin\theta\,d\theta \tag{2-2}$$

where $\theta = \omega t$ is any angle on the waveform.

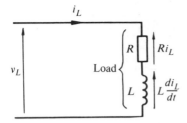

Figure 2-3 Load equivalent circuit.

The majority of d.c. loads (such as d.c. motors) respond to the mean (d.c.) value of the voltage, hence the r.m.s. value of the output voltage is generally of little interest. However, the a.c. ripple content in the direct voltage waveform, that is, the instantaneous variation of the load voltage relative to the mean value, is often a source of unwanted losses and is part of the characteristic of the circuit.

To select a suitable diode for the circuit, both the diode current and voltage waveforms must be studied. The diode voltage v_D shows a peak reverse voltage value of V_{max}.

Almost all d.c. loads contain some inductance; the waveforms shown in Fig. 2-2c are for an inductive load having the equivalent circuit shown in Fig. 2-3. Current flow will commence directly the supply voltage goes positive, but the presence of the inductance will delay the current change, the current still flowing at the end of the half cycle, the diode remains on, and the load sees the negative supply voltage until the current drops to zero.

Reference to Fig. 2-3 shows that the instantaneous load voltage

$$v_L = Ri_L + L\,di_L/dt \tag{2-3}$$

which enables the waveshape of the load current i_L to be determined. A guide to the required current rating of the diode would be given by determining the r.m.s. value of its current waveform.

The mean voltage for Fig. 2-2c is given by

$$V_{mean} = (1/2\pi)\int_{\theta=0}^{\theta-\phi} V_{max}\sin\theta\,d\theta \tag{2-4}$$

and is lower than the case of no inductance.

The single-phase half-wave circuit can be controlled by the use of a thyristor as shown in Fig. 2-4a. The thyristor will only conduct when its voltage v_T is positive and it has received a gate firing pulse i_g. Figures 2-4b and c show the conduction of the thyristor delayed by an angle α beyond that position where a diode would naturally conduct (or commutate); in this case, the firing delay angle α is expressed relative to the supply voltage zero.

Without the commutating diode, the waveforms would be similar to those of Fig. 2-2 with the exception of a delayed start. The waveforms of Fig. 2-4 do, how-

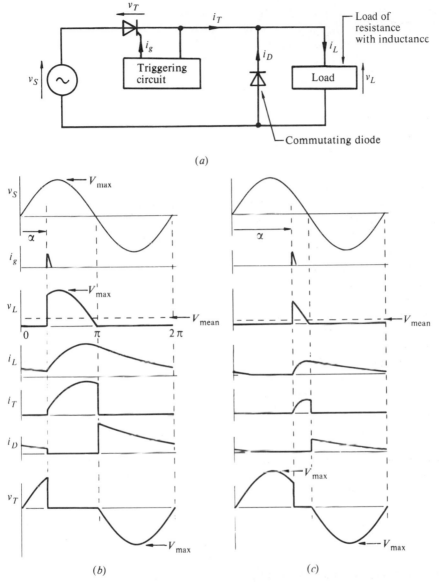

Figure 2-4 Single-phase half-wave controlled circuit with commutating diode. (*a*) Connection. (*b*) Small firing delay angle, and continuous current. (*c*) Large firing delay angle, and discontinuous current.

ever, assume the presence of a commutating diode which prevents the load voltage reversing beyond the diode volt-drop value, resulting in the waveforms shown.

During the thyristor on-period, the current waveform is dictated by Eq. (2-3), but once the voltage reverses, v_L is effectively zero and the load current follows an exponential decay. If the current level decays below the diode holding level, then the load current is discontinuous as shown in Fig. 2-4c. Figure 2-4b shows a continuous load-current condition, where the decaying load current is still flowing when the thyristor is fired in the next cycle. Analysis of the load-voltage waveform gives a mean value of

$$V_{mean} = (1/2\pi) \int_{\alpha}^{\pi} V_{max} \sin \theta \, d\theta = \frac{V_{max}}{2\pi} (1 + \cos \alpha) \qquad (2\text{-}5)$$

Inspection of the waveforms shows clearly that the greater the firing delay angle α, the lower is the mean load voltage, Eq. (2-5) confirming that it falls to zero when $\alpha = 180°$.

The thyristor voltage v_T waveform shows a positive voltage during the delay period, and also that both the peak forward and peak reverse voltages are equal to V_{max} of the supply.

Inspection of the waveforms in Fig. 2-4 clearly shows the two roles of the commutating diode, one to prevent negative load voltage and the other to allow the thyristor to regain its blocking state at the voltage zero by transferring (or commutating) the load current away from the thyristor.

2-4 BI-PHASE HALF-WAVE (OR SINGLE-WAY)

The bi-phase connection of Fig. 2-5a provides two voltages v_1 and v_2 in anti-phase relative to the mid-point neutral N. In this half-wave connection, the load is fed via a thyristor in each supply line, the current being returned to the supply neutral N.

In any simple half-wave connection, only one rectifying device (thyristor or diode) will conduct at any given time, in the diode case this being that one connected to the phase having the highest voltage at that instant. In the controlled circuit, a given thyristor can be fired during any time that its anode voltage is positive relative to the cathode.

With reference to Fig. 2-5, thyristor T_1 can be fired into the on-state at any time after v_1 goes positive. The firing circuits are omitted from the connection diagram to avoid unnecessary confusion to the basic circuit, but can be assumed to produce a firing pulse into the respective thyristor gates as shown in the waveforms. Each pulse is shown delayed by a phase angle α relative to the instant where diodes would conduct, that is, if the thyristors were replaced by diodes, α would be zero.

Once thyristor T_1 is turned on, current builds up in the inductive load, maintaining thyristor T_1 in the on-state into the period when v_1 goes negative. However, once v_1 goes negative, v_2 becomes positive, and the firing of thyristor T_2 immediately turns on thyristor T_2 which takes up the load current, placing a reverse voltage on thyristor T_1, its current being commutated (transferred) to thyristor T_2. The thyristor voltage v_T waveform in Fig. 2-5b shows that it can be fired into conduction at any time when v_T is positive. The peak reverse (and forward) voltage that appears across the thyristor is $2V_{max}$, that is, the maximum value of the com-

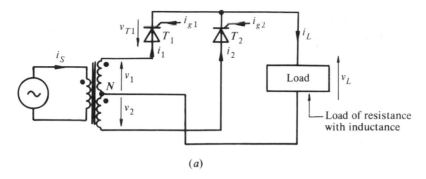

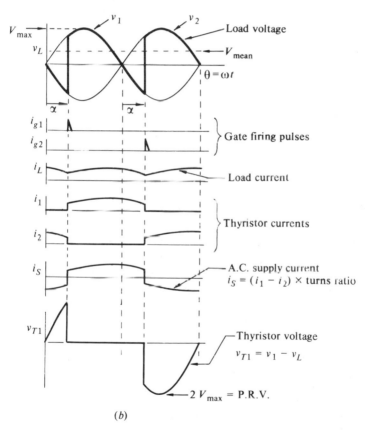

Figure 2-5 Bi-phase half-wave circuit. (*a*) Connection. (*b*) Waveforms.

plete transformer secondary voltage. Figure 2-6 illustrates this fact more clearly in that when thyristor T_2 is on and effectively a short-circuit, the entire transformer voltage appears across the off-state thyristor T_1.

Inspection of the load voltage waveform in Fig. 2-5*b* reveals that it has a mean

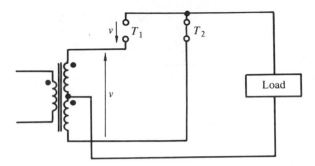

Figure 2-6 Illustrating instantaneous circuit condition.

value of

$$V_{\text{mean}} = (1/\pi)\int_{\alpha}^{\pi+\alpha} V_{\text{max}} \sin \theta \, d\theta = \frac{2V_{\text{max}}}{\pi} \cos \alpha \qquad (2\text{-}6)$$

In practice, the load voltage will be reduced by the volt-drop across one thyristor, as at all times there is one thyristor in series with the supply to the load. Also, this equation assumes high enough load inductance to ensure continuous load current. The highest value of the mean voltage will be when the firing delay angle α is zero, that is, the diode case. When the firing delay angle is $90°$, the load voltage will contain equal positive and negative areas, giving zero output voltage, a value confirmed by Eq. (2-6) which shows the mean voltage follows a cosine variation with firing delay angle. As the load voltage waveform repeats itself twice in the time of one supply cycle, the output has a two-pulse characteristic.

The current waveforms in Fig. 2-5b show a continuous load current, but the ripple will increase as the mean voltage is reduced until, given a light load inductance, the load current will become discontinuous, the load voltage then having zero periods. The thyristor currents are of half-cycle duration and tend to be square-shaped for continuous load current. The a.c. supply current can be seen to be non-sinusoidal and delayed relative to the voltage.

2-5   SINGLE-PHASE BRIDGE (OR DOUBLE-WAY)

The bridge (full-wave or double-way) connection can be arranged to be either uncontrolled, fully-controlled, or half-controlled configurations, and this section will describe each connection in turn.

2-5-1   Uncontrolled

The single-phase bridge circuit connection is shown in Fig. 2-7a in its simplest diagrammatic layout. This layout, whilst almost self-explanatory and widely used in electronic circuit layouts, does not at a glance demonstrate that it is two half-wave

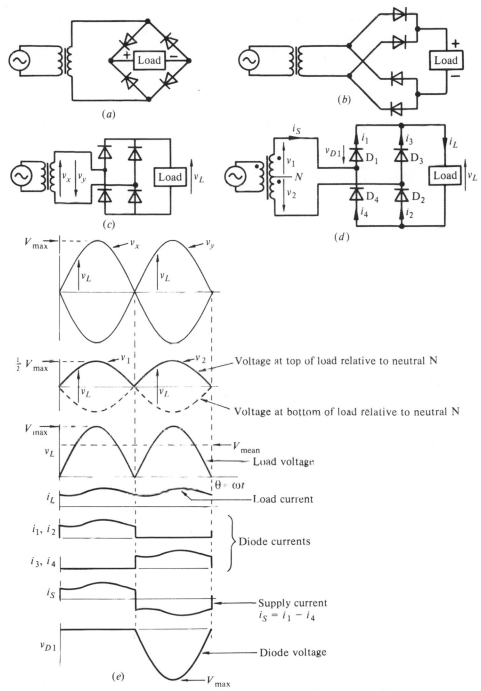

Figure 2-7 Single-phase bridge circuit. (*a*) – (*d*) Connection drawn to differing diagram layouts. (*e*) Waveforms.

circuits in series, nor is it possible to draw a similar layout for the three-phase circuits. The same circuit drawn to a different diagrammatic layout as in Fig. 2-7b shows clearly the concept of two half-wave circuits in series making the full-wave connection, two diodes with common cathodes feeding into the load, two diodes with common anodes returning the load current to the other supply line. However, the layout of Fig. 2-7b is rather cumbersome, and for power applications the layout of Fig. 2-7c is used.

In constructing the voltage waveforms, some circuit reference must be used, and in this respect one can construct the supply waveforms with reference to a midpoint neutral N as shown in Fig. 2-7d, thus enabling a comparison to be made to the half-wave circuit of Fig. 2-5. As only a simple two-winding transformer is required, the mid-point is neither required nor available in practice, and in this respect it is useful to look at Fig. 2-7c where the supply is given two labels v_x and v_y shown in the waveforms of Fig. 2-7e.

The load voltage shown in Fig. 2-7e can be constructed either by taking the waveforms of v_x and v_y when each is positive, or by constructing the voltages on each side of the load relative to the neutral N, the difference between them being the load voltage v_L. The use of the neutral N does demonstrate that the load voltage is the addition of two half-wave circuit voltages in series, making a full-wave connection. The diode voltage v_{D1} has a peak reverse value of the maximum value of the supply voltage, this being only half the value in the half-wave connection of Fig. 2-5 for the same load voltage; however, two diodes are always conducting at any given instant, giving a double volt-drop.

The diode and supply current waveforms shown in Fig. 2-7e are identical in shape to the half-wave connection of Fig. 2-5. The output characteristic is two-pulse, hence as regards the load response and supply requirements the bridge connection is similar to the bi-phase half-wave circuit.

2-5-2 Fully-controlled

The fully controlled circuit shown in Fig. 2-8 has thyristors in place of the diodes of Fig. 2-7. Conduction does not take place until the thyristors are fired and, in order for current to flow, thyristors T_1 and T_2 must be fired together, as must thyristors T_3 and T_4 in the next half cycle. To ensure simultaneous firing, both thyristors T_1 and T_2 are fired from the same firing circuit as shown in Fig. 2-9, the output being via a pulse transformer as the cathodes of the respective thyristors are at differing voltages in the main circuit.

The load voltage is the same as that described for the bi-phase half-wave connection with a mean value as for Eq. (2-6) of

$$V_{\text{mean}} = \frac{2V_{\text{max}}}{\pi} \cos \alpha \qquad (2-7)$$

less in this case by two thyristor volt-drops. This equation will not apply if the load current is not continuous.

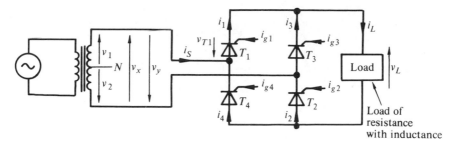

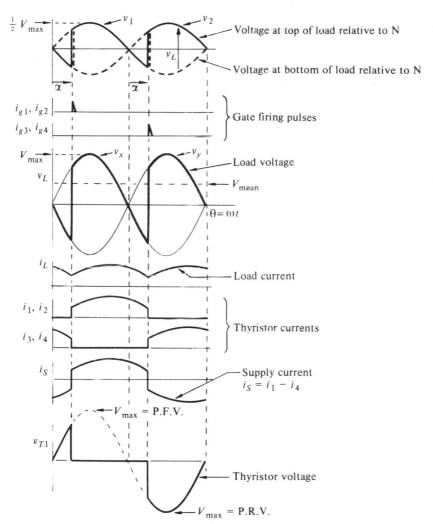

Figure 2-8 Fully-controlled single-phase bridge. (*a*) Connection. (*b*) Waveforms.

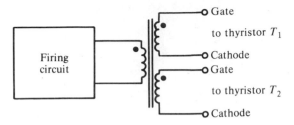

Figure 2-9 Firing circuit output connections.

2-5-3 Half-controlled

It is possible to control the mean d.c. load voltage by using only two thyristors and two diodes, as shown in the half-controlled connection of Fig. 2-10a. Drawing the circuit to a different diagrammatic layout, as shown in Fig. 2-11, shows clearly that the full-wave connection is the addition of two half-wave circuits, the input current to the load being via the thyristors, whilst the diodes provide the return path. The approach to determining the load-voltage waveform in Fig. 2-10b is to follow previous reasoning by plotting the voltage at each end of the load relative to the supply neutral N. The thyristors commutate when fired, the diodes commutating at the supply voltage zeros. The load voltage so constructed never goes negative and follows a shape as if the load were pure resistance, reaching a zero mean value when the firing delay angle α is $180°$.

The presence of the commutating diode obviously prevents a negative load voltage, but this would in any case be the situation, even without the commutating diode. After the supply voltage zero and before (say) thyristor T_3 is fired, thyristor T_1 would continue conducting, but the return load-current path would have been commutated from diode D_2 to diode D_4, hence a free-wheeling path for the load current would be provided via T_1 and D_4, resulting in zero supply current. The commutating diode will provide a preferential parallel path for this free-wheeling load current compared to the series combination of a thyristor and diode, and hence enable the thyristor to turn off and regain its blocking state.

The mean value of the load voltage will be given by

$$V_{\text{mean}} = (1/\pi) \int_{\alpha}^{\pi} V_{\text{max}} \sin \theta \, d\theta = \frac{V_{\text{max}}}{\pi} (1 + \cos \alpha) \qquad (2\text{-}8)$$

different by the various thyristor and diode volt-drops.

The current duration in the thyristors and main diodes is less than $180°$ by the firing delay angle α, leading to an a.c. supply current which has zero periods. The commutating diode conducts the decaying load current during the zero voltage periods.

Compared to the fully-controlled circuit, the half-controlled circuit is cheaper, but the a.c. supply current is more distorted due to its zero periods. The half-controlled connection cannot be used in the inversion mode described in Sec. 3-3, only the fully-controlled connection allows a reversal of the mean direct voltage.

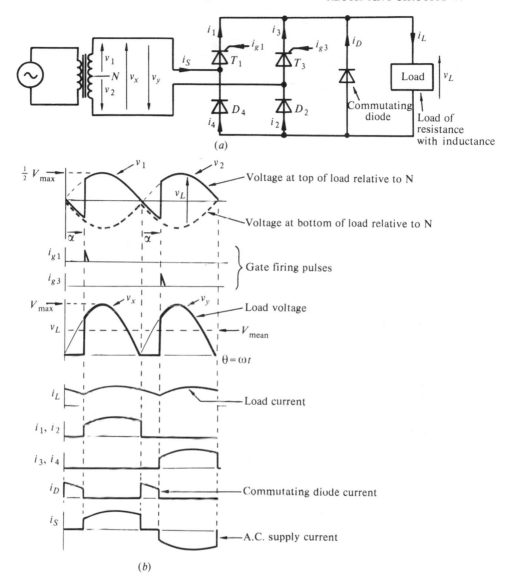

Figure 2-10 Half-controlled single-phase bridge. (*a*) Connection. (*b*) Waveforms.

2-6 THREE-PHASE HALF-WAVE (OR SINGLE-WAY)

The three-phase half-wave connection is the basic element in most of the polyphase rectifier circuits, although its use in its own right is limited, due in part to it requiring a supply transformer having an interconnected-star (zig-zag) secondary. However, to ease the explanation, the supply will initially be assumed as a simple star-connected winding.

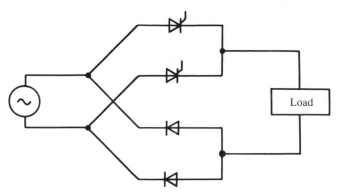

Figure 2-11 Half-controlled full-wave (bridge) connection.

With the polyphase connections, the time intervals between the repetitions in the d.c. load waveforms are shorter than for single-phase connections, and also in practice they will be supplying larger loads having heavier inductance. The net result is for the ripple content of the load current to be less, and it is reasonable to assume the current to be continuous and level. Hence, in developing the current waveforms for the polyphase circuits, it will be assumed that the load current is continuous and level, that is, it has negligible ripple.

The connection of the three-phase half-wave circuit is shown in Fig. 2-12a, each supply phase being connected to the load via a diode and, as in all half-wave connections, the load current being returned to the supply neutral.

The circuit functions in a manner such that only one diode is conducting at any given instant, that one which is connected to the phase having the highest instantaneous value. This results in the load voltage v_L having the waveform shown in Fig. 2-12b, which is the top of the successive phase voltages. While v_1 is the most positive phase, diode D_1 conducts but, directly v_2 becomes more positive than v_1, the load current commutates (transfers) from diode D_1 to diode D_2. Confirmation of the instant of commutation can be seen by examining the diode voltage wave-form v_D, which goes negative directly v_1 has an instantaneous value below v_2, hence diode D_1 turns off.

The instantaneous d.c. load voltage varies between the maximum value of the phase voltage and half this value, and it also repeats itself three times per cycle, thus having a three-pulse characteristic. Comparison of the load voltage in Fig. 2-12b to the two-pulse load voltage of Fig. 2-5 or 2.7 shows the three-pulse connection has a much smaller ripple.

The mean value of the load voltage is given by

$$V_{\text{mean}} = \frac{1}{2\pi/3} \int_{\pi/6}^{5\pi/6} V_{\text{max}} \sin \theta \, d\theta = \frac{3\sqrt{3}}{2\pi} V_{\text{max}} \tag{2-9}$$

less by the single diode volt-drop.

Assuming level d.c. load current I_L, the diode currents shown in Fig. 2-12b

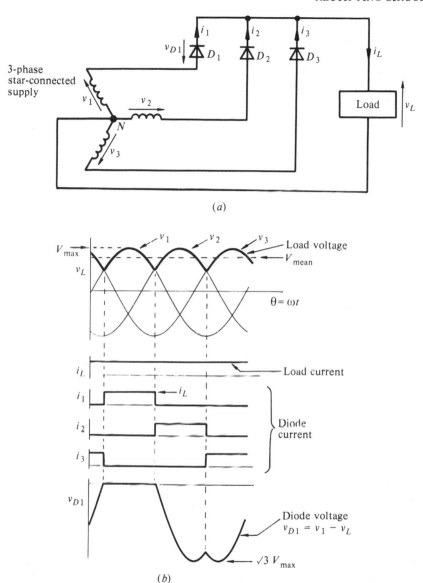

Figure 2-12 Three-phase half-wave circuit using diodes. (*a*) Connection. (*b*) Waveforms.

are each blocks one-third of a cycle in duration. Using the r.m.s. value of the diode current for the required rating purposes for each diode,

$$I_{\mathrm{rms}} = I_L/\sqrt{3} \qquad (2\text{-}10)$$

an expression which can be calculated by using calculus, or more simply by taking the square root of the mean of the sum of the (current)^2 over three equal intervals

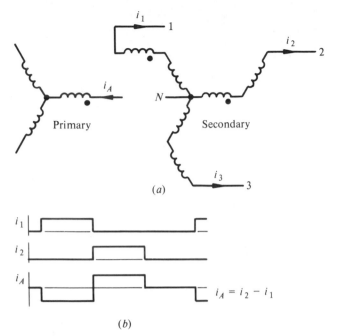

Figure 2-13 Interconnected-star connection of secondary. (*a*) Transformer circuit. (*b*) Current waveform.

in the cycle, that is,

$$I_{\text{rms}} = \left(\frac{I_L^2 + 0^2 + 0^2}{3}\right)^{1/2} = I_L/\sqrt{3}$$

Examination of the diode voltage waveform shows the peak reverse voltage to be $\sqrt{3}V_{\text{max}}$, which is the maximum value of the voltage between any two phases, that is, the maximum value of the line voltage.

It was stated earlier that the simple star connection of the supply was not appropriate, the reason being that the unidirectional current in each phase will lead to possible d.c. magnetization of the transformer core. To avoid this problem the interconnected-star (sometimes called zig-zag) winding shown in Fig. 2-13*a* is used as the secondary of the supply transformer. The current which is reflected into the primary is now a.c. as shown in Fig. 2-13*b*, being as much positive as negative, hence avoiding any d.c. component in the core m.m.f.

When the diodes of Fig. 2-12*a* are replaced by thyristors as shown in Fig. 2.14*a*, the circuit becomes fully controllable, with the mean load voltage being adjustable by control of the firing delay angle α. Again the firing circuits are not shown, but it may be assumed that each thyristor has a firing circuit connected to its gate and cathode, producing a firing pulse relative in position to its own phase voltage. A master control will ensure that the three gate pulses are displaced by 120° relative to each other, giving the same firing delay angle to each thyristor.

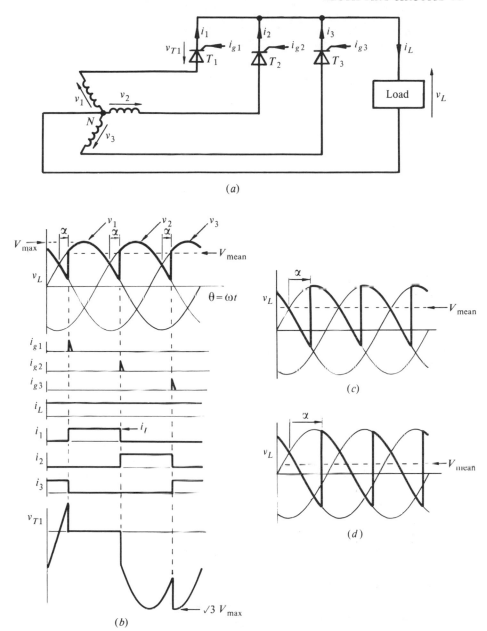

Figure 2-14 Three-phase half-wave controlled circuit using thyristors. (*a*) Connection. (*b*) Waveforms with samll firing delay angle. (*c*) and (*d*) Load voltage waveform with large firing delay angle.

The firing delay angle α is defined such that it is zero when the output mean voltage is a maximum, that is, the diode case. Hence, the firing delay angle α shown in Fig. 2-14 is defined relative to the instant when the supply phase voltages cross and diodes would commutate naturally, not the supply voltage zero.

Reference to Fig. 2-14b shows that the thyristors will not take up conduction until turned on by the gate pulse, thereby allowing the previous phase voltage to continue at the load, so giving an overall lower mean load voltage. The ripple content of the load voltage is increased, but it still has the three-pulse characteristic. The current waveform shapes have not changed to, but are delayed by, the angle α relative to the diode case. The thyristor voltage v_T shows that the thyristor anode voltage is positive relative to the cathode after the position of zero firing delay angle.

The load voltage waveforms of Fig. 2-14c and d show the effect of a larger delay angle, the voltage having instantaneous negative periods after the firing delay angle $\alpha = 30°$. The mean load voltage is given by

$$V_{\text{mean}} = \frac{1}{2\pi/3} \int_{\frac{\pi}{6}+\alpha}^{\frac{5\pi}{6}+\alpha} V_{\text{max}} \sin\theta \, d\theta = \frac{3\sqrt{3}}{2\pi} V_{\text{max}} \cos\alpha \qquad (2\text{-}11)$$

less by the single thyristor volt-drop.

As in the two-pulse Eq. (2-6), the mean voltage is proportional to the cosine of the firing delay angle α, being zero at 90° delay. The assumption of level continuous load current will be less valid as the mean voltage approaches zero, due to the greatly increased ripple content of the load voltage.

2-7 SIX-PHASE HALF-WAVE (OR SINGLE-WAY)

The connections, with waveforms, of the six-phase half-wave circuit using a simple star supply are shown in Fig. 2-15. The theory of the connection is an extension of the three-phase half-wave circuit, each thyristor conducting for one sixth cycle.

The load-voltage waveform is the top of the six-phase voltages for the diode case, delayed by the firing delay angle α as shown in Fig. 2-15b when thyristors are used. The load-voltage waveform is of six-pulse characteristic, having a small ripple in the diode case at a frequency of six times the supply frequency. Including firing delay, the mean value of the load voltage is

$$V_{\text{mean}} = \frac{1}{2\pi/6} \int_{\frac{\pi}{3}+\alpha}^{\frac{2\pi}{3}+\alpha} V_{\text{max}} \sin\theta \, d\theta = \frac{3}{\pi} V_{\text{max}} \cos\alpha \qquad (2\text{-}12)$$

The thyristor voltage waveform of Fig. 2-15b shows the peak reverse (and forward) voltage to be twice the maximum value of the phase voltage. The diode is inefficiently used as it conducts for only one-sixth of the cycle, giving an r.m.s. value of

$$I_{\text{rms}} = I_L/\sqrt{6} \qquad (2\text{-}13)$$

for a level load current I_L.

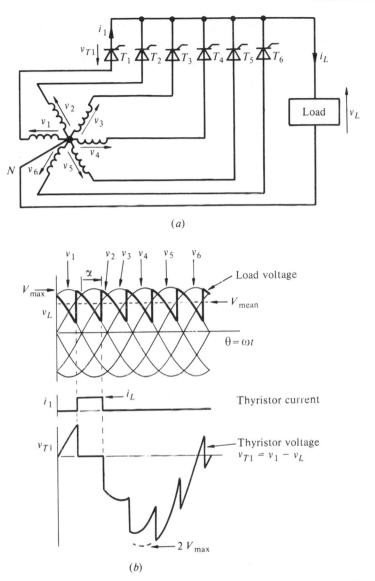

Figure 2-15 Simple six-phase half-wave circuit. (*a*) Connection. (*b*) Waveforms.

The simple star connection of Fig. 2-15*a* is not used in practice as the currents reflected into the primary winding have a large third-harmonic component. To eliminate the third-harmonic component, the fork connection shown in Fig. 2-16 can be used, but more frequently the double-star connection shown in Fig. 2-17 is used.

The double-star connection is essentially two independent three-phase half-wave circuits operating in parallel to give a six-pulse output. Fig. 2-17*a* shows the

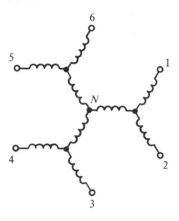

Figure 2-16 The six-phase fork connection.

two star groups supplied at 180° to each other, that is, if the two star points were rigidly connected, we would have a simple six-phase system. The two star points are linked via an interphase transformer, which is best considered as a reactor rather than a transformer. The load current is returned to the centre of the reactor.

Reference to the load-voltage waveform in Fig. 2-17b shows the two three-pulse waveforms of each star group relative to its own star point. The reactor allows each star group to conduct at the same time by taking up the voltage difference between the two star points, the load voltage then being midway between the two three-pulse groups. The load voltage has a six-pulse characteristic with a maximum instantaneous value of $(\sqrt{3}/2)V_{\max}$ occurring where the phase voltages cross.

For the diode case shown in Fig. 2-17b, the load voltage can be calculated by finding the mean value of either three-pulse group, or directly from the actual six-pulse load-voltage waveform, giving

$$V_{\mathrm{mean}} = \frac{3\sqrt{3}}{2\pi} V_{\max} \tag{2-14}$$

less by one diode volt-drop as the two groups are in parallel.

Because the two groups act independently, each diode conducts for one third of each cycle, hence at any instant one diode in each group is conducting, each carrying one half of the load current. The current waveforms in Fig. 2-17b are developed to show that in a delta-connected primary a stepped current waveform is drawn from the three-phase a.c. supply. Compared to the simple six-pulse connection of Fig. 2-15, the current utilization of the diode and the input a.c. waveform are both much superior.

The reactor voltage v_R waveform shown in Fig. 2-17b is the difference between the two star groups, having an approximately triangular shape with a maximum value of one half that of the phases, and is at a frequency of three times that of the supply. In order for a voltage to be developed across the reactor, there must be a changing magnetic flux, which can only be developed with a magnetizing current.

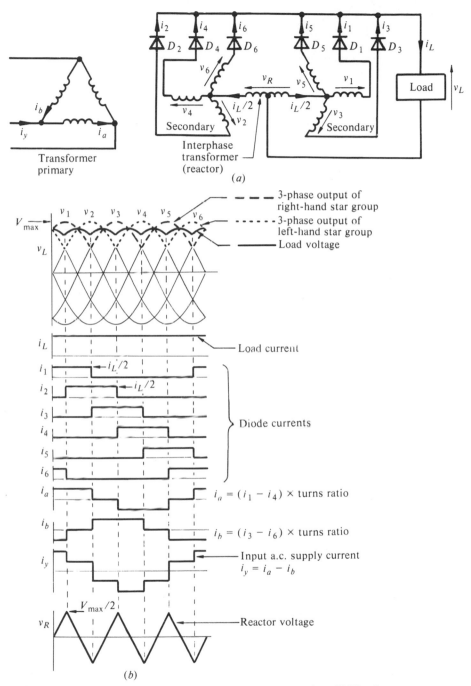

Figure 2-17 Double-star six-phase half-wave circuit. (*a*) Connection. (*b*) Waveforms.

To star point

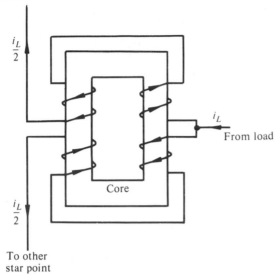

Core

To other
star point

Figure 2-18 Construction of interphase transformer (reactor).

The path for the magnetizing current must be through the diodes, which is only possible when there is load current flowing — the magnetizing-current path being through the reverse diode as a component slightly reducing the diode forward current. The level of the load current must exceed the level of the magnetizing current for it to exist. If the load is disconnected, no magnetizing current can flow; hence no voltage can be developed across the reactor, so the star points are electrically common, and the circuit acts as a simple six-phase half-wave connection. To guarantee correct functioning of the circuit under all load conditions, a small permanent load in excess of the magnetizing current must be connected across the rectifier.

A typical construction of the interphase transformer (reactor) is shown in Fig. 2-18, which shows a two-legged core with two closely coupled windings on each leg. The close coupling of the windings will ensure an m.m.f. balance as in a transformer, forcing the load current to divide equally between the windings. The magnetizing current flowing from one star point to the other star point will act in the same direction in all coils to satisfy the flux requirement. As in the normal transformer, the magnetizing current represents the slight unbalance between the total current in the two windings on the same leg.

Each diode will have a peak reverse voltage requirement of $2V_{\mathrm{max}}$, as they will be required to withstand this voltage if the interphase transformer fails to excite, the circuit then acting as the simple six-phase half-wave connection shown in Fig. 2-15.

Replacing the diodes shown in Fig. 2-17a by thyristors converts the double-star connection into a fully-controlled circuit. The load-voltage waveform with a

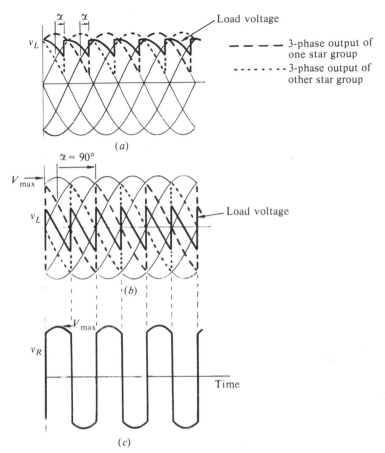

Figure 2-19 Voltage waveforms with controlled double-star connection. (*a*) Load voltage with small firing delay angle α. (*b*) Load voltage waveform with α = 90°, V_{mean} = zero. (*c*) Interphase transformer (reactor) voltage at α = 90°.

small firing delay angle α is shown in Fig. 2-19*a*, developed in the same manner as the diode case with the six-pulse load-voltage waveform midway between the two three-pulse groups. The mean voltage is, as in the earlier circuits, proportional to cos α when the load current is continuous.

When the firing delay angle is 90°, the mean voltage is zero, and the load-voltage waveform is as shown in Fig. 2-19*b*. At this condition of zero mean load voltage, the interphase transformer (reactor) voltage is approximately rectangular as shown in Fig. 2-19*c*. The flux change in the reactor is proportional to the area of the voltage-time curve (from $v = d\phi/dt$ giving $\delta\phi = \int v \, dt$). Comparison of the rectangular waveform of Fig. 2-19*c* to the triangular waveform in the diode case shows an area three times larger; hence, with a flux change three times greater, the interphase transformer will be physically three times larger in the fully-controlled circuit as compared to the diode circuit.

2-8 THREE-PHASE BRIDGE (OR DOUBLE-WAY)

The three-phase bridge connection is most readily seen as a full-wave or double-way connection by reference to the circuit layout shown in Fig. 2-20. The load is fed via a three-phase half-wave connection, the return current path being via another half-wave connection to one of the three supply lines, no neutral being required. However, the circuit connection layout is more usually drawn as shown in Fig. 2-21a.

The derivation of the load-voltage waveform for the all-diode connection of Fig. 2-21 can be made in two ways. Firstly, one can consider the load voltage to be the addition of the two three-phase half-wave voltages, relative to the supply neutral N, appearing at the positive and negative sides of the load respectively. As the voltage waveforms of Fig. 2-21b show, the resultant load voltage is six-pulse in characteristic, having as its maximum instantaneous value that of the line voltage. An alternative approach to deriving the load-voltage waveform is to consider that the two diodes which are conducting are those connected to the two lines with the highest voltage between them at that instant. This means that when v_a is the most positive phase diode D_1 conducts, and during this period first v_b is the most negative with diode D_6 conducting, until v_c becomes more negative when the current in diode D_6 commutates to diode D_2. The load voltage follows in turn six sinusoidal voltages during one cycle, these being $v_a - v_b$, $v_a - v_c$, $v_b - v_c$, $v_b - v_a$, $v_c - v_a$, $v_c - v_b$, all having the maximum value of the line voltage, that is, $\sqrt{3}$ times the phase voltage. Although the supply is shown as star-connected in Fig. 2-21, a delta connection can equally well be used.

The mean value of the load voltage can either be calculated from the sum of the two three-pulse waveforms which, using Eq. (2-9), gives

$$V_{\text{mean}} = 2 \times \frac{3\sqrt{3}}{2\pi} V_{\text{ph(max)}} = \frac{3}{\pi} V_{\text{line(max)}} \tag{2-15}$$

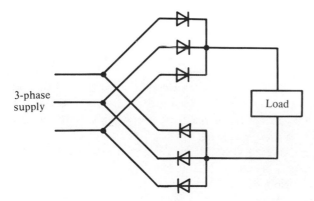

Figure 2-20 Three-phase full-wave circuit.

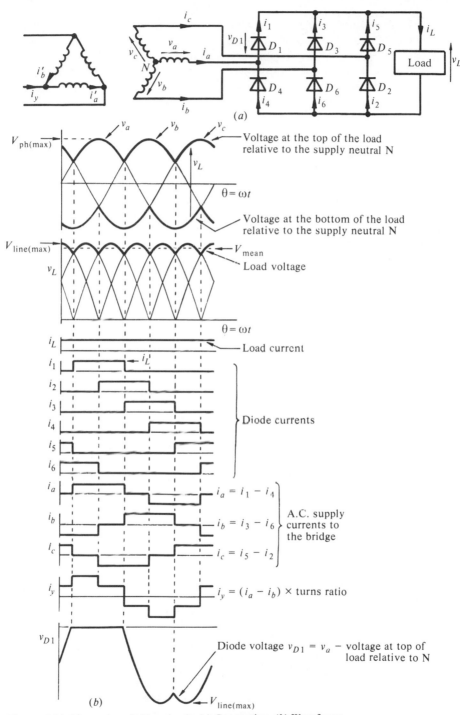

Figure 2-21 Three-phase bridge circuit. (*a*) Connection. (*b*) Waveforms.

or it can be calculated directly from the six-pulse load-voltage waveform which from Eq. (2-12) yields the same as Eq. (2-15) above. As two diodes are in series with the load, the mean value is reduced by two diode volt-drops.

The diode current waveforms shown in Fig. 2-21b reveal that each diode conducts the full-load current for one third of a cycle, the order of commutation determining the numbering of the diodes in the circuit. The diode voltage v_{D1} waveform can be determined as the difference between the phase voltage v_a and the voltage at the top of the load relative to the supply neutral N. The peak reverse voltage appearing across the diode is the maximum value of the line voltage.

Figure 2-21b shows the a.c. supply current to be symmetrical, but of a quasi-square shape. However, the current waveforms are closer to a sinusoidal shape than those in the single-phase bridge connection.

The three-phase bridge can be made into a fully-controlled connection by making all six rectifying elements thyristors, as shown in Fig. 2-22a. As in previous circuits, the mean load voltage is now controllable by delaying the commutation of the thyristors by the firing delay angle α.

With a small firing delay angle as shown in Fig. 2-22b, the waveshapes can be readily understood by reference to earlier circuits. The two three-pulse waveforms add to give the six-pulse load-voltage waveform. The current-waveform shapes are similar to the diode case, except they are delayed by the angle α.

A problem does arise with the bridge circuit that was not present in the earlier circuits, and that is the one of starting. When connected to the a.c. supply, firing gate pulses will be delivered to the thyristors in the correct sequence but, if only a single firing gate pulse is used, no current will flow, as the other thyristor in the current path will be in the off-state. Hence, in order to start the circuit functioning, two thyristors must be fired at the same time in order to commence current flow. With reference to Fig. 2-22b (say), the supply is connected when v_a is at its peak value, the next firing pulse will be to thyristor T_2. However, thyristor T_2 will not conduct unless at the same time thyristor T_1 is pulsed, as reference to the waveforms shows these are the two thyristors conducting at that instant. Hence, for starting purposes, the firing circuit must produce a firing pulse 60° after its first pulse. Once the circuit is running normally, the second pulse will have no effect, as the thyristor will already be in the on-state.

The starting pulse can be fed to the thyristor by each firing circuit having two isolated outputs, one to its own thyristor and the other to the previous thyristor. Alternatively, the firing circuits can be electronically linked so that, when each firing circuit initiates a pulse to its own thyristor, it also does likewise to the previous firing circuit.

When the firing delay is large, with the load voltage having negative periods, it is difficult to visualize the load-voltage waveform from the two three-pulse pictures; hence, as shown in Fig. 2-22c, the six line voltages $v_a - v_b$, $v_a - v_c$, $v_b - v_c$, $v_b - v_a$, $v_c - v_a$, $v_c - v_b$ give a direct picture of the load-voltage waveform and clearly show that zero mean voltage is reached when the firing delay angle is 90°.

The value of the mean load voltage is given by

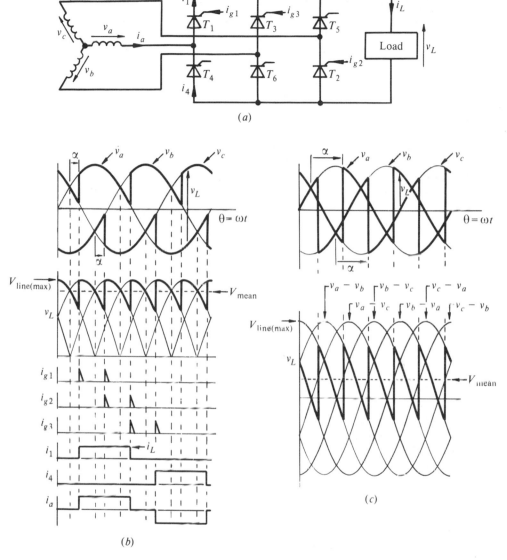

Figure 2-22 Fully-controlled three-phase bridge. (*a*) Connection. (*b*) Waveforms with small firing delay angle. (*c*) Voltage waveforms with large firing delay angle.

$$V_{\text{mean}} = \frac{3}{\pi} V_{\text{line(max)}} \cos \alpha \qquad (2\text{-}16)$$

less by the two thyristor volt-drops.

Reference to Fig. 2-20 would indicate quite correctly that control of the load voltage is possible if three thyristors were used in the half-wave connection feeding

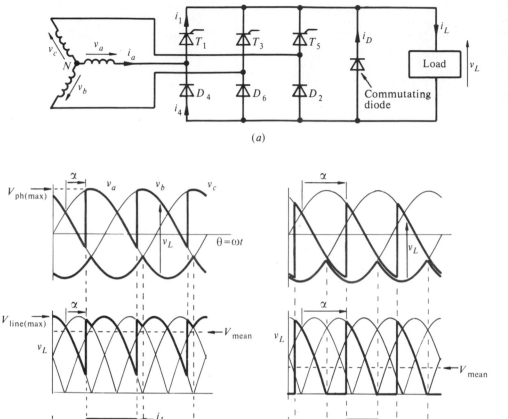

Figure 2-23 Half-controlled three-phase bridge. (*a*) Connection. (*b*) Waveforms with small firing delay angle. (*c*) Waveforms with larger firing delay angle.

the load, and diodes only used to return the current to the supply. Such a connection is shown in Fig. 2-23*a*, with the addition of a commutating diode whose function is similar to that in the single-phase half-controlled bridge.

The action of the circuit is most clearly explained by the two three-pulse voltage waveforms shown in Fig. 2-23*b*, where the upper waveform shows a small firing delay, whereas the lower waveform is that of the diode case. The addition of these two waveforms gives the load voltage v_L, with only three notches of voltage removed per cycle, not the six notches of the fully-controlled circuit. The wave-form is now three-pulse, having a higher harmonic ripple component than the fully-controlled connection.

The current waveforms of Fig. 2-23b show that the thyristor current i_1 is delayed, but i_4 remains in phase with its voltage, resulting in an unsymmetrical a.c. supply-line current which will contain even harmonics.

Figure 2-23c shows a firing delay angle above 90°, making the upper waveform more negative than positive [to get to (c) from (b) think of the vertical line at the firing position moving to the right]. The load voltage now has periods of zero volt-age, the commutating diode taking the freewheeling load current in preference to the series arm (of a thyristor plus diode) in a like manner to that described for the single-phase half-controlled bridge.

Inspection of the load-voltage waveforms in Fig. 2-23 shows that zero mean load voltage is reached when the firing delay angle α reaches 180°. The mean volt-age can be considered as the addition of the two half-wave three-pulse voltages giving from Eqs. (2-9) and (2-11)

$$V_{\text{mean}} = \frac{3\sqrt{3}}{2\pi} V_{\text{ph(max)}} (1 + \cos \alpha) = \frac{3}{2\pi} V_{\text{line(max)}} (1 + \cos \alpha) \qquad (2\text{-}17)$$

Compared to the fully-controlled circuit, the half-controlled circuit is cheaper, has no starting problems, but has a higher harmonic content in its load-voltage and supply-current waveforms.

2-9 TWELVE-PULSE CIRCUITS

Figure 2-24 illustrates the twelve-pulse voltage waveform for an uncontrolled, that is, diode, connection, where it is clearly close to a smooth direct voltage. The as-sociated current shown is typical of the waveshape of the current drawn from a three-phase a.c. supply, this being closer to sinusoidal form than in the lower-pulse circuits. It is possible to conclude that the higher the pulse number of a rectifier, the closer it comes to the ideal of giving a level direct voltage and drawing a sinu-soidal current from the a.c. supply.

Three of the most common connections which give a twelve-pulse characteristic are shown in Fig. 2-25. The half-wave connection of Fig. 2-25a is an extension of the double-star circuit described in Sec. 2-7. Here four star groups are displaced to give twelve phases 30° apart, linked via interphase transformers (reactors) to the load. Four diodes conduct simultaneously with only one diode volt-drop reducing the mean load voltage.

The full-wave connections involve linking two three-phase bridges as shown in Figs. 2-25b and c. The a.c. supply is from a transformer having two secondaries: one star-connected, the other delta-connected. In this manner the three-phase volt-ages supplying the two bridges are displaced by a phase angle of 30°, hence the two six-pulse outputs are symmetrically displaced to give an overall twelve-pulse output.

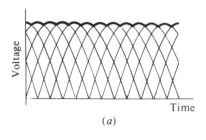

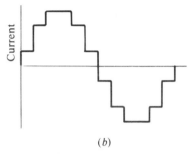

Figure 2-24 Twelve-phase waveforms. (*a*) Voltage of uncontrolled circuit. (*b*) Typical a.c. input current.

The series connection of Fig. 2-25*b* is for loads demanding a high voltage as the individual bridge outputs add, but the diode ratings relate to the individual bridge. The series connection also provides access to a centre point for earthing purposes. The two bridges may be joined in parallel as shown in Fig. 2-25*c*.

Higher pulse-number circuits may be built using the basic three-phase building blocks as for the twelve-pulse connections.

The circuits of Fig. 2-25 may be fully controlled by using thyristors, or partly controlled using a combination of thyristors and diodes.

2-10 TRANSFORMER RATING

It is evident from the rectifier circuits described that the supply transformers carry currents which are non-sinusoidal, and the secondary is sometimes a connection of windings from different legs of the transformer core. The rating (or size) of the transformer must take these factors into account.

The winding rating of a transformer is the sum of the products of the number of windings, times their r.m.s. voltage, times the winding r.m.s. current.

The primary rating may differ from the secondary, particularly in the half-wave circuits due to the better current waveform and the absence of phases composed of windings linked from different legs. The fork connection of Fig. 2-16 demonstrates a connection where the secondary winding has a higher rating than the primary.

In those transformers where there are two or more secondary windings linked

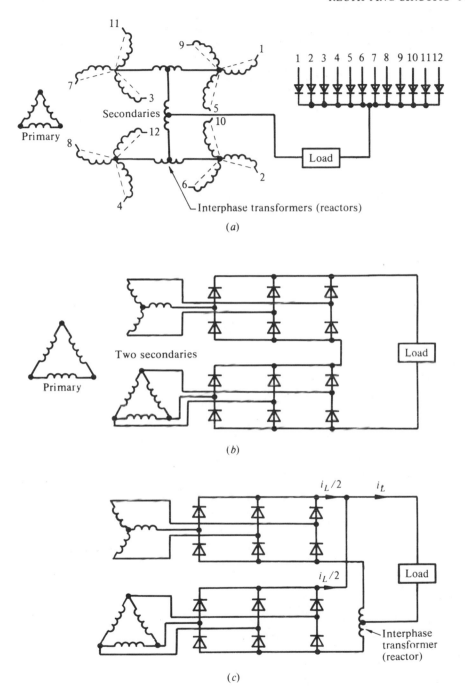

Figure 2-25 Typical twelve-pulse connections. (*a*) Half-wave (single-way). (*b*) Bridge, series connection. (*c*) Bridge, parallel connection.

to a single primary winding, such as in the bi-phase, interconnected-star, or double-star secondaries, the winding design must ensure that the mean distance between windings is the same. The secondaries are sectionalized and interlinked to give the same space and hence the same leakage flux between the primary and each secondary winding. Each secondary winding must extend for the same length as the primary, so that there is an m.m.f. balance, otherwise there would be excessive mechanical stresses.

2-11 SUMMARY

In this chapter several rectifying circuits have been described, so that in a given application one can be aware of the comparisons between the various circuits in order to make the right choice.

A low-voltage load, say 20 V, will impose no severe voltage stresses on the diode (or thyristor) voltage ratings but, at this low voltage, the difference between the one-diode volt-drop of the half-wave circuit and the double volt-drop of the full-wave bridge circuits is significant, suggesting a lower loss with the half-wave circuits.

A high-voltage load, say 2 kV, will indicate that the choice should be a bridge circuit, as the diode (or thyristor) voltage ratings would be excessive in a half-wave circuit. The double volt-drop in the bridge circuit would be insignificant with high-voltage loads.

In the medium-voltage range, the more complex transformer design would possibly rule out the half-wave circuits on cost considerations.

Single-phase circuits are limited to lower power applications, say 15 kW, because there is a limit to the distorted current which can be drawn from the supply, in addition to the usual reasons for using the three-phase supply for heavier loads.

Where applications require a reversal of the mean load voltage, the fully-controlled connection must be used. The half-controlled connections are cheaper where no load-voltage reversal is required, but the greater distortion in the voltage and current waveforms leads to technical limitations to their use.

In Chapter 7 it is shown that, in order to reduce the harmonic content of the waveforms, it is necessary to adopt the higher pulse-number connections. Restrictions imposed by the electricity supply authorities on the harmonic current which can be drawn by a load enforces the use of a higher pulse-number circuit to supply heavy loads. Where an application requires a very smooth direct voltage, the use of a high pulse-number connection may be the most economical solution.

2-12 WORKED EXAMPLES

Example 2-1

A circuit is connected as shown in Fig. 2-2 to a 240 V 50 Hz supply. Neglecting the diode volt-drop, determine the current waveform, the mean load voltage, and the mean load current for a load of (i) a pure resistor of 10 Ω, (ii) an inductance of 0.1 H in series with a 10 Ω resistor.

SOLUTION The supply voltage quoted is an r.m.s. value relating to a sinewave, hence referring to Fig. 2-2b,

$$V_{max} = 240\sqrt{2} = 339.4 \text{ V}$$

(i) For a load of pure resistance R, the load current will be a half sinewave of maximum value

$$I_{max} = V_{max}/R = 339.4/10 = 33.94 \text{ A}.$$

From Eq. (2-2), $V_{mean} = 339.4/\pi = 108 \text{ V}$

$$I_{mean} = V_{mean}/R = 108/10 = 10.8 \text{ A}.$$

(ii) The current waveform may be determined by considering the circuit to be a series combination of $R = 10 \Omega$, $L = 0.1$ H, being switched to an a.c. supply at its voltage zero, current ceasing when it falls to zero.
The equation to determine the current i is

$$V_{max} \sin \omega t = L \, di/dt + Ri, \text{ with } i = 0 \text{ at } t = 0,$$

$$339.4 \sin 2\pi50t = 0.1 \, di/dt + 10i.$$

Using the Laplace transform method of solution where $\bar{i}(s)$ is the transform of $i(t)$, then

$$339.4 \times \frac{2\pi50}{s^2 + (2\pi50)^2} = 0.1(s\bar{i} - i_0) + 10\bar{i}, \text{ where } i_0 = 0,$$

$$\bar{i} = \frac{1\,066\,000}{[s^2 + (2\pi50)^2](s + 100)}$$

Transforming this equation yields

$$i = 9.81 \, e^{-100t} + 10.29 \sin(2\pi50t - 1.262) \text{ A}$$

Alternatively the current expression may be determined by considering the current to be composed of the steady-state a.c. value plus a decaying d.c. transient of initial value such as to satisfy the initial condition of zero current.
The a.c. impedance is 10 Ω resistance in series with a reactance of $2\pi50 \times 0.1 = 31.4 \Omega$ giving an impedance of 32.97 Ω. The a.c. component of the current is therefore $339.4/32.97 = 10.29$ A peak, lagging the voltage by arctan $(31.4/10) = 72.3° = 1.262$ rad.
The a.c. component of the current is $10.29 \sin(2\pi50t - 1.262)$ A, which has a value of -9.81 A at $t = 0$.
The time constant of the circuit is $0.1/10 = 1/100$ second, hence the d.c. component of the current is $9.81 \, e^{-100t}$ A.

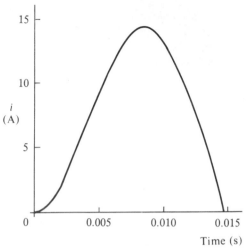

Figure 2-26

The total current equation is

$$i = 9.81 \, e^{-100t} + 10.29 \sin(2\pi 50t - 1.262) \text{ A}$$

A sketch of the current waveform is shown in Fig. 2-26, current ceasing when it attempts to reverse.

The current will cease when $i = 0$ yielding a time $t = 0.01472$ s which gives an angular duration of current of 265°.

Hence
$$V_{\text{mean}} = \frac{1}{2\pi} \int_{0°}^{265°} 339.4 \sin\theta \, d\theta = 58.8 \text{ V}$$

The mean current can be calculated with the aid of calculus, but it is easier to find it by dividing the mean voltage by the d.c. impedance, that is, the resistance:

$$I_{\text{mean}} = 58.8/10 = 5.88 \text{ A}.$$

Example 2-2

If the diode of Fig. 2-2 is replaced by a thyristor, determine the mean load voltage and current if the load is 10 Ω in series with an inductor of 0.1 H, and the firing of the thyristor is delayed by 90°. The a.c. supply is 240 V, 50 Hz, and the thyristor volt-drop is to be neglected.

SOLUTION This problem is similar to that of Example 2-1(ii) except for the delayed start, the waveshapes being as shown in Fig. 2-27.

Assuming that time $t = 0$ at the instant of firing, then the steady-state a.c. component of current is $10.29 \sin(2\pi 50t - 1.262 + 1.571)$ A, which has a value of 3.12 A at $t = 0$.

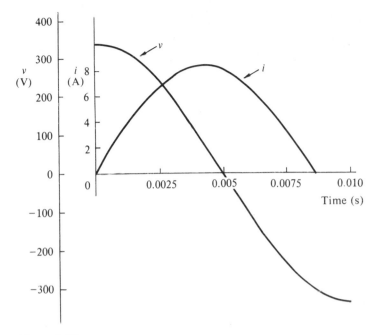

Figure 2-27

Hence the equation to the current is

$$i = 10.29 \sin(2\pi 50t + 0.309) - 3.12\, e^{-100t} \text{ A.}$$

The current will cease when $i = 0$ which occurs at $t = 0.0086$ s which is equivalent to 155°.

$$V_{\text{mean}} = \frac{1}{2\pi} \int_{90°}^{90°+155°} 339.4 \sin\theta\; d\theta = 22.8 \text{ V}$$

$$I_{\text{mean}} = 22.8/10 = 2.28 \text{ A}$$

Example 2-3

The single-phase half-wave circuit with a commutating diode as shown in Fig. 2-4 is used to supply a heavily inductive load of up to 15 A from a 240 V a.c. supply. Determine the mean load voltage for firing delay angles of 0°, 45°, 90°, 135°, and 180°, neglecting the thyristor and diode volt-drops. Specify the required rating of the thyristor and diode.

SOLUTION Using Eq. (2-5), the mean load voltage is

$$V_{\text{mean}} = \frac{240\sqrt{2}}{2\pi}(1 + \cos\alpha)$$

which yields the following values

α	0°	45°	90°	135°	180°
V_{mean}	108 V	92 V	54 V	16 V	0 V

Thyristor rating:

Peak forward (or reverse) voltage $= V_{max}$

$$\text{P.F.V.} = \text{P.R.V.} = 240\sqrt{2} = 340 \text{ V.}$$

The thyristor will conduct for a maximum duration at $\alpha = 0°$ of one half cycle and, if one assumes level current, then using two equal time intervals the r.m.s. current rating can be calculated as

$$I_{rms} = \left(\frac{15^2 + 0^2}{2}\right)^{1/2} = 10.6 \text{ A}$$

Diode rating:

$$\text{P.R.V.} = V_{max} = 340 \text{ V}$$

As the firing delay approaches 180°, the diode will conduct for almost the whole cycle; hence the required current rating would be 15 A; however, in practice, some decay in this current would occur.

Example 2-4

Using the single-phase half-wave circuit of Fig. 2-4, a low-voltage load is supplied by a 20 V a.c. supply. Assuming continuous load current, calculate the mean load voltage when the firing delay angle is 60°, assuming forward volt-drops of 1.5 V and 0.7 V across the thyristor and diode respectively.

SOLUTION Using Eq. (2-5) and neglecting volt-drops,

$$V_{mean} = \frac{20\sqrt{2}}{2\pi}(1 + \cos 60°) = 6.752 \text{ V.}$$

The thyristor will conduct for $(180 - 60)°$, giving an average volt-drop over the cycle of $\frac{120}{360} \times 1.5 = 0.5$ V.

The diode when conducting imposes a 0.7 V drop across the load; in this case, it averages over the cycle to $0.7 \times \frac{180 + 60}{360} = 0.467$ V. Therefore the mean load voltage is $6.752 - 0.5 - 0.467 = 5.78$ V.

It can be seen that at a low voltage the device volt-drops are not negligible.

Example 2-5

A load of 10 Ω resistance, 0.1 H inductance, is supplied via the circuit of Fig. 2-4 from a 70.7 V, 50 Hz a.c. supply. If the thyristor is fired at a delay angle of 90°,

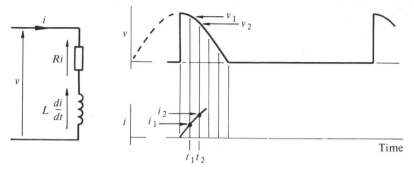

Figure 2-28

determine by a graphical method the waveform of the current during the first two cycles of operation. Neglect device volt-drops.

SOLUTION Figure 2-28 relates to a graphical method where use is made of $v = Ri + L\, di/dt$.

Now taking any time (say) t_1, when the current is i_1, then $v_1 = Ri + L\delta i/\delta t$, that is, $\delta i = \dfrac{v_1 - Ri_1}{L}\,\delta t$, where the time interval $\delta t = t_2 - t_1$. Substitute the values from the graph (Fig. 2-28) for v_1, i_1 and δt into the equation, and δi is obtained. Plot i_2 as $i_1 + \delta i$, and the next point on the current graph is obtained. Proceed in a like manner to find the next current value i_3 at t_3 by using values at t_2.

At 50 Hz the cycle time is 20 ms so, if 20 intervals are selected, then $\delta t = 1$ ms. With $R = 10$ and $L = 0.1$, $\delta i = 0.01 v_n - 0.1 i_n$.

The first calculation is when the load is switched to the voltage peak of 100 V (i.e. $70.7\sqrt{2}$) and no current, that is, $i_0 = 0$, giving

$$\delta i = (0.01 \times 100 \sin 90°) - (0.1 \times 0) = 1 \text{ A}$$

$$i_1 = 1 \text{ A}$$

The second calculation gives

$$\delta i = (0.01 \times 100 \sin 108°) - (0.1 \times 1) = 0.85 \text{ A}$$

$$i_2 = 1 + 0.85 = 1.85 \text{ A}$$

Continuing in a like manner, at the end of 5 ms ($90°$) the load voltage is zero, $i_5 = 2.83$ A.

For the next 15 ms, the commutating diode is conducting, with v in Fig. 2-28 being zero, hence, to find i_6,

$$\delta i = 0 - (0.1 \times 2.83) = -0.283 \text{ A}$$

giving $i_6 = 2.83 - 0.28 = 2.55$ A.

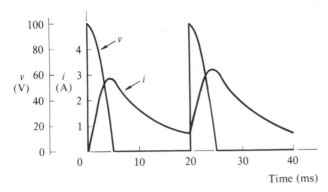

Figure 2-29

Continuing in a like manner $i_{20} = 0.59$ A when the thyristor is again fired, so to calculate i_{21},

$$\delta i = (0.01 \times 100 \sin 90°) - (0.1 \times 0.59) = 0.94 \text{ A}$$

$i_{21} = 0.59 + 0.94 = 1.53$ A until at the next voltage zero, $i_{25} = 3.17$ A and after a further 15 ms, $i_{40} = 0.65$ A.

The current is shown plotted in Fig. 2-29, from which it is possible to conclude that by the end of the third or fourth cycle steady-state conditions will have been reached. To improve the accuracy of the plot, shorter time intervals could be taken. An exact calculation could be made along the lines used in Example 2-1. In the steady state one would have to assume a current I_1 at the start of the commutating diode conduction period falling exponentially to I_2 when the thyristor is fired, then

$$I_2 = I_1 e^{-100t} \qquad (t = 15 \text{ ms})$$

During the thyristor on-period,

$$i = \frac{100}{32.97} \sin (\omega t - 1.262) + I_x e^{-100t},$$

I_x being the d.c. transient component found by substituting at $t = 0, i = I_2$ and at $t = 5$ ms, $i = I_1$, giving $I_1 = 3.09$ A and $I_2 = 0.69$ A.

Example 2-6

The bi-phase half-wave circuit shown in Fig. 2-5 is supplied at 120 V line to neutral. Determine the mean load voltage for firing delay angles of $0°$, $30°$, $60°$, and $90°$, assuming the load current to be continuous and level with a constant 1.5 V drop on each thyristor.

Determine the required thyristor ratings given that the load current is 15 A.

SOLUTION Using Eq. (2-6), $V_{\text{mean}} = \frac{2}{\pi} 120\sqrt{2} \cos \alpha - 1.5$, giving these values:

α	$0°$	$30°$	$60°$	$90°$
V_{mean}	106.5 V	92.1 V	52.5 V	0 V

Thyristor rating: from Fig. 2-5b, P.R.V. = P.F.V. = $2V_{\text{max}} = 2 \times 120\sqrt{2} = 340$ V; and for half cycle conduction, $I_{\text{rms}} = 15/\sqrt{2} = 10.6$ A.

Example 2-7

A single-phase diode bridge is supplied at 120 V. Determine the mean load voltage assuming each diode to have a volt-drop of 0.7 V.

SOLUTION Figure 2-7 relates to this circuit and, using Eq. (2-7) with $\alpha = 0$,

$$V_{\text{mean}} = \frac{2}{\pi} 120\sqrt{2} - (2 \times 0.7) = 106.6 \text{ V.}$$

Example 2-8

A single-phase fully-controlled bridge is supplied at 120 V. Determine the mean load voltage for firing delay angles of $0°$, $45°$, and $90°$, assuming continuous load current. Allow a thyristor volt-drop of 1.5 V. Determine also the required peak voltage of each thyristor.

SOLUTION Figure 2-8 and Eq. (2-7) relate to this question.

$$V_{\text{mean}} = \frac{2}{\pi} 120\sqrt{2} \cos \alpha - (2 \times 1.5), \text{ giving values:}$$

α	$0°$	$45°$	$90°$
V_{mean}	105 V	73.4 V	0 V

The peak voltage across each thyristor = $V_{\text{max}} = 120\sqrt{2} = 170$ V.

Example 2-9

The half-controlled single-phase bridge circuit shown in Fig. 2-10 is supplied at 120 V. Neglecting volt-drops, determine the mean load voltage at firing delay angles of $0°$, $60°$, $90°$, $135°$, and $180°$. If the load is highly inductive taking 25 A, determine the required device ratings.

SOLUTION Using Eq. (2-8), $V_{\text{mean}} = \dfrac{120\sqrt{2}}{\pi}(1 + \cos \alpha)$, giving these values:

α	$0°$	$60°$	$90°$	$135°$	$180°$
V_{mean}	108 V	81 V	54 V	16 V	0 V

Each thyristor and diode must withstand $V_{\text{max}} = 120\sqrt{2} = 170$ V.
The bridge components conduct for a maximum of one half-cycle, hence for level current $I_{\text{rms}} = 25/\sqrt{2} = 17.7$ A.
The commutating diode will conduct for almost the complete cycle when $\alpha \rightarrow 180°$,

therefore it must be rated to 25 A. In practice, the current would decay somewhat at low load voltages.

Example 2-10

Repeat the calculation for the mean voltage in Example 2-9 at $\alpha = 90°$, assuming thyristor and diode volt-drops of 1.5 V and 0.7 V respectively.

SOLUTION At $\alpha = 90°$ the bridge components conduct for half the time, hence over one cycle their volt-drops will reduce the mean voltage by $(1.5 + 0.7)/2 = 1.1$ V. The commutating diode imposes a 0.7 V negative voltage on the load for the other half of the time, averaging over the cycle to $0.7/2 = 0.35$ V.

$$V_{mean} = \frac{120\sqrt{2}}{\pi}(1 + \cos 90°) - 1.1 - 0.35 = 52.6 \text{ V}.$$

Example 2-11

A highly inductive d.c. load requires 12 A at 150 V from a 240 V single-phase a.c. supply. Give design details for this requirement using (i) bi-phase half-wave, and (ii) bridge connection. Assume each diode to have a volt-drop of 0.7 V. Make comparisons between the two designs.

SOLUTION (i) Referring to Fig. 2-5 and Eq. (2-6), using diodes,

$$V_{mean} = 150 = \frac{2}{\pi}V_{max} - 0.7$$

$$V_{max} = \frac{\pi}{2}(150 + 0.7) = 236.7 \text{ V}$$

Hence each section of the transformer secondary requires an r.m.s. voltage of $236.7/\sqrt{2} = 167.4$ V, and carries an r.m.s. current of $12/\sqrt{2} = 8.5$ A.
Transformer secondary rating $= 2 \times 167.4 \times 8.5 = 2.84$ kVA.
Transformer voltage ratio $= 240/167.4$.
Transformer primary current is square wave of r.m.s. value $12(167.4/240) = 8.4$ A.
P.R.V. for each diode $= 2V_{max} = 474$ V, with $I_{rms} = 12/\sqrt{2} = 8.5$ A.
(ii) Referring to Fig. 2-7,

$$V_{mean} = \frac{2}{\pi}V_{max} - (2 \times 0.7)$$

$$V_{max} = \frac{\pi}{2}(150 + 1.4) = 237.8 \text{ V}$$

Transformer secondary r.m.s. voltage $= 237.8/\sqrt{2} = 168.2$ V.
The secondary-current waveform is square wave 12 A and, as the current is level, its r.m.s. value $= 12$ A.
Transformer secondary rating $= 168.2 \times 12 = 2.02$ kVA.

Transformer voltage ratio = 240/168.2.

Transformer primary r.m.s. current = $12(168.2/240) = 8.4$ A.

For each diode, P.R.V. = $V_{max} = 238$ V; $I_{rms} = 12/\sqrt{2} = 8.5$ A.

Comparing the two circuits, the bridge is superior on transformer size and diode voltage rating requirements.

The diode loss in (i) is $0.7/150 = 0.47\%$ of the load power, compared to $1.4/150 = 0.93\%$ in circuit (ii).

Example 2-12

Repeat Example 2-11 using a load voltage of 15 V.

SOLUTION Calculations now give:

	(i)	(ii)
transformer secondary voltage	17.4 V	18.2 V
transformer secondary rating	296 VA	219 VA
P.R.V. of diode	50 V	26 V
% loss to load power	4.7%	9.3%

The comparison now favours the bi-phase, half-wave circuit, as the P.R.V. requirement is inconsequential and the relative losses are important.

Example 2-13

An alternative connection for the half-controlled single-phase bridge is shown in Fig. 2-30. Assuming level load current, sketch the current waveshapes in the thyristor and diodes at a firing delay angle of 90°.

SOLUTION Figure 2-31 shows the waveforms. When thyristor T_1 is fired, it and diode D_2 will conduct. During the periods of zero load voltage, both diodes will conduct, carrying the freewheeling load current.

Example 2-14

Draw the load-voltage waveform which would occur with the half-controlled single-phase bridge circuit of Fig. 2-10 if there were no commutating diode, the firing delay angle were 150°, and one thyristor missed firing in one cycle. Assume continuous load current.

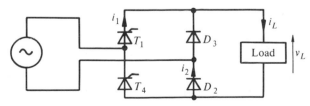

Figure 2-30

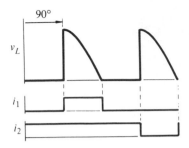

Figure 2-31

SOLUTION The normal operation without a commutating diode is shown in the first two half-cycles of Fig. 2-32 where, during the zero load-voltage period, the inductive load current freewheels via the series combination of a thyristor and diode.

However, if thyristor T_1 fails to fire, thyristor T_3 remains conducting to the start of the next half-cycle, when the supply voltage reverses and thyristor T_3, together with diode D_4, conducts a complete half-cycle into the load. The sudden burst of energy into the load could result in dangerous conditions, particularly with a motor load.
The presence of a commutating diode would prevent this danger, as thyristor T_3 would have been turned off at the end of the previous half-cycle. Misfiring would have merely resulted in the absence of one output voltage period.

Example 2-15

A three-phase half-wave rectifier as shown in Fig. 2-12 has a supply of 150 V/phase. Determine the mean load voltage and the required diode rating, assuming the load current is level at 25 A. Assume each diode has a volt-drop of 0.7 V.

SOLUTION From Eq. (2-9), $V_{\mathrm{mean}} = \dfrac{3\sqrt{3}}{2\pi} 150\sqrt{2} - 0.7 = 174.7$ V.

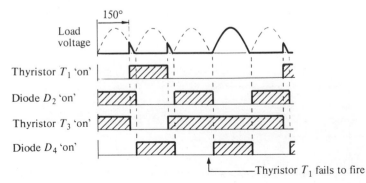

Figure 2-32

Diode rating from Fig. 2-12, and Eq. (2-10):

$$\text{P.R.V.} = \sqrt{3}V_{\text{max}} = \sqrt{3} \times 150\sqrt{2} = 386 \text{ V}, \quad I_{\text{rms}} = \frac{25}{\sqrt{3}} = 14.4 \text{ A}.$$

Example 2-16

A three-phase half-wave rectifier circuit is fed from an interconnected star transformer (shown in Fig. 2-13). If the load (highly inductive) is 200 V at 30 A, specify the transformer, given the a.c. supply to be 415 V and each diode to have a volt-drop of 0.7 V.

SOLUTION Using Eq. (2-9) to find the secondary voltage,

$$200 = \frac{3\sqrt{3}}{2\pi} V_{\text{max}} - 0.7, \text{ giving}$$

$V_{\text{rms}} = 171.6$ V.

This voltage is the phasor addition of two windings equal in voltage magnitude displaced $60°$; voltage of each section $= \dfrac{171.6}{2 \cos 30°} = 99.1$ V.

For a star-connected primary, the winding voltage $= 415/\sqrt{3} = 239.6$ V.

Each secondary winding carries a block of 30 A for a one-third cycle giving $I_{\text{rms}} = 30/\sqrt{3} = 17.3$ A.

The 30 A is reflected into the primary to be $30(99.1/239.6) = 12.4$ A and, referring to Fig. 2-13b, the primary winding r.m.s. current value is

$$I_{\text{rms}} = \left(\frac{12.4^2 + 12.4^2 + 0^2}{3}\right)^{1/2} = 10.1 \text{ A}.$$

Secondary rating $= 6 \times 99.1 \times 17.3 = 10.3$ kVA.

Primary rating $= 3 \times 239.6 \times 10.1 = 7.3$ kVA.

Example 2-17

A three-phase half-wave controlled rectifier has a supply of 150 V/phase. Determine the mean load voltage for firing delay angles of $0°$, $30°$, $60°$, and $90°$, assuming a thyristor volt-drop of 1.5 V and continuous load current.

SOLUTION From Eq. (2-11), $V_{\text{mean}} = \dfrac{3\sqrt{3}}{2\pi} 150\sqrt{2} \cos \alpha - 1.5$, giving these values:

α	$0°$	$30°$	$60°$	$90°$
V_{mean}	173.9 V	150.4 V	86.2 V	0 V

Example 2-18

If the circuit of Fig. 2-14a for the three-phase half-wave controlled rectifier is modified by placing a commutating diode across the load, plot a curve of mean load

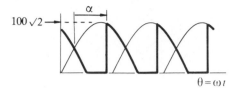

Figure 2-33

voltage against firing delay angle. Take the supply to be 100 V/phase and neglect device volt-drops.

SOLUTION The effect of the commutating diode is to prevent reversal of the load voltage, leading to the waveform shown in Fig. 2-33. From inspection of the waveform,

$$V_{\text{mean}} = \frac{1}{2\pi/3} \int_{30°+\alpha}^{180°} 100\sqrt{2} \sin\theta \, d\theta \quad \text{when } \alpha \gtrless 30°$$

but for firing delay angles below 30°, Eq. (2-11) applies.
The values of mean load voltages against firing delay angle are shown in the plot of Fig. 2-34, together with (for comparison) the fully controlled values with continuous current and no commutating diode.

Example 2-19

The simple six-phase half-wave connection using diodes supplies a d.c. load of 40 V, 50 A. Determine the required diode rating, and specify the transformer for (i) simple six-phase (Fig. 2-15), (ii) fork connection (Fig. 2-16). Assume level load current, a diode volt-drop of 0.7 V, and that the transformer primary is delta-connected, fed from a 415 V supply.

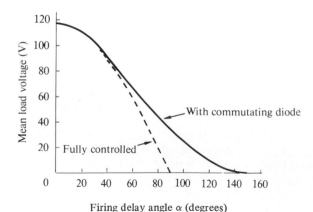

Figure 2-34

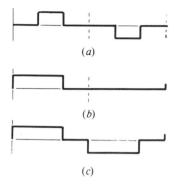

(a)

(b)

(c)

Figure 2-35 (a) Primary winding current, simple star secondary. (b) Inner winding current in fork secondary. (c) Primary winding current, fork secondary.

SOLUTION Using Eq. (2-12), $V_{\text{mean}} = \dfrac{3}{\pi} V_{\text{max}} - 0.7$, giving the r.m.s. voltage/phase as 30.14 V.

Referring to Fig. 2-15, and Eq. (2-13) for the diode rating, P.R.V. $= 2 \times 30.14\sqrt{2} = 86$ V, $I_{\text{rms}} = 50/\sqrt{6} = 20.4$ A.

To determine the transformer ratings, the current waveforms in the sections must be known. Figure 2-35 shows the various waveforms which can be derived from the outgoing phase currents.

(i) R.M.S. secondary current $= 50/\sqrt{6} = 20.4$ A

Primary/secondary voltage $= 415/30.14$

Magnitude of primary current $= 50(30.14/415) = 3.63$ A

R.M.S. value of primary current $= 3.63/\sqrt{3} = 2.1$ A

Secondary rating $= 6 \times 30.14 \times 20.4 = 3.69$ kVA

Primary rating $= 3 \times 415 \times 2.1 = 2.61$ kVA

(ii) R.M.S. voltage of each section $= \dfrac{30.14}{2 \cos 30°} = 17.40$ V

R.M.S. current in outer winding section $= 20.4$ A

R.M.S. current in inner winding section $= 50/\sqrt{3} = 28.9$ A

Magnitude of primary current $= 50(17.4/415) = 2.1$ A

R.M.S. value of primary current $= \left(\dfrac{2.1^2 + 2.1^2}{3} \right)^{1/2} = 1.71$ A

Secondary rating $= (6 \times 17.4 \times 20.4) + (3 \times 17.4 \times 28.9) = 3.64$ kVA

Primary rating $= 3 \times 415 \times 1.71 = 2.13$ kVA

Example 2-20

Using the same data as for Example 2-19, determine diode, transformer and interphase transformer specifications using the double-star connection shown in Fig. 2-17.

SOLUTION Using Eq. (2-14), r.m.s. voltage/phase $= \dfrac{(40 + 0.7)2\pi}{3\sqrt{3}\sqrt{2}} = 34.8$ V. Referring to Fig. 2-17, each diode carries $50/2 = 25$ A for a one-third cycle, hence diode ratings are: P.R.V. $= 2 \times 34.8\sqrt{2} = 99$ V, $I_{rms} = 25/\sqrt{3} = 14.4$ A.

Transformer rating:

R.M.S. secondary current $= 14.4$ A

Primary/secondary current $= 415/34.8$

Magnitude of primary current $= 25(34.8/415) = 2.1$ A

R.M.S. value of primary current $= \left(\dfrac{2.1^2 + 2.1^2}{3} \right)^{1/2} = 1.71$ A

Secondary rating $= 6 \times 34.8 \times 14.4 = 3.01$ kVA

Primary rating $= 3 \times 415 \times 1.71 = 2.13$ kVA

Interphase transformer rating: The voltage rating is a function of the total flux change $\delta\phi$ which from $v = d\phi/dt$ gives $\delta\phi = \int v\, dt$, the area under the voltage-time curve.

As transformers are normally rated to sinewaves, comparison of the area under the curve of v_R (Fig. 2-17b) to a sinewave will yield an r.m.s value for rating (size) purposes; v_R is the first $30°$ of the voltage between two phase voltages $60°$ apart, that is, $v_R = 34.8\sqrt{2} \sin \omega t$.

Area under v_R curve for the first quarter cycle

$$= \int_0^{\pi/6} 34.8\sqrt{2} \sin \omega t\, d\omega t = 6.593 \text{ units}$$

Area under sine curve of r.m.s. value V

$$= \int_0^{\pi/6} \sqrt{2}V \sin 3\omega t\, d\omega t = 0.472V \text{ units,}$$

$6.593 = 0.472V$, giving $V = 13.98$ volts.

The current rating $= (1/2) \times$ load current $= 50/2 = 25$ A.

Interphase transformer rating $= 13.98 \times 25 = 350$ VA.

Example 2-21

A double-star fully-controlled rectifier is supplied by a transformer having a secondary voltage of 200 V/phase. Determine the load voltage with firing delay angles of $0°$, $30°$, $45°$, and $90°$, assuming continuous load current. Determine the maximum mean load voltage if the interphase transformer failed to excite. Neglect thyristor volt-drops.

Determine the rating of the interphase transformer if the load current is 40 A, and what rating it would have if the circuit were uncontrolled.

SOLUTION Using Eq. (2-14) times $\cos \alpha$ for the controlled case,

$$V_{mean} = \frac{3\sqrt{3}}{2\pi} 200\sqrt{2} \cos \alpha, \text{ giving these values:}$$

α	$0°$	$30°$	$45°$	$90°$
V_{mean}	234 V	203 V	165 V	0 V

If the load current is insufficient to excite the interphase transformer, then the circuit operates as the simple six-phase circuit of Fig. 2-15. Maximum mean load voltage occurs at zero firing delay angle, giving

$$V_{\text{mean(max)}} = \frac{1}{2\pi/6} \int_{60°}^{120°} 200\sqrt{2} \sin\theta \, d\theta = 270 \, \text{V}.$$

Following the reasoning developed in Example 2-20 for the equivalent r.m.s. voltage V across the interphase transformer, the area under the v_R curve of Fig. 2-19c for the first quarter cycle is

$$\int_{\pi/3}^{\pi/2} 200\sqrt{2} \sin \omega t \, d\omega t = 141.4 \, \text{units}$$

Sinewave area $= 0.472V$ units, hence $V = 141.4/0.472 = 300$ volts.
Rating $= 300(40/2) = 6$ kVA.
If the circuit were uncontrolled, the r.m.s. voltage is 80.4 V, giving the rating as 80.4(40/2) $= 1.61$ kVA.

Example 2-22

Derive a general expression for the mean load voltage of a p-pulse fully-controlled rectifier.

SOLUTION Figure 2-36 defines the general waveform where p is the pulse-number of the output. The angles are defined relative to the peak of the voltage waveform. From the waveform,

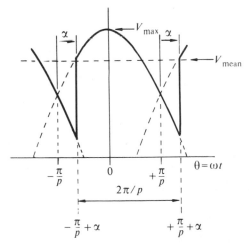

Figure 2-36

$$V_{mean} = \frac{1}{2\pi/p} \int_{-\frac{\pi}{p}+\alpha}^{+\frac{\pi}{p}+\alpha} V_{max} \cos\theta \, d\theta$$

$$= \frac{pV_{max}}{2\pi} \left[\sin\left(\frac{\pi}{p}+\alpha\right) - \sin\left(-\frac{\pi}{p}+\alpha\right) \right]$$

$$= \frac{pV_{max}}{2\pi} \left[\sin\frac{\pi}{p}\cos\alpha + \cos\frac{\pi}{p}\sin\alpha - \sin\left(-\frac{\pi}{p}\right)\cos\alpha + \cos\left(-\frac{\pi}{p}\right)\sin\alpha \right]$$

$$= \frac{pV_{max}}{\pi} \sin\frac{\pi}{p} \cos\alpha$$

Example 2-23

Determine the r.m.s. value of a level current I that flows for $1/p$ of each cycle.

SOLUTION Divide the cycle into p intervals, then the current in one interval is I, and zero in the other intervals.
The sum of the squares for each interval $= I^2$
Mean value of the sum of the squares $= I^2/p$
The r.m.s. value of the current $= \left(\frac{I^2}{p}\right)^{1/2} = \frac{I}{\sqrt{p}}$

Example 2-24

A three-phase bridge rectifier supplies a d.c. load of 300 V, 60 A from a 415 V, 3-phase, a.c. supply via a delta-star transformer. Determine the required diode and transformer specification. Assume a diode volt-drop of 0.7 V, and level load current.

SOLUTION Divide the cycle into p intervals; then the current in one interval is I,
$\frac{3}{\pi}V_{line(max)} - (2 \times 0.7)$, giving $V_{line(max)} = 315.6$ V, giving an r.m.s. phase voltage
of $315.6/(\sqrt{3} \times \sqrt{2}) = 128.9$ V
Diode rating: P.R.V. $= V_{line(max)} = 315.6$ V; $I_{rms} = 60/\sqrt{3} = 34.6$ A

Secondary phase current r.m.s. value $= \left(\frac{60^2 + 60^2}{3}\right)^{1/2} = 49$ A

Transformer rating $= 3 \times 128.9 \times 49 = 18.9$ kVA
The primary and secondary ratings are the same, because each winding current has the same waveform.
Turns ratio primary/secondary $= 415/128.9$
Primary phase r.m.s. current $= 49(128.9/415) = 15.2$ A

Example 2-25

A fully-controlled three-phase rectifier bridge circuit is supplied by a line voltage of

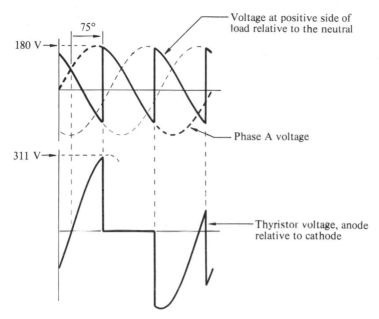

Figure 2-37

220 V. Assuming continuous load current and a thyristor volt-drop of 1.5 V, determine the mean load voltage at firing delay angles of $0°, 30°, 45°, 60°,$ and $90°$.
Plot the waveform of the thyristor voltage at a firing delay angle of $75°$.

SOLUTION Using Eq. (2-16), $V_{mean} = \dfrac{3}{\pi} 220\sqrt{2} \cos \alpha - (2 \times 1.5)$, giving these values:

α	$0°$	$30°$	$45°$	$60°$	$90°$
V_{mean}	294 V	254 V	207 V	146 V	0 V

Referring to Fig. 2-22, the voltage across (say) thyristor T_1 is the difference between the phase A voltage and the voltage at the top of the load relative to the neutral. Figure 2-37 shows the thyristor voltage at a firing delay angle of $75°$. It can be seen that the thyristor voltage is positive throughout the delayed firing period.

Example 2-26

If a commutating diode is placed across the load in the fully-controlled three-phase bridge circuit of Fig. 2-22, explain how the load-voltage waveform will be affected.
Given the a.c. line voltage to be 200 V, plot values of mean load voltage against firing delay angle, neglecting any thyristor or diode volt-drops.
Sketch the a.c. line-current waveform at a firing delay angle of $90°$.

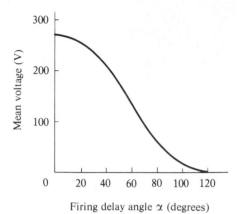

Figure 2-38

SOLUTION The effect of placing a diode across the load is to prevent voltage reversal, and so commutate the load current away from the thyristors.

Referring to the load-voltage waveforms in Fig. 2-22, up to $\alpha = 60°$ the voltage is always positive and Eq. (2-16) applies. Above $\alpha = 60°$ the load voltage loses its negative component, so the mean load voltage is given by

$$V_{\text{mean}} = \frac{1}{2\pi/6} \int_{60°+\alpha}^{180°} 200\sqrt{2} \sin \theta \, d\theta$$

being zero at $\alpha = 120°$.

The values of mean load voltage against firing delay angle are shown plotted in Fig. 2-38.

Assuming level load current, the presence of the diode modifies the waveforms at $\alpha = 90°$ to those shown in Fig. 2-39, drawn to the references as given in Fig. 2-22a. Note that the current waveforms while being further removed from a sinewave are still symmetrical. The commutating diode conducts during the zero load-voltage periods.

Example 2-27

A three-phase half-controlled rectifier bridge as shown in Fig. 2-23 is supplied at a line voltage of 415 V. Plot a curve relating mean load voltage to firing delay angle, and sketch the load-voltage waveform at firing delay angles of $0°$, $30°$, $60°$, $90°$, $120°$, and $150°$. Neglect thyristor and diode volt-drops.

SOLUTION Using Eq. (2-17),

$$V_{\text{mean}} = \frac{3}{2\pi} 415\sqrt{2}(1 + \cos \alpha)$$

the values of which are shown plotted in Fig. 2-40.

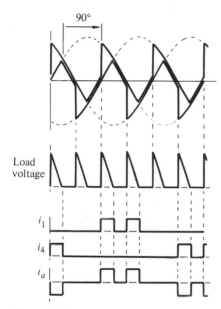

Figure 2-39

The waveforms of the load voltage are shown in Fig. 2-41. These waveforms may be most easily constructed by allowing the vertical line at commutation to move to the right as the firing delay increases.

Example 2-28

A d.c. load requires control of voltage from its maximum down to one quarter of that value. Using the half-controlled three-phase bridge, determine the current rating required for the commutating diode if the load current is level at 20 A.

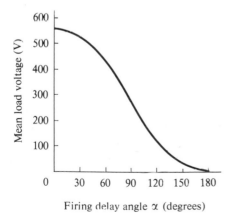

Figure 2-40

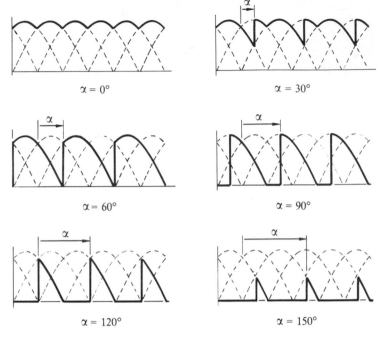

$\alpha = 0°$ 　　　　　　　　　　 $\alpha = 30°$

$\alpha = 60°$ 　　　　　　　　　　 $\alpha = 90°$

$\alpha = 120°$ 　　　　　　　　　 $\alpha = 150°$

Figure 2-41

SOLUTION From Eq. (2-17), the mean load voltage is proportional to $(1 + \cos \alpha)$; therefore at one quarter voltage we have $\dfrac{1}{4} = \dfrac{1 + \cos \alpha}{1 + \cos 0}$, giving $\alpha = 120°$.

The sketch in Fig. 2-41 shows that at $\alpha = 120°$ the commutating diode will conduct for $60°$ every $120°$ (during the zero voltage periods); therefore r.m.s. current in diode $= \left(\dfrac{20^2 + 0^2}{2}\right)^{1/2} = 14.14$ A.

Example 2-29

Determine the percentage value of the peak-to-peak ripple voltage compared to the mean voltage for uncontrolled rectifiers having pulse numbers of $2, 3, 6, 12,$ and 24.

SOLUTION Taking the waveform to have a peak value of V, then for a p-pulse rectifier the minimum value of the waveform will be $V \cos \dfrac{\pi}{p}$ (see Example 2-22 with $\alpha = 0$).

Peak-to-peak value $= V - V \cos \dfrac{\pi}{p}$ and from Example 2-22, $V_{\text{mean}} = \dfrac{pV}{\pi} \sin \dfrac{\pi}{p}$

Hence relative percentage values are:

Pulse-number	2	3	6	12	24
% ripple	157.1	60.46	14.03	3.447	0.858

Example 2-30

A d.c. load of 2000 V, 300 A is to be supplied by a twelve-pulse bridge rectifier. Determine the required thyristor (or diode) ratings and transformer secondary voltages for both the series and parallel connections. Neglect thyristor or diode volt-drops.

SOLUTION The series connection is shown in Fig. 2-25b.
Mean voltage of each bridge = 2000/2 = 1000 V.
Each bridge has a six-pulse characteristic, hence Eq. (2-15) applies, giving a maximum line voltage of 1047 V.
Each thyristor (or diode) carries the total load current for one third cycle.
Thyristor (or diode) ratings are: P.R.V. = 1047 V, I_{rms} = 300/$\sqrt{3}$ = 173 A.
The star secondary winding r.m.s. voltage = 1047/($\sqrt{3} \times \sqrt{2}$) = 428 V.
The delta secondary winding r.m.s. voltage = 1047/$\sqrt{2}$ = 740 V.
The parallel connection is shown in Fig. 2-25c. Compared to the series connection, voltages are doubled but currents halved. Hence thyristor (or diode) ratings are 2094 V, 87 A; winding voltages, star 855 V, delta 1481 V.

Example 2-31

Estimate the total thyristor losses compared to the load power for each of the twelve-pulse circuits shown in Fig. 2-25, when the mean voltage is 60 V, and thyristors are used having a volt-drop of 1.5 V.

SOLUTION Half-wave connection, one thyristor volt-drop, loss = 1.5/60 = 2.5%.
Bridge, parallel connection, two thyristor volt-drops, loss = (2 × 1.5)/60 = 5%.
Bridge, series connection, four thyristor volt-drops, loss = (4 × 1.5)/60 = 10%.

Example 2-32

Derive the current waveshape in the primary winding of the twelve-pulse bridge circuit.

SOLUTION With reference to Fig. 2-25b, the current in the star and delta secondary windings are as shown in Fig. 2-42. The shapes are as derived in Fig. 2-21, the delta connection converting the quasi-square shaped line current into the stepped waveform. Deriving directly the stepped waveform shape of the delta phase current is difficult but, by finding the difference between two stepped phase currents 120° apart, one easily arrives at the quasi-square wave for the line current.
The turns ratio for the delta is different from the star by $\sqrt{3}$, hence $I_{primary}$ = $I_{star} + \sqrt{3}I_{delta}$, giving the stepped wave shown in Fig. 2-42.

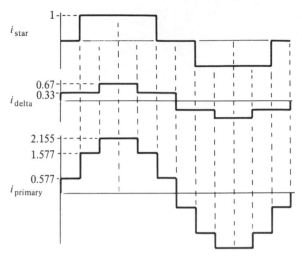

Figure 2-42

Example 2-33

A connection of three single-phase bridges fed from a three-phase supply as shown in Fig. 2-43 will give a six-pulse output. Sketch the waveform of the a.c. line current i_y, and determine the required diode and transformer specifications, if the load is

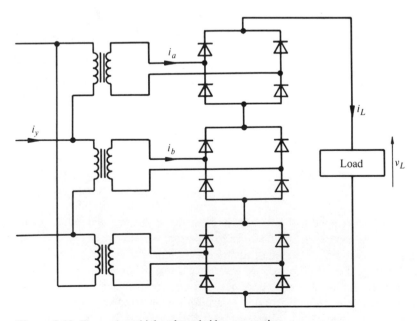

Figure 2-43 Three-phase high-voltage bridge connection.

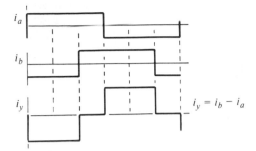

Figure 2-44

60 A at 300 V. Assume a diode volt-drop of 0.7 V and level load current.
Compare this circuit to the normal three-phase bridge.

SOLUTION The output voltage is the addition of three two-pulse voltages displaced so as to give a six-pulse output. The mean voltage of each separate bridge is
one third of the mean load voltage.

Let V = r.m.s. voltage of supply to each bridge, then

$$V_{mean} = \frac{300}{3} = \frac{2}{\pi} V\sqrt{2} - (2 \times 0.7)$$

for each two-pulse bridge, giving $V = 112.6$ volts.
Each diode conducts the complete load current for one half cycle as shown in Fig.
2-44. The figure also shows the input line current i_y.
Diode rating: P.R.V. $= \sqrt{2}V - 159$ volts, $I_{rms} = 60/\sqrt{2} = 42.4$ A.
Transformer rating $= 3 \times 112.6 \times 60 = 20.3$ kVA.
Comparison with the normal three-phase bridge can be examined by looking at
Example 2-24, as the load specification is the same.
The P.R.V. is halved, but the current rating is higher for each diode by a factor of
1.22.
The total transformer rating (size) is greater. The input current waveform is the
same, but a third-harmonic component does circulate in the transformer primary.
The volt-drop is higher, as at all times six diodes are conducting in series.
In conclusion, this circuit would only be used for high-voltage loads, because the
peak reverse-voltage rating of each diode is less than for the normal bridge circuit.

THREE

CONVERTER OPERATION

In Chapter 2, the basic characteristics of the common rectifying circuits were introduced, ignoring the effect of the a.c. supply impedance and concentrating only on the characteristics of the circuits as rectifiers. In this chapter, the analysis of those circuits will be widened to include the effect of the supply impedance, the power factor of the current drawn from the supply, and extend the study to reverse power flow.

The word *rectification* implies conversion of energy from an a.c. source to a d.c. load. In practice, under certain conditions, the power flow can be reversed, when the circuit is said to be *operating in the inverting mode*. As the circuit can be operated in either direction of power flow, the word *converter* better describes the circuit, the words *rectifier* and *inverter* being retained when the converter operates in those particular modes.

3-1 OVERLAP

In Chapter 2, the assumption was made that the transfer or commutation of the current from one diode (or thyristor) to the next took place instantaneously. In practice, inductance and resistance must be present in the supply source, and time is required for a current change to take place. The net result is that the current commutation is delayed, as it takes a finite time for the current to decay to zero in the outgoing diode (or thyristor), whilst the current will rise at the same rate in the incoming diode.

The inductive reactance of the a.c. supply is normally much greater than its resistance and, as it is the inductance which delays the current change, it is reasonable to neglect the supply resistance. The a.c. supply may be represented by its Thévenin equivalent circuit, each phase being a voltage source in series with its inductance. The major contributor to the supply impedance is the transformer leakage reactance.

To explain the phenomenon associated with the current transfer, the three-phase half-wave rectifier connection will be used, as once the explanation with this circuit has been understood, it can be readily transferred to the other connections. Figure 3-1a shows the three-phase supply to be three voltages, each in series with an inductance L. Reference to the waveforms in Fig. 3-1b shows that at commutation there is an angular period γ during which both the outgoing diode and incoming

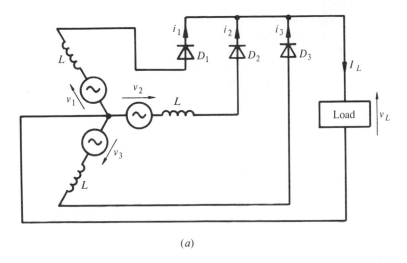

(a)

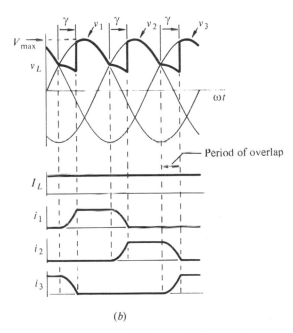

(b)

Figure 3-1 Overlap in the three-phase half-wave rectifier. (a) Circuit reference. (b) Waveforms.

diode are conducting. This period is known as the *overlap period*, and γ is defined as the *commutation angle* or alternatively the *angle of overlap*. During the overlap period, the load current is the addition of the two diode currents, the assumption being made that the load is inductive enough to give a sensibly level load current. The load voltage is the mean of the two conducting phases, the effect of overlap being to reduce the mean level.

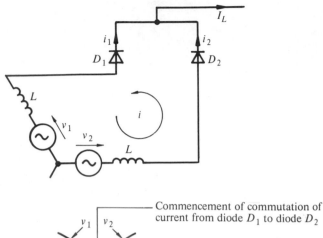

Figure 3-2 Conditions during the overlap period.

The overlap is complete when the current level in the incoming diode reaches the load-current value. To determine the factors on which the overlap depends, and to derive an expression for the diode current, a circulating current i can be considered to flow in the closed path formed by the two conducting diodes D_1 and D_2 as shown in Fig. 3-2. Ignoring the diode volt-drops,

$$v_2 - v_1 = L \, di/dt + L \, di/dt \tag{3-1}$$

The voltage $v_2 - v_1$ is the difference between the two phases, having a zero value at $t = 0$, the time at which commutation commences. The voltage difference between two phases is the *line voltage* having a maximum value $\sqrt{3} V_{max}$ where V_{max} is of the phase voltage.

Using Eq. (3-1),

$$\sqrt{3} V_{max} \sin \omega t = 2L \frac{di}{dt}$$

$$di = \frac{\sqrt{3} V_{max}}{2L} \sin \omega t \, dt$$

Integrating both sides,

$$i = \frac{\sqrt{3} V_{max}}{2L} \left(-\frac{\cos \omega t}{\omega} \right) + C$$

at $t = 0$, $i = 0$, $\therefore C = \dfrac{\sqrt{3}V_{max}}{2\omega L}$

$$\therefore i = \frac{\sqrt{3}V_{max}}{2\omega L}(1 - \cos \omega t) \tag{3-2}$$

The overlap is complete when $i = I_L$, at which instant $\omega t = \gamma$, the overlap angle. Also $\omega L = X$, the supply source reactance. Hence,

$$I_L = \frac{\sqrt{3}V_{max}}{2X}(1 - \cos \gamma) \tag{3-3}$$

or

$$\cos \gamma = 1 - \frac{2I_L X}{\sqrt{3}V_{max}} \tag{3-4}$$

From Eq. (3-2), the current change in the diodes during overlap is cosinusoidal, as illustrated in Fig. 3-1b.

It is worth noting that for commutation involving two phases of a three-phase group, conditions during overlap are as a line-line short-circuit fault. As the positive and negative phase sequence reactance values of a transformer are equal, then the commutating reactance value is the normal short-circuit reactance.

To determine the mean voltage of the waveform shown in Fig. 3-1a, one can use calculus to find the area under the two sections of the curve, one based on the sinewave shape after overlap is complete and the other during overlap. During overlap, the load voltage is the mean between two sinewaves, that is, the shape is sinusoidal, but if we consider the curve as a cosine wave, then the integration limits will be 0 to γ on a cosine wave of peak value $V_{max} \sin \dfrac{\pi}{6}$, giving

$$V_{mean} = \frac{1}{2\pi/3}\left[\int_{\frac{\pi}{6}+\gamma}^{\frac{5\pi}{6}} V_{max} \sin \theta \; d\theta + \int_0^{\gamma} V_{max} \sin \frac{\pi}{6} \cos \phi \; d\phi\right]$$

$$= \frac{3\sqrt{3}V_{max}}{4\pi}(1 + \cos \gamma) \tag{3-5}$$

If one neglects overlap, that is, let $\gamma = 0$, then Eq. (3-5) is identical to Eq. (2-9).

An alternative approach to analysing the effect of the supply inductance is to consider the relationship of the inductor voltage to its current as the current rises from zero to the load value (or collapses from the load value to zero).

$$L \; di/dt = v, \text{hence} \int L \; di = \int v \; dt$$

therefore

$$LI = \int v \; dt = \text{volt-seconds} \tag{3-6}$$

where I is the change in current and $\int v \; dt$ is the area under a curve representing the instantaneous voltage across the inductance during overlap. Therefore, if the mean value of the load voltage is found via terms in volt-seconds, we can subtract LI_L to take account of overlap.

For example, using the waveform in Fig. 3.1,

$$V_{\text{mean}} = \frac{1}{2\pi/3\omega} \left[\int_{t=\pi/6\omega}^{t=5\pi/6\omega} V_{\text{max}} \sin \omega t \, dt - LI_L \right] = \frac{3\sqrt{3}\,V_{\text{max}}}{2\pi} - \frac{3\omega}{2\pi}LI_L$$

(3-7)

From Eq. (3-3) we can substitute for I_L, giving

$$V_{\text{mean}} = \frac{3\sqrt{3}\,V_{\text{max}}}{2\pi} - \frac{3\omega L \times \sqrt{3}\,V_{\text{max}}}{2\pi \times 2X}(1 - \cos \gamma) = \frac{3\sqrt{3}\,V_{\text{max}}}{4\pi}(1 + \cos \gamma)$$

an expression which is identical to Eq. (3-5).

In the controlled 3-pulse circuit, the overlap will lead to the waveform shown in Fig. 3-3 (circuit reference Fig. 3-1a using thyristors), where it can be seen that with a firing delay angle α a finite voltage is present from the start of commutation. Using Eq. (3-1), $v_2 - v_1 = \sqrt{3}\,V_{\text{max}} \sin(\omega t + \alpha)$, where t is the time from the start of commutation, when $i = 0$.

Therefore $\qquad \sqrt{3}\,V_{\text{max}} \sin(\omega t + \alpha) = 2L\, di/dt$

which yields $\qquad i = \dfrac{\sqrt{3}\,V_{\text{max}}}{2\omega L}[\cos \alpha - \cos(\omega t + \alpha)]$ (3-8)

overlap being complete when $i = I_L$ and $\omega t = \gamma$.

$$I_L = \frac{\sqrt{3}\,V_{\text{max}}}{2\omega L}[\cos \alpha - \cos(\gamma + \alpha)]$$ (3-9)

Compared to the uncontrolled case ($\alpha = 0$), the overlap angle γ will be shorter and the current change during commutation will be towards a linear variation. The

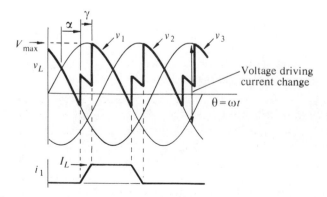

Figure 3-3 Overlap in a controlled 3-pulse rectifier.

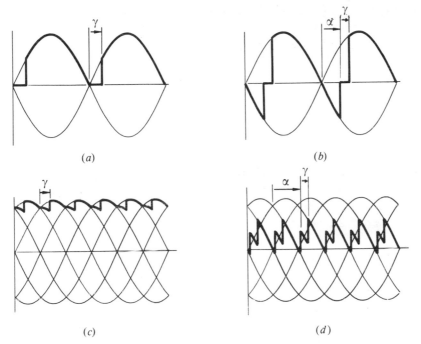

Figure 3-4 Typical load-voltage waveforms showing overlap. (*a*) 2-pulse uncontrolled. (*b*) 2-pulse controlled. (*c*) 6-pulse uncontrolled. (*d*) 6-pulse controlled.

mean load voltage is given by

$$V_{mean} = \frac{1}{2\pi/3} \int_{\frac{\pi}{6}+\alpha+\gamma}^{\frac{5\pi}{6}+\alpha} V_{max} \sin \theta \, d\theta + \int_{\alpha}^{\alpha+\gamma} V_{max} \sin \frac{\pi}{6} \cos \phi \, d\phi$$

$$= \frac{3\sqrt{3}V_{max}}{4\pi} [\cos \alpha + \cos(\alpha + \gamma)] \tag{3-10}$$

The effect of overlap is present in all rectifiers and Fig. 3-4 shows typical wave-forms with pulse-numbers other than that discussed in detail above. The location of the waveform during overlap is at a position midway between the outgoing and incoming voltages. With the 2-pulse waveform, the load voltage is zero during the overlap period.

Circuits with a commutating diode across the load will experience the overlap effect in so far as time is required to transfer the load current away from the supply and into the diode. Typically, the condition can be represented at overlap by the circuit shown in Fig. 3-5. When the supply voltage v reverses, that is, to the direction shown in Fig. 3-5, then a circulating current i will be set up in the closed path formed with the diode. Commutation is complete when i equals the load current I_L. The load will not influence the commutating conditions if I_L is assumed to remain at a constant level.

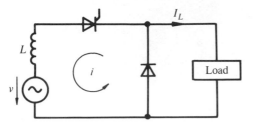

Figure 3-5 Circuit conditions at overlap when the load current is being transferred to the commutating diode.

Using similar reasoning to that developed in the derivation of Eq. (3-2), then from Fig. 3-5, and ignoring thyristor and diode volt-drops, $v = L\,di/dt$, where $v = V_{max} \sin \omega t$, and $i = 0$ at $t = 0$, starting from the instant when the load voltage attempts to reverse; hence we have the equations

$$i = \frac{V_{max}}{\omega L}(1 - \cos \omega t) \tag{3-11}$$

$$I_L = \frac{V_{max}}{\omega L}(1 - \cos \gamma) \tag{3-12}$$

The neglect of the device volt-drops could lead to considerable errors, particularly in the bridge circuits where two devices are concerned.

Following the commutation diode conduction period, the next thyristor is fired and the load current reverts back to the supply, giving another overlap period during which time the load voltage remains effectively zero. Figure 3-6 shows the circuit conditions which apply but, unlike those relating to Fig. 3-5, the voltage v this time will be above zero at the instant of thyristor firing, hence giving a shorter overlap period.

It is possible, given high enough supply reactance, for the overlap period to continue into the time when the next commutation is due to occur, say if, for example, overlap exceeds 60° in a 6-pulse connection. Conditions when this happens must be analysed with reference to the particular connection being used. In practice, it is rare for overlap to continue through to the next commutation, although it does occur, for example, when starting d.c. motors at low voltage.

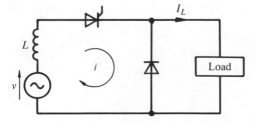

Figure 3-6 Circuit conditions at overlap when the load current is being transferred away from the commutating diode.

3-2 POWER FACTOR

The power factor of a load fed from an a.c. supply is defined as

$$\text{power factor} = \frac{\text{mean power}}{V_{\text{rms}} I_{\text{rms}}} \tag{3-13}$$

In the usual a.c. system where the current is sinusoidal, the power factor is the cosine of the angle between current and voltage. The rectifier circuit, however, draws non-sinusoidal current from the a.c. system, hence the power factor cannot be defined simply as the cosine of the displacement angle.

Inspection of the waveforms of the various controlled rectifiers in Chapter 2 shows that firing delay has the effect of delaying the supply current relative to its phase voltage. The current does contain harmonic components which result in its overall r.m.s. value being higher than the r.m.s. value of its fundamental component, therefore the power factor is less than that calculated from the cosine of its displacement angle.

Normally, the supply phase voltage can be taken as being sinusoidal, hence there will be no power associated with the harmonic current, which therefore results in

$$\text{power} = V_{1(\text{rms})} I_{1(\text{rms})} \cos \phi_1 \tag{3-14}$$

where the suffix 1 relates to the fundamental component of the current, ϕ_1 being the phase angle between the voltage and the fundamental component of the current.

For a sinusoidal voltage supply, substituting Eq. (3-14) into Eq. (3-13) yields

$$\text{power factor} = \frac{I_{1(\text{rms})}}{I_{\text{rms}}} \cos \phi_1 \tag{3-15}$$

where
$$\frac{I_{1(\text{rms})}}{I_{\text{rms}}} = \text{input distortion factor} \tag{3-16}$$

and
$$\cos \phi_1 = \text{input displacement factor.} \tag{3-17}$$

ϕ_1 will equal the firing delay angle α in the fully-controlled connections that have a continuous level load current.

The power factor will always be less than unity when there are harmonic components in the supply current, even when the current is in phase with the voltage, as in the diode case.

3-3 INVERSION

The 3-pulse converter has been chosen to demonstrate the phenomenon of inversion, although any fully-controlled converter could be used. Assuming continuous load current, consider the firing delay angle to be extended from a small value to almost $180°$ as shown in Fig. 3-7b to f. Up to a delay of $90°$ the converter is rectifying, but at $90°$ the voltage is as much negative as positive, resulting in a

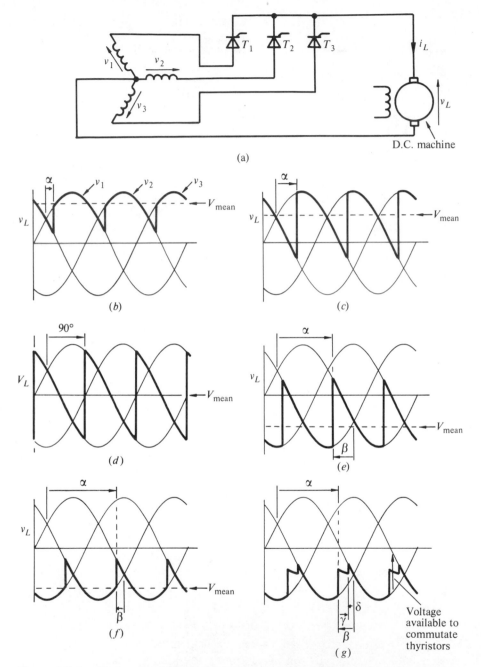

Figure 3-7 3-pulse waveform showing the effect on the load voltage when firing delay is extended towards 180°. (*a*) 3-pulse connection with a d.c. machine as the load. (*b*) Rectifying, small firing delay angle. (*c*) Rectifying, but instantaneous voltage partly negative. (*d*) Firing delay = 90°, V_{mean} is zero. (*e*) Inverting, V_{mean} is negative. (*f*) Inverting, approaching limit as $\beta \rightarrow 0$. (*g*) Inverting waveform including overlap effect.

zero mean value. Further delay beyond $90°$ will result in the waveform having an overall negative mean value, until, as α approaches $180°$, the waveform shape is similar to that at $\alpha = 0$, except it is reversed. To construct these waveforms, let the vertical line at the instants of commutation move to the right as the firing delay is extended.

The circuit connection of Fig. 3-7a shows a d.c. machine as the load element, which acts as a motor while the converter is rectifying. However, once the load voltage v_L reverses, the d.c. machine acts as a generator, and the converter is now said to be operating in the inverting mode. The current direction cannot reverse, as it is constrained by the thyristor direction, hence, if the machine runs in the same direction of rotation, it can only generate by having its armature or field connections reversed.

The reversal of the direct voltage means that current is flowing in each phase when the phase voltage is negative, that is, power is being fed back into the a.c. system from the d.c. generator.

In order for the thyristors to commute, the converter must be connected to a large a.c. synchronous system, such as the public supply, so that the a.c. voltages are of a defined waveform. The energy fed back into the a.c. system will be absorbed by the many other loads on the system.

It is only possible to commutate current from (say) thyristor T_1 to thyristor T_2 while the instantaneous voltage of phase 2 is higher than phase 1 (that is, while v_2 is less negative than v_1). At $\alpha = 180°$, $v_2 = v_1$ and the relative voltage between the two phases after this reverses, making commutation impossible, hence $\alpha = 180°$ is the limit of operation. When in the inverting mode, it is more usual to designate the firing position as firing advance angle β as shown in Fig. 3-7e and f, the relationship between β and α being

$$\beta = 180° - \alpha \tag{3-18}$$

The limit of $180°$ and the relation between β and α apply whatever the pulse-number of the converter.

In deriving the waveforms of Fig. 3-7b to f, the effect of overlap has been ignored, so as to simplify the explanation. In Fig. 3-7g the overlap is shown, the effect being to delay the commutation, the waveform having a voltage midway between the incoming and outgoing voltages. If the commutation is not complete before the two commutating phases reach equal voltage values, then transfer of current is impossible as the load (generator) current will revert to the outgoing thyristor. Therefore, the overlap angle γ must be less than the angle of firing advance β. In practice, β can never be reduced to zero. In Fig. 3-7g, an angle

$$\delta = \beta - \gamma \tag{3-19}$$

is shown where δ represents as an angle the time available to the outgoing thyristor to regain its blocking state after commutation. δ is known as the *recovery* or *extinction angle* and would typically be required to be not less than $5°$.

The firing circuits to the thyristors are designed so that, irrespective of any other control, a firing pulse is delivered to the thyristor early enough to ensure complete commutation. For example, at (say) $\beta = 20°$, a firing (end-stop) pulse will always be delivered to the thyristor gate.

A more detailed study of converter operation in the inverting mode can be explained by reference to Fig. 3-8. Taking the generator as the source of power feeding into the a.c. system, it is convenient to reverse the voltage references as compared to the rectifying mode. Compared to Fig. 3-7a, the d.c. machine shown in Fig. 3-8a has its armature connections reversed, so as to emphasize that for a given direction of rotation the voltage direction at the brushes is unchanged, whether the machine is generating or motoring, only the current flow in the brushes being reversed. Further, the references to the three-phase voltages are reversed to emphasize that the a.c. system is absorbing power, that is, the current is flowing into the positive (arrowhead) end of the voltage reference.

The generator voltage waveform in Fig. 3-8b now becomes the inverse of that shown in Fig. 3-7g because of the reversal of the frame of reference voltages. It can now be seen that the angle of firing advance β is an angle in advance of the commutation limit, a similar reasoning to the delay meaning in the rectifier case. Because the frame of reference has been reversed, it needs to be emphasized that inversion will take place only if the firing delay angle α is extended beyond 90°; one must not be misled from Fig. 3-8b into a belief that a movement of the firing pulse forward in time will give inversion.

The generator mean voltage can be calculated from the waveform in a similar manner to the derivation of Eqs. (3-5) and (3-10), hence giving

$$V_{\text{mean}} = \frac{1}{2\pi/3}\left[\int_{\frac{\pi}{6}-\beta+\gamma}^{\frac{5\pi}{6}-\beta} V_{\text{max}} \sin\theta\, d\theta + \int_{-\beta}^{-\beta+\gamma} V_{\text{max}} \sin\frac{\pi}{6}\cos\phi\, d\phi\right]$$

$$= \frac{3\sqrt{3}\, V_{\text{max}}}{4\pi}[\cos\beta + \cos(\beta-\gamma)] \tag{3-20}$$

The generator voltage will be higher than V_{mean} by the thyristor volt-drop.

Assuming continuous-level direct current in the generator, the individual thyristor currents are as shown in Fig. 3-8b. It can be seen that these currents are in advance of their respective phase voltages, hence the power is being fed back into the a.c. system at a leading power factor.

The thyristor voltage waveform in Fig. 3-8b shows that the anode-cathode voltage is reversed for only the short time represented by δ, which gives the turn-off time available to the thyristor to regain its blocking state. The waveform does demonstrate that the anode voltage is positive with respect to the cathode for most of its off-time.

3-4 REGULATION

The term *regulation* is used to describe the characteristic of equipment as it is

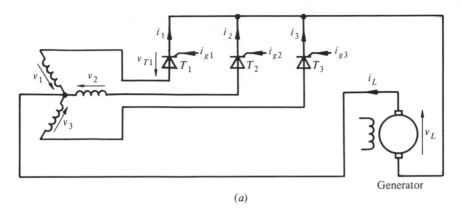

(a)

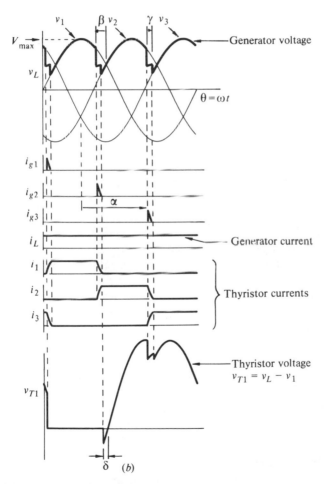

(b)

Figure 3-8 3-pulse inverter operation. (a) Connection and circuit reference. (b) Waveforms.

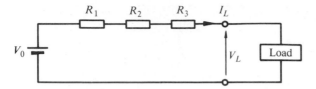

Figure 3-9 Equivalent circuit of a loaded rectifier.

loaded. In the case of the rectifier, regulation describes the drop in mean voltage with load relative to the no-load or open-circuit condition.

There are three main sources contributing to loss of output voltage:

1. The voltage drop across the diodes and/or thyristors.
2. The resistance of the a.c. supply source and conductors.
3. The a.c. supply source inductance.

These three voltage drops can be represented respectively by the three resistors R_1, R_2 and R_3 in the equivalent circuit of Fig. 3-9. The open-circuit voltage is given by V_0 and the actual load voltage by V_L. If the load current I_L is taken to be level, that is, a pure direct current, then any voltage drop can only be represented by resistors.

The voltage drop across the thyristors and diodes can in the first instance be taken as a constant value, or secondly can more accurately be represented by a smaller constant volt-drop (junction potential) plus a resistance value for the bulk of the silicon. In circuits containing a mixture of thyristors and diodes, the volt-drop and equivalent resistance attributed to this cause may depend on the degree of firing delay.

The resistance of the leads and a.c. source resistance can be considered constant in most cases. If (say) throughout a cycle the current is always flowing in two of the supply phases, then the resistance per phase can be doubled and added to the d.c. lead resistance to give the value for the equivalent circuit.

The voltage drop due to the a.c. supply source inductance is the overlap effect, and was calculated for the three-pulse uncontrolled circuit in Eq. (3-7). If the fully controlled case is considered, then the limits in the integral will be

$$\left(\frac{\pi}{6\omega} + \frac{\alpha}{\omega}\right) \text{ to } \left(\frac{5\pi}{6\omega} + \frac{\alpha}{\omega}\right),$$

giving

$$V_{\text{mean}} = \frac{3\sqrt{3}\,V_{\text{max}}}{2\pi}\cos\alpha - \frac{3\omega}{2\pi}LI_L \qquad (3\text{-}21)$$

Examination of Eq. (3-21) shows that, independent of whether the rectifier is controlled or not, the load voltage is reduced by $(3\omega L/2\pi)I_L$ for the three-pulse output, hence this voltage can be represented in the equivalent circuit of Fig. 3-9 as a resistance of value $3\omega L/2\pi$. Unlike the other equivalent circuit resistances, this value does not represent any power loss, but merely represents the voltage drop due to overlap.

The voltage drop due to overlap is changed to a higher value if overlap continues into the period of the next commutation. Simple overlap is known as a *mode* 1 condition, whilst overlap involving three elements is known as a *mode* 2 condition. The paper by Jones, V. H. (see Bibliography) analyses in detail these conditions.

In the inverting mode, the voltage drop due to overlap can be determined for the 3-pulse connection in a like manner to the method used for deriving Eqs. (3-7) and (3-21), and with reference to Fig. 3-8*b* gives

$$V_{\text{mean}} = \frac{1}{2\pi/3\omega}\left[\int_{\frac{\pi}{6\omega}-\beta}^{\frac{5\pi}{6\omega}-\beta} V_{\text{max}} \sin \omega t \, dt + LI_L\right]$$

$$= \frac{3\sqrt{3}V_{\text{max}}}{2\pi}\cos\beta + \frac{3\omega L}{2\pi}I_L \tag{3-22}$$

3-5 EQUATIONS FOR p-PULSE CONVERTER

A general expression for the mean load voltage of a p-pulse fully-controlled rectifier, including the effects of the overlap angle γ, can be determined by reference to Fig. 3-10, where the mean voltage is given by

$$V_{\text{mean}} = \frac{1}{2\pi/p}\left[\int_{-\frac{\pi}{p}+\alpha+\gamma}^{\frac{\pi}{p}+\alpha} V_{\text{max}} \cos\theta \, d\theta + \int_{\alpha}^{\alpha+\gamma} V_{\text{max}} \cos\frac{\pi}{p}\cos\phi \, d\phi\right]$$

$$-\frac{pV_{\text{max}}}{2\pi}\left[\sin\left(\frac{\pi}{p}+\alpha\right) - \sin\left\{-\frac{\pi}{p}+(\alpha+\gamma)\right\}\right.$$

$$\left. + \cos\frac{\pi}{p}\sin(\alpha+\gamma) - \cos\frac{\pi}{p}\sin\alpha\right]$$

$$\therefore V_{\text{mean}} = \frac{pV_{\text{max}}}{2\pi}\sin\frac{\pi}{p}[\cos\alpha + \cos(\alpha+\gamma)] \tag{3-23}$$

less the device volt-drops.

By substituting $\alpha = \pi - \beta$ and calling the mean voltage positive then, for the inverting mode,

$$V_{\text{mean}} = \frac{pV_{\text{max}}}{2\pi}\sin\frac{\pi}{p}[\cos\beta + \cos(\beta-\gamma)] \tag{3-24}$$

The voltage drop due to the supply commutating reactance X Ω/phase when a p-pulse rectifier supplies a load of current value I_L can be determined from

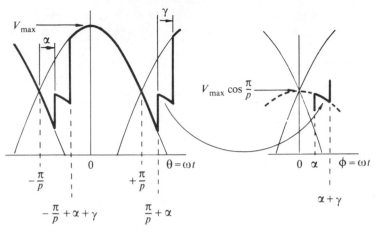

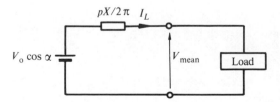

Figure 3-10 General waveform of a p-pulse rectifier.

Fig. 3-10. Taking the base as time, but allowing for the loss of area due to overlap as LI_L as shown in developing Eq. (3-7), then

$$V_{\text{mean}} = \frac{1}{2\pi/p\omega}\left[\int_{-\frac{\pi}{p\omega}+\frac{\alpha}{\omega}}^{\frac{\pi}{p\omega}+\frac{\alpha}{\omega}} V_{\text{max}} \cos \omega t\, dt - LI_L\right]$$

$$= \frac{pV_{\text{max}}}{\pi}\sin\frac{\pi}{p}\cos\alpha - \frac{pX}{2\pi}I_L \tag{3-25}$$

where $X = \omega L$.

This equation represents a voltage $V_0 \cos \alpha$, the mean open-circuit voltage, minus a voltage drop $(pX/2\pi)I_L$, giving an equivalent circuit as shown in Fig. 3-11, ignoring device and true resistance volt-drops. V_0 is the open-circuit voltage at zero firing delay angle.

The relationship between the overlap angle γ, load current I_L, supply voltage V_{max}, and commutating reactance X for a p-pulse rectifier operating at any firing delay angle α, can be determined by equating Eqs. (3-23) and (3-25), yielding

$$XI_L = V_{\text{max}}\sin\frac{\pi}{p}[\cos\alpha - \cos(\alpha + \gamma)] \tag{3-26}$$

Figure 3-11 Equivalent circuit of a p-pulse fully-controlled rectifier, allowing for the commutating reactance.

3-6 WORKED EXAMPLES

Example 3-1

A single-phase diode bridge circuit is supplied at 50 V, 50 Hz to feed a d.c. load taking a level current of 60 A. Determine the volt-drops due to: (i) the supply having an inductance of 0.1 mH, (ii) each diode having a forward volt-drop of $(0.6 + 0.002i)$ volts, (iii) supply and lead resistance of 0.002 Ω.
Draw an equivalent circuit to represent the rectifier.

SOLUTION From Eq. (3-25) using $p = 2$, voltage drop due to inductance

$$\text{(overlap)} = \frac{2 \times 2\pi 50 \times 2 \times 0.1 \times 10^{-3} \times 60}{2\pi} = 1.2 \text{ V}$$

Note that in this circuit the current at commutation is reversed in the supply, hence in Eq. (3-25) the value of inductance used is twice that of the supply.
There are two diodes conducting at any given time, therefore the total diode volt-drop $= 2 \times [0.6 + (0.002 \times 60)] = 1.44$ V.
Resistance volt-drop $= 0.002 \times 60 = 0.12$ V.
Mean voltage from Eq. (3-25), ignoring all drops, is

$$\frac{2 \times 50\sqrt{2}}{\pi} \sin \frac{\pi}{2} = 45.02 \text{ V}$$

The equivalent circuit is shown in Fig. 3-12.

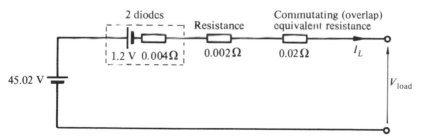

Figure 3-12

Example 3-2

A single-phase fully-controlled bridge rectifier has an overlap angle of 30° with a given load current at zero firing delay angle. Determine the overlap angle when the firing delay angle is 45°, the load current being unchanged. Sketch the voltage waveform and the current waveforms for both conditions, neglecting other volt-drops.

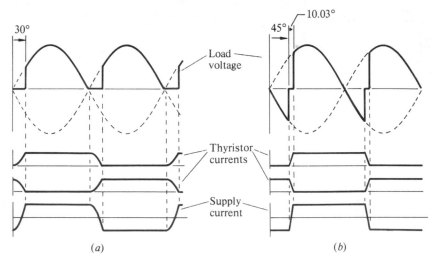

Figure 3-13

SOLUTION Using Eq. (3-9), $\cos \alpha - \cos(\alpha + \gamma) = \dfrac{2\omega L I_L}{\sqrt{3} V_{max}}$

which for a given current is constant at all delay angles.

$$\cos 0° - \cos(0° + 30°) = \cos 45° - \cos(45° + \gamma)$$

giving $\gamma = 10.03°$ at a firing delay angle of 45°.

Figure 3-13 shows the waveforms. In (a) with zero firing delay angle, the current during overlap follows a cosinusoidal form. From Eq. (3-8), $i = k(1 - \cos \omega t)$, where $k = $ a constant. In (b) at $\alpha = 45°$, $i = k[\cos 45° - \cos(\omega t - 45°)]$, and the current variation (although cosinusoidal) is over that section where the current change is almost linear.

Example 3-3

A half-controlled single-phase bridge with a commutating diode is fed at 120 V, 50 Hz. If the load is highly inductive, taking 10 A, determine the load voltage and current waveforms in the circuit at a firing delay angle of 90°. Take the supply to have an inductance of 3 mH, and ignore device and other volt-drops.

SOLUTION Figure 2-10 refers to the circuit, and it can be assumed that the load current is level at 10 A.

The commutating diode will start to conduct when the supply voltage reverses, conditions being as shown in Fig. 3-14 where $v = 120\sqrt{2} \sin \omega t = L\, di/dt$, giving

$$120\sqrt{2} \sin 2\pi 50 t = 3 \times 10^{-3} di/dt$$

$$i = -180 \cos 2\pi 50 t + C$$

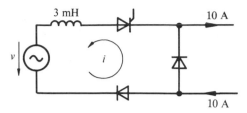

Figure 3-14

at $t = 0, i = 0, C = 180$,

$$i = 180(1 - \cos 2\pi50t) = 180(1 - \cos \omega t)\,\text{A}$$

The overlap is complete when $i = 10\,\text{A}$, giving $\omega t = 19.2°$.

When the thyristors are fired at $\alpha = 90°$, the commutating voltage (v and i reversed in Fig. 3-14) is $120\sqrt{2} \sin (\omega t + 90°) = 120\sqrt{2} \cos \omega t$.

From $120\sqrt{2} \cos \omega t = L\,di/dt$, and $i = 0$ at $t = 0$, we get for the circulating current $i = 180 \sin \omega t$ amperes, the overlap being complete when $i = 10\,\text{A}$, giving $\omega t = 3.2°$.

The waveforms are shown in Fig. 3-15. In practice, although the load current may be continuous, it is unlikely to be level with a 2-pulse rectifier supplying relatively low power loads. However, it would be valid to assume level load current over the short overlap period, making the waveform shapes derived above correct during overlap.

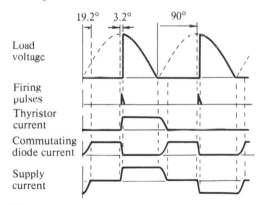

Figure 3-15

Example 3-4

Describe overlap conditions for the circuit given in Example 3-3 when the firing delay angle is 15°.

SOLUTION Using the data of Example 3-3, the commutating diode will start to take up the load current, but after 15° will have reached only $180(1 - \cos 15°) =$

6.13 A. However, before the load sees the supply voltage, this current will have to be transferred back to the thyristors. Using $120\sqrt{2} \sin(\omega t + 15°) = L\, di/dt$, it can be calculated that it will take a further 12.2° to transfer the 6.13 A away from the commutating diode and reverse the 10 A in the supply.

From Eq. (2-8), the no-load (i.e., no overlap) mean voltage is

$$\frac{120\sqrt{2}}{\pi}(1 + \cos 15°) = 106.2 \text{ V, which overlap reduces to}$$

$$\frac{120\sqrt{2}}{\pi}[1 + \cos(15° + 12.2°)] = 102.1 \text{ V, a 3.9\% reduction.}$$

Example 3-5

A three-phase controlled half-wave rectifier is supplied at 150 V/ph, 50 Hz, the source inductance and resistance being 1.2 mH and 0.07 Ω respectively per phase. Assuming a thyristor volt-drop of 1.5 V and a continuous load current of 30 A, determine the mean load voltage at firing delay angles of 0°, 30°, and 60°.

SOLUTION At all times there will be a constant thyristor volt-drop of 1.5 V, and a resistance volt-drop of $30 \times 0.07 = 2.1$ V. From Eq. (3-25), the source reactance will, at 30 A, lead to a volt-drop of $(3 \times 2\pi 50 \times 1.2 \times 10^{-3} \times 30)/2\pi = 5.4$ V.

From Eq. (2-11) or (3-25)

$$\text{the mean voltage ignoring drops} = \frac{3\sqrt{3} \times 150\sqrt{2}}{2\pi} \cos\alpha$$

$$= 175.4 \cos\alpha$$

At $\alpha = 0°$, $V_{\text{mean}} = 175.4 \cos 0° - 1.5 - 2.1 - 5.4 = 166.4$ V

$\alpha = 30°$, $= 175.4 \cos 30° - 1.5 - 2.1 - 5.4 = 142.9$ V

$\alpha = 60°$, $= 175.4 \cos 60° - 1.5 - 2.1 - 5.4 = 78.7$ V

Example 3-6

The supply to a three-phase half-wave diode rectifier has a source impedance of 0.8 p.u. reactance and 0.01 p.u. resistance. Determine the required voltage of the transformer secondary to feed a load of 30 V, 50 A, assuming a diode volt-drop of 0.7 V.

SOLUTION To include the volt-drop due to the supply source impedance it is necessary to convert the given per-unit values to ohmic values, but to do this we must know the secondary r.m.s. voltage and current rating. Hence, in the first instance calculate the voltage ignoring source impedance, and then correct this solution.

Ignoring source impedance and using Eq. (2-9), $30 = (3\sqrt{3}\, V_{\text{max}}/2\pi) - 0.7$, giving the r.m.s. voltage as 26.25 V; from Eq. (2-10) the r.m.s. phase current $= 50/\sqrt{3} = 28.87$ A. Based on these values, the supply resistance $= 0.01(26.25/28.87) =$

0.0091 Ω, and the supply reactance $= 0.08(26.25/28.87) = 0.073\ \Omega$; volt-drop due to supply resistance $= 50 \times 0.0091 = 0.45$ V; from Eq. (3.25) overlap volt-drop $= (3 \times 0.073 \times 50)/2\pi = 1.74$ V.

Recalculating the supply voltage now yields $30 = (3\sqrt{3}V_{max}/2\pi) - 0.7 - 0.45 - 1.74$, giving the supply r.m.s. voltage as 28.12 V.

A further calculation to find more accurately the supply ohmic impedances gives $R = 0.0097\ \Omega$, $X = 0.078\ \Omega$, and $V_{rms} = 28.3$ V/ph.

Substituting values in Eq. (3-26) reveals the overlap angle to be approximately $27°$.

Example 3-7

A fully-controlled three-phase bridge rectifier is supplied at 415 V, 50 Hz, the supply source inductance being 0.9 mH. Neglecting resistance and thyristor volt-drops, (i) plot a graph of the mean load voltage against firing delay angle for a level load current of 60 A, (ii) at a firing delay angle of $30°$ plot a graph of mean load voltage against load current up to 60 A, (iii) plot waveforms of load, thyristor, and line-to-line voltages, at a firing delay angle of $30°$ with a load current of 60 A.

SOLUTION From Eq. (3-25),

$$V_{mean} = \frac{6 \times 415\sqrt{2}}{\pi} \sin \frac{\pi}{6} \cos \alpha - \frac{6 \times 2\pi 50 \times 0.9 \times 10^{-3}}{2\pi} I_L$$

for (i) $V_{mean} = 560.45 \cos \alpha - 16.2$

for (ii) $V_{mean} = 485.4 - 0.27 I_L$

These equations are plotted in Fig. 3-16.

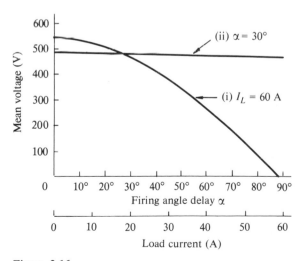

Figure 3-16

(iii) At $\alpha = 30°$, $I_L = 60$ A, from Eq. (3-26),

$$2\pi 50 \times 0.9 \times 10^{-3} \times 60 = 415\sqrt{2} \sin \frac{\pi}{6}[\cos 30° - \cos(30° + \gamma)]$$

giving $\gamma = 6.08°$.

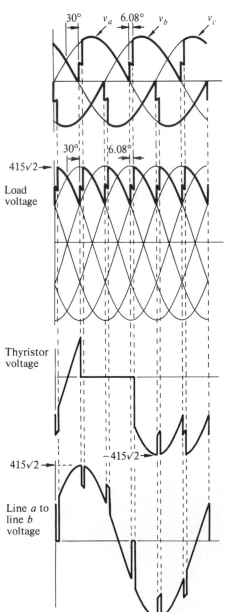

Figure 3-17

Using Fig. 2-22 for the circuit reference, the voltage waveforms are shown in Fig. 3-17. During commutation, a short-circuit appears across the two commutating lines giving in the line voltage waveform two zero periods when thyristor T_1 commutates to T_3 and thyristor T_4 to T_6. The other notches in the waveform occur when commutations occur between other lines, say line b to line c, when the voltage between line a and line b is to a voltage from v_a to midway between v_b and v_c.

Example 3-8

A three-phase diode bridge is supplied at 220 V from a supply having a short-circuit kVA rating of 75 kVA. Plot the load voltage and supply current waveforms when the load current is level at (i) 80 A, (ii) 140 A.

SOLUTION   A short-circuit across the supply will result in $75\,000/(\sqrt{3} \times 220) =$ 196.8 A, which shows the supply to have a high impedance for this load, hence on load a high voltage drop can be expected.

Assuming the supply impedance to be purely reactive, the supply reactance $X = 220/(\sqrt{3} \times 196.8) = 0.645\ \Omega/\text{phase}$.

(i) Using Eq. (3-26) to find the overlap,

$$0.645 \times 80 = 220\sqrt{2}\ \sin\frac{\pi}{6}(1 - \cos\gamma), \text{ giving } \gamma = 48.1°.$$

Following the reasoning developed in deriving Eq. (3-2), the current growth in the diode is $i = [220\sqrt{2}/(2 \times 0.645)](1 - \cos\omega t) = 241(1 - \cos\omega t)$ A.

The waveforms are shown in Fig. 3-18.

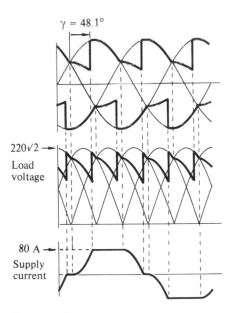

Figure 3-18

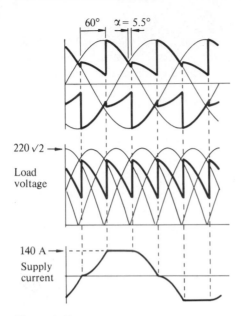

Figure 3-19

(ii) Using Eq. (3-26) yields $\gamma = 65.2°$ for $I_L = 140\,\text{A}$. As this calculation indicates an overlap above 60°, then the overlap will not be complete before the next commutation is due to commence, as can be seen by looking at Fig. 3-18.

With Fig. 2-21a as reference, consider the situation when diode D_2 is due to commutate to diode D_4 but diode D_1 is still commutating the load current to diode D_3. While diode D_1 is conducting, diode D_4 will be reverse-biased for as long as v_L is positive. The net result is that commutation 2 to 4 will not commence until 1 to 3 is complete. Each overlap will be $360/6 = 60°$, the circuit acting as if it were phase-delayed thyristors. The waveforms are shown in Fig. 3-19. To find α, Eq. (3-26) can be used with $\gamma = 60°$, hence

$$0.645 \times 140 = 220\sqrt{2}\, \sin\frac{\pi}{6}\,[\cos\alpha - \cos(\alpha + 60°)], \text{ giving } \alpha = 5.5°.$$

It is to be noted that the above reasoning only holds true up to $\alpha = 30°$, after which the load voltage has zero periods and another set of conditions apply. In practice, at this condition the assumption of level load current is becoming much less tenable.

Example 3-9

Using the data of Example 3-6, determine the a.c. supply voltage requirement for a bridge circuit. Compare the full-load losses of the two connections.

SOLUTION Taking the supply to be star-connected, ignoring source impedance yields from Eq. (2-15), $30 = (3/\pi)V_{\text{line(rms)}} - (0.7 \times 2)$, giving the r.m.s. phase voltage as 13.42 V.

The r.m.s. line current $= [(50^2 + 50^2)/3]^{1/2} = 40.82$ A.
Based on these values, the supply resistance $= 0.01(13.42/40.82) = 0.0033$ Ω, and supply reactance $= 0.08(13.42/40.82) = 0.0263$ Ω.
Voltage drop due to supply resistance $= 50 \times 0.0033 \times 2 = 0.33$ V (two lines conducting at all times).
From Eq. (3-25), overlap volt-drop $= (6 \times 0.0263 \times 50)/2\pi = 1.26$ V.
Recalculating the supply voltage from $30 = (3/\pi)V_{\text{line(max)}} - (0.7 \times 2) - 0.33 - 1.26$ gives final values as $R = 0.0035$ Ω, $X = 0.0277$ Ω, and $V_{\text{ph(rms)}} = 14.14$ V.
Loss in half-wave connection $= (50 \times 0.7) + (50^2 \times 0.0097) = 59.25$ W.
Loss in bridge connection $= (50 \times 0.7 \times 2) + (50^2 \times 0.0035 \times 2) = 87.5$ W.
Load power $= 30 \times 50 = 1500$ W.
The half-wave connection is the more efficient because of the single diode volt-drop; but at a higher output load voltage, it would be found that the lower I^2R loss would make the bridge connection the more efficient.

Example 3-10

A half-controlled three-phase bridge rectifier is supplied at 220 V from a source of reactance 0.24 Ω/ph. Neglecting resistance and device volt-drops, determine the mean load voltage, and plot the load-voltage and supply-current waveforms for a level load current of 40 A, at firing delay angles of (i) 45°, (ii) 90°.

SOLUTION (i) The overlap angle γ associated with the diodes will be given by Eq. (3-4) where $\sqrt{3}V_{\text{max}}$ is the line voltage

$$\cos \gamma = 1 - (2 \times 40 \times 0.24)/220\sqrt{2}, \quad \text{giving } \gamma = 20.2°$$

The equation to the commutating current from Eq. (3-2) is

$$i = \frac{220\sqrt{2}}{2 \times 0.24}(1 - \cos \omega t) = 648(1 - \cos \omega t) \text{ A}$$

The overlap γ associated with the thyristors will be given from Eq. (3-9)

$$40 = \frac{220\sqrt{2}}{2 \times 0.24}[\cos 45° - \cos(45° + \gamma)]$$

giving $\gamma = 4.8°$. From Eq. (3.8) the commutating current change $di/dt = (648/\omega)$ $\sin (45° + \omega t)$, which is almost linear from $\omega t = 0$ to $\omega t = \gamma = 4.8°$.
The waveforms are shown in Fig. 3-20.
The mean voltage is most easily calculated by the addition of the two 3-pulse waveforms, hence using Eq. (3-23)

$$V_{\text{mean}} = \frac{3 \times 220\sqrt{2}}{\sqrt{3} \times 2\pi}\sin \frac{\pi}{3} \{[1 + \cos 20.2°] + [\cos 45° + \cos(45° + 4.8°)]\}$$

$$= 244.4 \text{ V}$$

(ii) Reference to Fig. 2-41 shows that a firing delay angle over 60° will result in zero load-voltage periods.

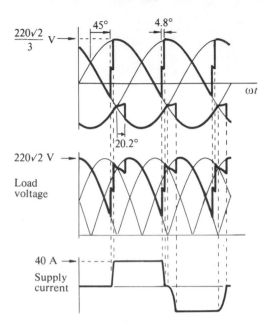

Figure 3-20

When the load voltage reaches zero, the load current transfers into the commutating diode; the voltage v shown in Fig. 3-21a starts from zero, hence i will be identical to the change from diode to diode of (i) above, hence $i = 648(1 - \cos \omega t)$ with an overlap of 20.2°. This overlap will be complete before the thyristor is fired at the firing delay angle of 90°, when the voltage commutating the current back to the supply will be $220\sqrt{2} \sin(90° + \omega t)\,V$, giving $i = 648 \sin \omega t$ A (from $v = 2L\,di/dt$) and an overlap angle of 3.5°(ωt when $i = 40$), during which time the load voltage will remain zero.

The waveforms are shown in Fig. 3-21b, from which the mean load voltage can be directly calculated as

$$V_{\text{mean}} = \frac{1}{2\pi/3} \int_{3.5°}^{90°} 220\sqrt{2} \cos \theta \; d\theta = 139.4\,V$$

Example 3-11

Determine the input power factor and distortion factor of the supply current to (i) single-phase diode bridge, and (ii) three-phase diode bridge. Ignore overlap and diode volt-drops, and assume the load current is level.

SOLUTION (i) The input current will be a square wave (see Fig. 2-7) of r.m.s. value I_L, and also I_L is the mean load current.
From Eq. (2-7), $V_{\text{mean}} = (2/\pi)\sqrt{2}V_{\text{rms}}$. Hence, from Eq. (3-13),

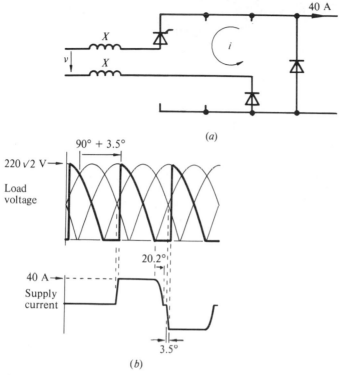

Figure 3-21

$$\text{the power factor} = \frac{V_{\text{mean}} I_{\text{mean}}}{V_{\text{rms}} I_{\text{rms}}} = (2/\pi)\sqrt{2} = 0.9$$

The current is not displaced to the voltage, hence from Eqs. (3-15) and (3-16) the distortion factor = power factor = 0.9.

(ii) The input current will be quasi-square wave (see Fig. 2-21) of r.m.s. value $[(I_L^2 + I_L^2)/3]^{1/2} = \sqrt{2}I_L/\sqrt{3}$.

From Eq. (2-15), $V_{\text{mean}} = (3/\pi)\sqrt{2}V_{\text{rms}}$.

Hence, from Eq. (3-13),

$$\text{power factor} = \frac{V_{\text{mean}} I_{\text{mean}}}{\sqrt{3}V_{\text{line(rms)}} I_{\text{line(rms)}}} = \frac{3 \times \sqrt{2}}{\sqrt{3} \times \pi} \times \frac{\sqrt{3}}{\sqrt{2}} = 0.955$$

Again the distortion factor equals the power factor.

Example 3-12

Plot the power factor of the supply current against mean load voltage as the firing delay angle is varied, for a single-phase bridge, for (i) fully-controlled and (ii) half-controlled connections. Neglect overlap and device volt-drops, and assume level constant load current.

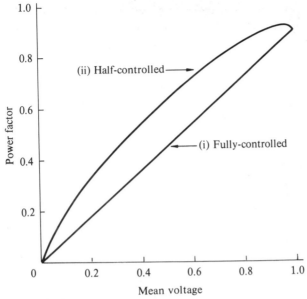

Figure 3-22

SOLUTION (i) With reference to Fig. 2-8 and Eq. (2-7), $I_{rms} = I_L$ (square wave), and $V_{mean} = (2/\pi)\sqrt{2}V_{rms} \cos \alpha$.
From Eq. (3.13), power factor $= (2/\pi)\sqrt{2} \cos \alpha$.
Normalize values by letting V_{mean} at $\alpha = 0$ be 1 unit, then $V_{mean} = \cos \alpha$ units. Calculating values for a range of values of α from $0°$ to $90°$ enables the curves shown in Fig. 3-22 to be plotted.
(ii) With reference to Fig. 2-10 and Eq. (2-8), the supply current is level, of length $(\pi - \alpha)$, for each half cycle at a firing delay angle α.

$$I_{rms} = \left(\frac{1}{\pi}\int_\alpha^\pi I_L^2 \, d\theta\right)^{1/2} = I_L\left(\frac{\pi - \alpha}{\pi}\right)^{1/2}$$

$$V_{mean} = \frac{\sqrt{2}V_{rms}}{\pi}(1 + \cos \alpha)$$

$$\text{power factor} = \frac{\sqrt{2}}{\pi}(1 + \cos \alpha)\left(\frac{\pi}{\pi - \alpha}\right)^{1/2}$$

$V_{mean} = (1 + \cos \alpha)/2$ units normalized as in (i) to give 1 unit at $\alpha = 0°$.
Calculating power and power factor values for a range of α from $0°$ to $180°$ enables the curves shown in Fig. 3-22 to be plotted.

Example 3-13

Calculate the power factor of the supply current to a three-phase rectifier bridge when it is phase-delayed to give half its maximum output voltage, for (i) fully-controlled, (ii) half-controlled connections. Neglect device volt-drops and overlap, and assume level load current.

SOLUTION (i) Let the phase voltage be V_{rms} then, from Eq. (2-16),
$V_{mean} = (3/\pi)\sqrt{3} \times \sqrt{2}V_{rms} \cos \alpha$, with at half maximum $\cos \alpha = 0.5$.
From Fig. 2-22, I_{rms} of supply current $= [(I_L^2 + I_L^2)/3]^{1/2} = (\sqrt{2}/\sqrt{3})I_L$.
Using Eq. (3-13),

$$\text{power factor} = \frac{V_{mean} I_L}{3 V_{rms} I_{rms}} \qquad \text{(3 for 3 phases)}$$

$$= \frac{3 \times \sqrt{3} \times \sqrt{2} \times 0.5}{3 \times \pi} \frac{\sqrt{3}}{\sqrt{2}} = 0.477$$

(ii) Equation (2-17) gives $V_{mean} = [(3 \times \sqrt{3} \times \sqrt{2})/2\pi] V_{rms}(1 + \cos\alpha)$, with at half maximum output $\cos \alpha = 0$.
Reference to Fig. 2-23c shows that when $\alpha = 90°$, the supply current will have positive and negative durations of $90°$, therefore I_{rms} of the supply current $= I_L/\sqrt{2}$.

$$\text{Power factor} = \frac{3 \times \sqrt{3} \times \sqrt{2}}{3 \times 2\pi} \times \sqrt{2} = 0.551$$

Example 3-14

A single-phase fully-controlled converter is connected to a 240 V supply. Plot the d.c. generator and thyristor voltage waveforms when the firing advance angle is $45°$, overlap being $10°$. Calculate the mean d.c. generator voltage, taking the thyristor volt-drop to be 1.5 V. Assume the current is continuous.

SOLUTION Taking Fig. 2-8 as reference, the waveforms are shown in Fig. 3-23. From the waveform,

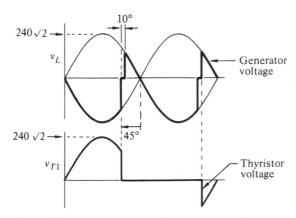

Figure 3-23

$$V_{\text{mean}} = \frac{1}{\pi} \int_{145°}^{315°} 240\sqrt{2} \sin \theta \, d\theta = -164.9 \text{ V}$$

or find, by using Eq. (3-24), with $p = 2$, $\beta = 45°$, $\gamma = 10°$.
The generator mean voltage will be $164.9 + (2 \times 1.5) = 167.9$ V.

Example 3-15

A three-phase fully-controlled bridge converter is connected to a 415 V supply, having a reactance of 0.3 Ω/phase and resistance of 0.05 Ω/phase. The converter is operating in the inverting mode at a firing advance angle of 35°. Determine the mean generator voltage, overlap angle, and recovery angle when the current is level at 60 A. Assume a thyristor volt-drop of 1.5 V.

SOLUTION Using Eq. (3-25),
the volt-drop due to overlap = $(6 \times 0.3 \times 60)/2\pi = 17.2$ V.
Volt-drop due to thyristors = $2 \times 1.5 = 3$ V.
Volt-drop due to the supply resistance = $2 \times 0.05 \times 60 = 6$ V.
At $\beta = 35°$, from Eq. (3-24) the no-load mean voltage ($\gamma = 0$) is

$$\frac{6 \times 415\sqrt{2}}{2\pi} \sin \frac{\pi}{6} (\cos 35° + \cos 35°) = 459.1 \text{ V}$$

The generator has to supply all the volt-drops, hence the mean generator voltage
= $459.1 + 17.2 + 3 + 6 = 485.3$ V.
Using Eq. (3-26), with $\alpha = 180 - 35 = 145°$, the overlap angle γ is 6.7°.
From Eq. (3-19), recovery angle = $35 - 6.7 = 28.3°$.

Example 3-16

Plot the direct voltage waveforms of the three-phase fully-controlled bridge at firing advance angles of 90° and 45°. Neglect overlap.

SOLUTION Using Fig. 2-22 as reference, the waveforms are shown in Fig. 3-24 for $\beta = 90°$, and in Fig. 3-25 for $\beta = 45°$.
Note that extending the delay angle of Fig. 2-22c to 90°, and then to 135°, results in the top and bottom waveforms crossing over.

Example 3-17

Plot the thyristor voltage waveform for a six-pulse converter when in the inverting mode for a firing advance angle of 30° and overlap 10°, for (i) six-phase half-wave, (ii) bridge connections.

SOLUTION (i) The waveform is shown in Fig. 3-26 using Fig. 2-15 as circuit reference.

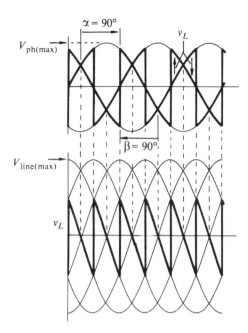

Figure 3-24

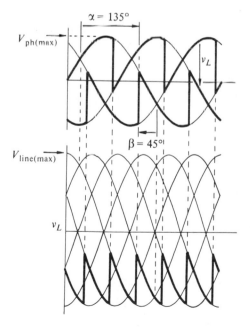

Figure 3-25

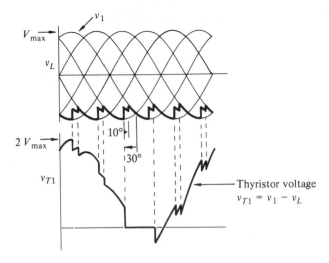

Figure 3-26

(ii) The waveform is shown in Fig. 3-27 using Fig. 2-22 as reference. The thyristor T_1 voltage is represented by the vertical lines shown in the upper waveforms. Note that during commutations the line-to-neutral voltage at the bridge terminals is midway between that line and the other commutating line.

Example 3-18

Using the data given in Example 3-15, determine the required angle of firing advance if the generator is set to a mean armature voltage of 300 V.

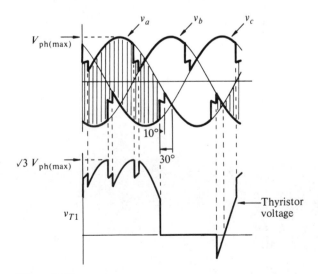

Figure 3-27

SOLUTION Subtracting the volt-drops calculated in Example 3-15 gives the no-load mean voltage as $300 - 17.2 - 3 - 6 = 273.8 \text{ V}$.

Using the general expression $V_{\text{mean}} = \dfrac{pV_{\text{max}}}{\pi} \sin \dfrac{\pi}{p} \cos \beta$,

$\cos \beta = 273.8 \times \dfrac{\pi}{6 \times 415\sqrt{2} \times \sin(\pi/6)}$, giving $\beta = 60.8°$.

Example 3-19

Plot the direct voltage waveform of two-, three-, and six-pulse converters operating in the inverting mode when the firing advance angle is such that a commutation failure occurs.

SOLUTION The waveforms are shown in Fig. 3-28. If commutation is not complete by the voltage cross-over point, then the current will revert back to the outgoing thyristor, resulting in a full half-cycle of rectification, giving a voltage

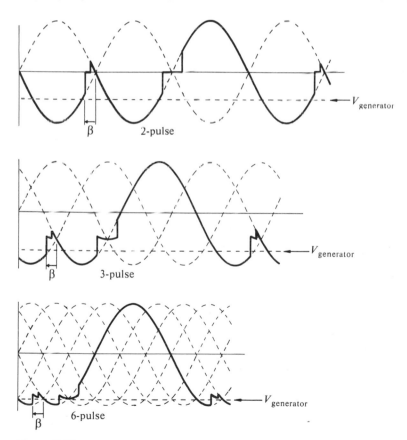

Figure 3-28

assisting (rather than the normal opposing) the generator voltage, resulting in very high load current and possible damage.

Example 3-20

A fully-controlled three-phase bridge converter is operating in the inverting mode with an angle of firing advance of 25°. If the a.c. supply is 220 V with a reactance of 0.1 Ω/phase, determine the maximum current that can be commutated, allowing a recovery angle of 5°. Neglect device volt-drops.

SOLUTION The overlap γ limit will be $25 - 5 = 20°$, so using Eq. (3-26) with $\alpha = 180° - \beta = 155°$ we get

$$0.1\,I_L \;=\; 220\sqrt{2}\,\sin\frac{\pi}{6}[\cos 155° - \cos(155 + 20)°],$$

giving $I_L = 140$ A.

FOUR

D.C. LINE COMMUTATION

In the previous two chapters, where the rectifying devices were linked to an a.c. supply, the devices turn off when the current naturally reaches its zero level in the a.c. cycle. Many applications of the devices are to be found where the supply is a d.c. source. In order to turn off the devices in the case of a d.c. supply, means must be employed whereby the current in the device is reduced to zero and a reverse voltage applied for sufficient time to enable the recovery of the device blocking state. The thyristor is the particular device in which external means must be employed to turn it off.

The power transistor can be turned off by reduction of its base current to effectively zero, without any need to employ extra external circuitry. Where the load is inductive, the load current must be diverted, otherwise the stored magnetic energy will be dissipated in the transistor resulting in gross overheating. Figure 4-1 shows that a commutating (freewheeling) diode is needed to divert the load current when the transistor turns off.

This chapter will outline the basic principles of the various turn-off circuits, the practical applications being left to the later chapters. Formulae relating to performance are not developed for each circuit, as they almost always relate to the exponential changes in a resistor-capacitor circuit or to the oscillation in a capacitor-inductor circuit. Appropriate general formulae are developed in the final summary section of the chapter.

4-1 PARALLEL CAPACITANCE

When the thyristor shown in Fig. 4-2a is turned on the current will be E/R. To turn off the thyristor, a capacitor charged as shown can be switched across the thyristor by closing switch S. A reverse voltage will appear across the thyristor, its current quenched, the current momentarily reverses to recover the thyristor stored charge. As the waveforms of Fig. 4-2b show, the capacitor and hence thyristor voltage will become positive after time elapses, the capacitor becoming oppositely charged to the level of the battery voltage E. With this simple arrangement, the capacitor C must be large enough to maintain a reverse voltage for the required turn-off time.

To convert the simple circuit of Fig. 4-2 into one with practical use, means

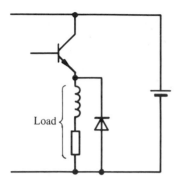

Figure 4-1 Transistor-controlled load.

must be devised to replace the mechanical switch S by an electronic switch, and ensure the charge on capacitor C is restored to its original direction ready for the next turn-off sequence. Such a circuit is shown in Fig. 4-3, a circuit frequently known as a *d.c. chopper* due to its ability to chop the battery voltage on or off to the load.

To understand the manner in which the circuit of Fig. 4-3*a* functions, it is best to take conditions at each switching sequence. Assuming ideal thyristors and lossless components, then the waveforms are as in Fig. 4-3*b*. T_1 is the main load thyristor, T_2 is the auxiliary thyristor to switch the capacitor charge across T_1 for turn-off, and the inductor L is necessary to ensure the correct polarity on capacitor C.

When the battery is connected, no current flows as both thyristors are off. For correct operation, capacitor C must first be charged by firing thyristor T_2, giving the simple equivalent circuit of Fig. 4-4*a*, except capacitor C has no initial charge on it; hence an exponential decay current of initial value E/R flows in the branches shown. After time has elapsed, the capacitor C will be charged ideally to the battery voltage E, but in practice when the charging current falls below thyristor T_2 holding level, current will cease.

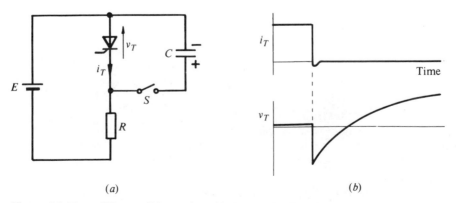

(*a*)   (*b*)

Figure 4-2 Turn-off by parallel capacitor. (*a*) Circuit. (*b*) Waveforms.

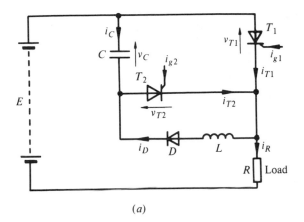

(a)

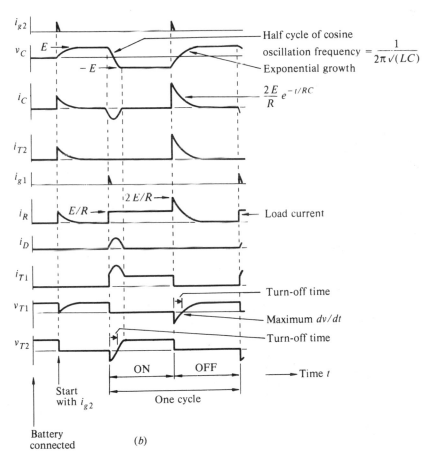

Half cycle of cosine

oscillation frequency $= \dfrac{1}{2\pi\sqrt{(LC)}}$

Exponential growth

$\dfrac{2E}{R}e^{-t/RC}$

Load current

Turn-off time

Maximum dv/dt

Turn-off time

Time t

ON   OFF

Start
with i_{g2} One cycle

Battery
connected   (b)

Figure 4-3 A chopper circuit employing parallel-capacitor commutation. (a) Circuit reference. (b) Waveforms.

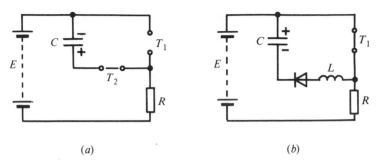

(a)   (b)

Figure 4-4 Equivalent circuit conditions for Fig. 4-3. (a) When T_2 initially fired. (b) When T_1 initially fired.

Firing thyristor T_1 connects the battery to the load, as is clear in the equivalent circuit of Fig. 4-4b. At the same time, an oscillation starts between the inductor L and capacitor C which continues for one half cycle as the diode prevents reverse current flow, hence the charge on capacitor C is reversed from that shown in Fig. 4-4b to that shown in Fig. 4-4a. Now firing T_2 clearly puts the charge on capacitor C across thyristor T_1, turning it off, and giving the conditions shown in Fig. 4-4a, where the current in thyristor T_2 and the load will have an initial value of $2E/R$.

To summarize conditions in the circuit of Fig. 4-3, firing thyristor T_1 connects the load to the battery; firing thyristor T_2 turns off thyristor T_1, disconnecting the load from the battery. This circuit suffers from the disadvantage that the capacitor charging current flows through the load. The diode D may in practice be replaced by a thyristor, which would be fired at the same time as thyristor T_1, as the presence of inductance in the battery-load circuit may result in a second oscillation after thyristor T_2 turns off, leading to a partial discharge of capacitor C via the battery and diode D. A thyristor instead of a diode D would prevent this secondary discharge.

The simple circuit shown in Fig. 4-5 avoids the use of an inductor. The principle is that firing thyristor T_1 connects the battery to the load R_1 and at the same time enables the capacitor C to charge via resistor R_2. Firing thyristor T_2 places the charge on capacitor C across thyristor T_1, turning it off. Thyristor T_2 will remain on

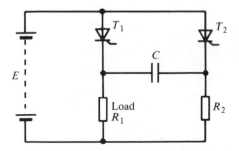

Figure 4-5 Simple parallel-capacitor chopper.

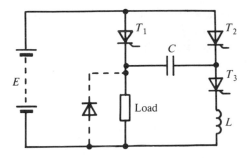

Figure 4-6 A further parallel-capacitance commutation circuit.

with current flow via R_2, capacitor C oppositely charging via R_1. Firing thyristor T_1 now connects the battery to the load R_1 and at the same time turns thyristor T_2 off by placing capacitor C across it. The disadvantage of this simple circuit is the loss in the resistor R_2, as it carries current throughout the load off-period. The loss can be minimized by making R_2 large compared to R_1, but this will lengthen the charging time of the capacitor, hence limiting the rate at which the load can be switched.

A circuit which avoids charging the capacitor via the load is shown in Fig. 4-6. In this circuit, firing thyristor T_1 connects the battery to the load. At the same time or later, thyristor T_3 can be fired, which places the LC circuit across the battery, setting up the start of an oscillation which will leave the capacitor C charged to a voltage of $2E$.

To understand how capacitor C in Fig. 4-6 is charged to $2E$, consider the waveforms shown in Fig. 4-7. Switching LC to the battery will set up an oscillation at a frequency of $1/2\pi\sqrt{(LC)}$ Hz if losses are neglected. However, assuming some loss, the oscillations will die away with capacitor C finally charged to voltage E. However,

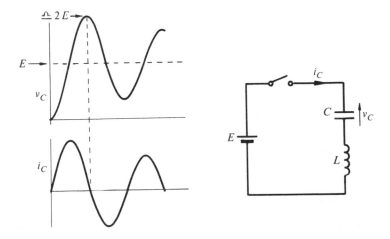

Figure 4-7 Waveforms when a series LC circuit is connected to a battery.

the thyristor T_3 will prevent any reverse current, so that only one half cycle of oscillation occurs, leaving capacitor C charged to a voltage of almost $2E$.

Referring back to Fig. 4-6, thyristor T_1 is turned off by the capacitor when thyristor T_2 is fired. Capacitor C will be oppositely charged to the battery voltage level, so that on the second firing of thyristors T_1 and T_3 a higher-energy oscillation will take place, the voltage on capacitor C possibly going to a level above $2E$. An advantage of this circuit is that the commutating capacitor is charged to a somewhat higher voltage than the battery value. A disadvantage is that inadvertent firing, such that both thyristors T_2 and T_3 conduct at the same time, will place a short-circuit across the battery.

If the load is inductive, then a commutating (freewheeling) diode is required across the load, as shown in outline in Fig. 4-6, to transfer the load current away from the thyristors during switching.

In choosing a circuit for a particular application, consideration must be given to: the type of load, the rate of switching, the size of components, the losses, and the cost. The worked Examples 4-1 to 4-6 demonstrate more clearly the relative merits of the circuits.

4-2 RESONANT TURN-OFF

The self-oscillating property of a capacitor-inductor combination can be utilized to turn off the load thyristor at a given time after turn-on without the need for a second or auxiliary thyristor.

The series resonant circuit shown in Fig. 4-8a must be underdamped for the thyristor current to attempt to reverse and turn-off to take place. The waveforms of Fig. 4-8b illustrate the manner in which the series resonance circuit functions. The first firing pulse after connecting the battery will switch the supply to an underdamped oscillating circuit, the thyristor current attempts to reverse after one half cycle, and the thyristor automatically turns off. During this half cycle the capacitor voltage will reach a value approaching $2E$ so that, at the cessation of thyristor current, a reverse voltage approaching the value of E appears across the thyristors. The capacitor will discharge into the load and, when its voltage drops below E, a forward voltage will appear at the thyristor.

The on-time of the thyristor in Fig. 4-8 is fixed by the time of the damped oscillation frequency. The off-time can be varied, but must be long enough for the capacitor voltage to drop considerably below E, so that oscillation can occur at the next firing of the thyristor. The second-cycle waveform shows that, because of the finite capacitor voltage at the start of the oscillation, a lower current is drawn from the battery.

In the parallel resonance circuit shown in Fig. 4-9a, the capacitor is charged to the level of the battery voltage when connected. Firing the thyristor connects the battery to the load, and at the same time sets up an oscillation in the LC circuit. Provided the oscillating current is greater than the load current (E/R), the thyristor current will attempt to reverse, and turn-off then takes place. For the first half cycle, the thyristor current will increase, but later be reduced to zero during the early part of the reverse half cycle of oscillation.

The waveforms shown in Fig. 4-9*b* are for the case where *R* has a value such that

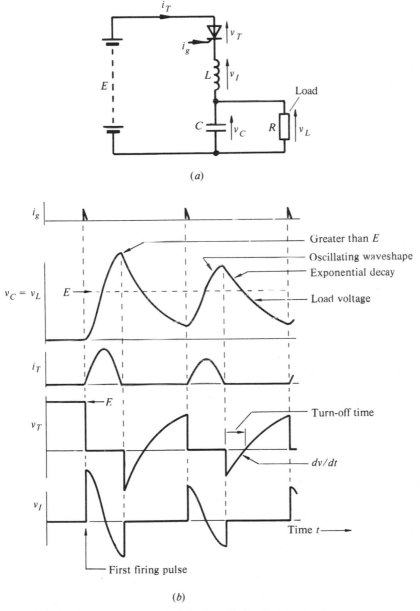

(*a*)

(*b*)

Figure 4-8 Series resonance turn-off. (*a*) Circuit. (*b*) Waveforms.

the LCR series circuit is critically damped ($R^2 = 4L/C$). If R were reduced, one rapidly reaches the case of the load current being greater than the oscillating LC current. If R is increased, then the charging of capacitor C will be slow, making the

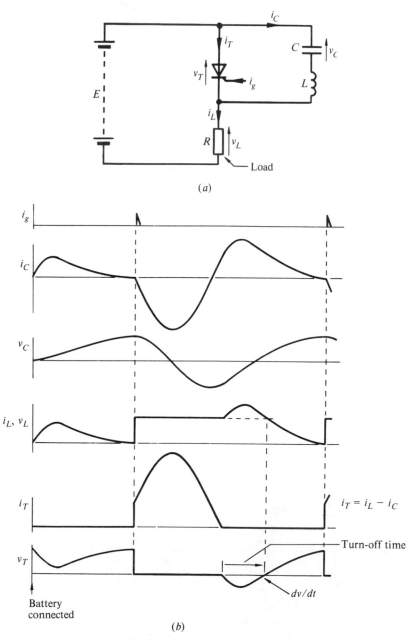

Figure 4-9 Parallel resonance turn-off. (a) Circuit. (b) Waveforms.

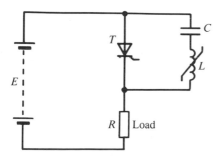

Figure 4-10 An improved parallel resonance turn-off using a saturable reactor.

minimum off-time long. Hence, this circuit is of limited value, being mostly re-stricted to loads of constant resistance.

The parallel resonance circuit can be improved by changing the linear inductor of Fig. 4-9 to a saturable reactor, as shown in Fig. 4-10. A saturable reactor is an inductor with an iron core designed to have a very high inductance; but the core cross-section is such that the flux can be taken into saturation, when it then tends to act as an air-cored coil of low inductance.

In the circuit of Fig. 4-10, firing the thyristor starts a very slow oscillation (be-cause L is large) until the flux change reduces L to a low value; the current then rapidly builds up, eventually doing a half cycle oscillation, and this current flows through the thyristor in the same direction as the load current. For the oscillating current to reverse, the saturable reactor must come out of saturation, its flux reverse to saturate in the opposite direction, and then the current builds up to a value which turns off the thyristor. Hence, the presence of the saturable reactor adds two time intervals, which increase the on-time of the load.

To help explain the action of the circuit with the saturable reactor, waveforms are shown in Fig. 4-11 to the same references as given in Fig. 4-9a. Inspection of the thyristor current i_T waveform shows quite clearly the extension of the on-time.

If the load is inductive (as is the normal case), then a commutating (free-wheeling) diode is required across the load as shown in Fig. 4-12. Further, a diode can be placed in reverse across the thyristor, so that at turn-off the oscillating LC current is diverted into the diode rather than into the load.

Where the load is highly inductive, having an inductance much greater than that of the resonant inductor L, the circuit of Fig. 4-13 has advantages over the previous circuits. Its advantages are that the rate of current rise in the thyristor is limited by inductor L, and the decay of current in thyristor T is relatively slow, thus avoiding excessive reverse storage charge current. The diode D prevents any disturbance to the capacitor charge by a live (motor) load. The relatively low charging rate of capacitor C via the load will not cause an excessive volt-drop in the d.c. supply, as the current level will be low, unlike the rapid charge rate in some other circuits.

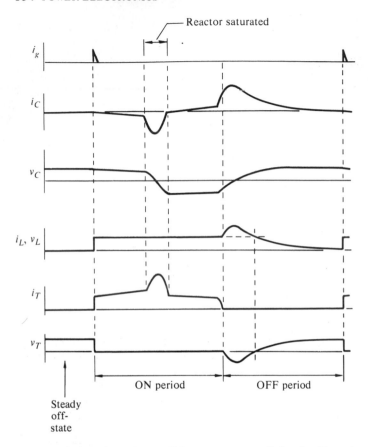

Figure 4-11 Waveforms for parallel resonance turn-off circuit with a saturable reactor.

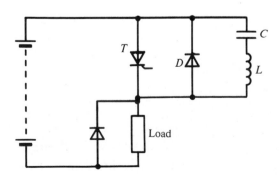

Figure 4-12 Parallel resonance turn-off circuit with the addition of diodes.

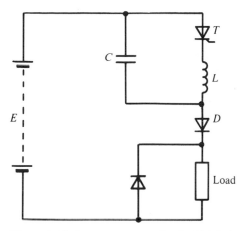

Figure 4-13 A resonant controller for high inductance loads.

4-3 COUPLED PULSE

The load-current-carrying thyristor can be turned off by placing in series with it an inductor, across which a voltage is developed which is greater than, and opposed to, the d.c. supply. In this manner, a reverse voltage is placed across the thyristor and, providing this voltage is held for a long enough time, the thyristor will turn off.

Reference to Fig. 4-14b shows how a charged capacitor can induce a voltage into the load circuit through a mutually coupled coil. The voltage induced into the load circuit must be greater than the battery supply voltage. In this particular circuit, there is no need for a mutually coupled coil; the connection can be direct to an inductor as shown in Fig. 4-14a.

When the circuit of Fig. 4-14a is connected to the battery, a current will flow via L_2, D_2 and C to charge the capacitor to approximately twice the battery voltage, diode D_2 preventing any reverse current. Turning on thyristor T_1 connects the load to the battery via inductor L_1. When thyristor T_2 is fired, the charge (voltage) on C is placed across inductor L_1, putting a reverse voltage on to thyristor T_1, thus turning it off. The resonant circuit formed by L_1 and C will oscillate for approximately half a cycle until stopped by thyristor T_2 blocking the reverse current. The capacitor now recharges via L_2 and D_2 ready for the next sequence. A condition that the events take place as described is that the resonant frequency of the $L_2 C$ link is much less than that of the $L_1 C$ link, so that when capacitor C discharges into inductor L_1 it is independent of inductor L_2. Hence L_2 must be much larger than L_1. If diode D_2 is replaced by a thyristor fired after thyristor T_2 turns off, then L_2 can be smaller, reducing the minimum off-time to the load.

Inspection of the waveforms shown in Fig. 4-14c shows that the load current is a square block without the capacitor current component which was present in some of the circuits described earlier. During turn-off, the current in inductor L_1

will rise from its load value (E/R) because the stored energy of capacitor C is transferred into inductor L_1. The current in thyristor T_2 will rise immediately to E/R and follow a sinusoidal shape. The time available for thyristor T_1 to turn off

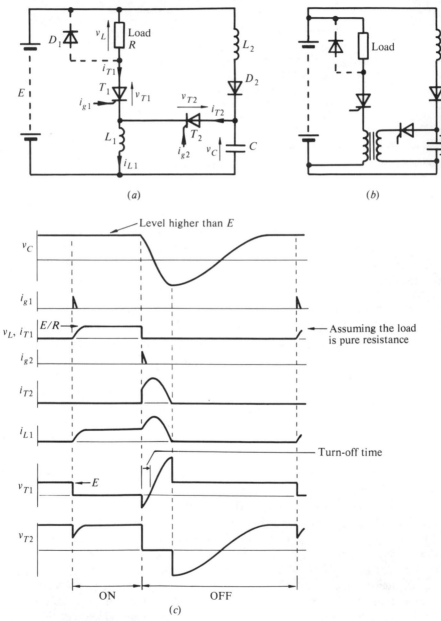

Figure 4-14 Commutation by external pulse. (a) Direct connection. (b) Through mutually coupled coil. (c) Waveforms.

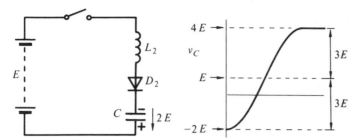

Figure 4-15 Conditions after first turn-off of T_1 in Fig. 4-14a.

is the time taken for the voltage on capacitor C to fall to the level E. The waveforms also show the much slower oscillation rate in the $L_2 C$ link.

A point not made in the explanation to the circuit of Fig. 4-14a is that, if the assumption is made that the components are lossless, then at each switching the voltage on capacitor C will build up by a value $2E$. Figure 4-15 illustrates this phenomenon in that, when thyristor T_2 turns off, the capacitor is left with a voltage of approximately $2E$ (bottom plate positive). Hence the oscillation within the $L_2 C$ battery circuit will be $\pm 3E$ about E, giving a final voltage of $4E$ on the capacitor C, at which time diode D_2 stops the reverse current flow. In practice, losses would limit the eventual voltage rise. A diode placed across thyristor T_2 in an anti-parallel connection would give a full cycle of $L_2 C$ oscillation, and the battery would then merely top up the charge to approximately $2E$.

The coupled-pulse circuit shown in Fig. 4-16a has the feature that, when the load is connected to the battery, the commutating capacitor is automatically charged in the correct sense ready for turn-off.

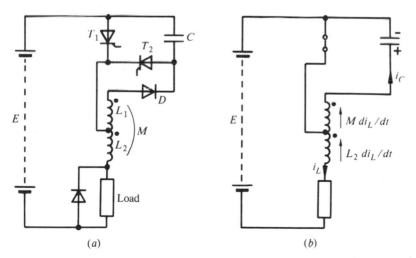

(a)   (b)

Figure 4-16 Coupled-pulse commutation. (a) Circuit. (b) Conditions immediately after thyristor T_1 is fired.

With reference to Fig. 4-16b, consider events when thyristor T_1 is fired. The growth of the load current will induce a voltage in L_2 which, in turn, induces a voltage into the mutually coupled coil L_1. The mutually induced voltage will set up a current flow in the capacitor, thus charging it in the direction shown. The diode D will prevent the capacitor discharging.

When thyristor T_2 is fired, the capacitor is placed across thyristor T_1, turning it off, the capacitor subsequently having its charge reversed. When next thyristor T_1 is fired, C, L_1 and D will form an oscillating circuit to bring the charge to the direction shown in Fig. 4-16b, the mutually induced voltage ($M\,di_L/dt$) adding additionally to the charge.

The advantage the coupled-pulse circuit of Fig. 4-16 has over other circuits is that the capacitor is charged via the load current, hence guaranteeing correct conditions for turn-off. Also, the heavier the load current, the higher will be the capacitor charge, hence compensating for the longer turn-off time required by the thyristor.

4-4 COMMUTATION BY ANOTHER LOAD-CARRYING THYRISTOR

Many circuits, particularly those employed in inverters (these generate an a.c. supply from a d.c. source), use a commutation technique of transferring the load current away from the thyristor being turned off into another load-current-carrying thyristor or diode.

A simple circuit which transfers current between two equal loads is illustrated with appropriate waveforms in Fig. 4-17. When thyristor T_1 is fired, one load is connected to the battery and at the same time the capacitor C is charged via the other load. When thyristor T_2 is fired, the charged capacitor is then placed across thyristor T_1, turning it off, the other load being then connected to the battery. The capacitor charge is reversed at each switching.

It is worth considering how the switching of current between two equal loads could form a simple inverter by reference to Fig. 4-18, where the loads of Fig. 4-17a are the respective primaries of a transformer whose secondary feeds the load. The alternative switching of the two thyristors at equal time intervals will give an alternating voltage to the load.

A complementary impulse-commutated bridge circuit due to McMurray-Bedford is shown in Fig. 4-19. If thyristors T_1 and T_2 are on, the battery is connected to the load. Alternatively, the battery may be connected to the load in the opposite sense by having thyristors T_3 and T_4 on. Hence, an alternating voltage can be made to appear at the load.

As regards commutation in Fig. 4-19, if (say) thyristor T_1 (and T_2) is on and carrying load current, then firing thyristor T_4 will automatically transfer the load current into T_4, turning T_1 off. If thyristor T_1 is turned off without any change in the other (T_2, T_3) side, then the load will be disconnected from the battery. If,

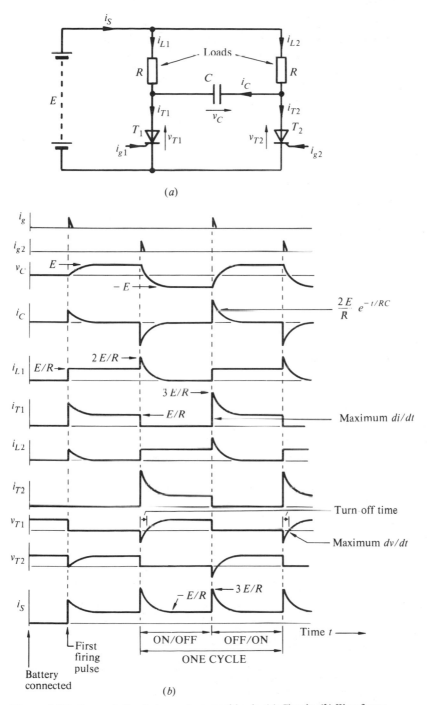

Figure 4-17 Commutation between two equal loads. (*a*) Circuit. (*b*) Waveforms.

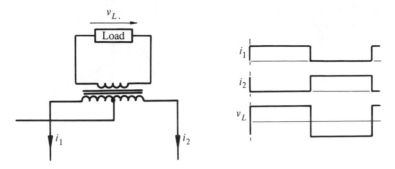

Figure 4-18 Conversion of Fig. 4-17*a* to a basic inverter.

however, thyristor T_3 is fired at the same time as thyristor T_4, then both thyristors T_1 and T_2 will be turned off, and the battery connected to the load in the reversed sense.

To aid the explanation of events during commutation, consider Fig. 4-20*a* which shows conditions prior to the firing of thyristor T_4. The capacitor C_4 is charged to the battery voltage E. The energy stored in the magnetic field of the inductor is $(\frac{1}{2})LI^2$. Immediately thyristor T_4 is fired into the on-state, conditions change to those shown in Fig. 4-20*b*. The capacitor C_4 voltage appears across coil L_4, which in turn induces a voltage E across its closely-coupled coil L_1. Inspection of the diagram shows thyristor T_1 to be reverse-biased by a voltage E, hence T_1 turns off.

The current distribution shown in Fig. 4-20*b* immediately after thyristor T_4 is fired arises from the condition that the stored magnetic field energy of L_1L_4 cannot change, hence the current I which was in L_1 switches to L_4, so maintaining the energy level at $(1/2)LI^2$. Assuming a resistive load with collapse of load current, then the current I will be drawn equally from the two capacitors C_1 and C_4 which

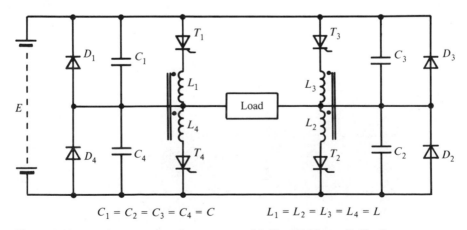

$$C_1 = C_2 = C_3 = C_4 = C \qquad L_1 = L_2 = L_3 = L_4 = L$$

Figure 4-19 Complementary impulse-commutated bridge (McMurray-Bedford).

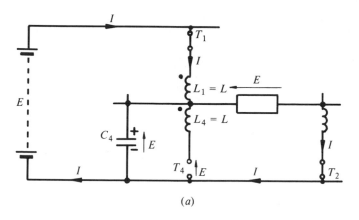

(a)

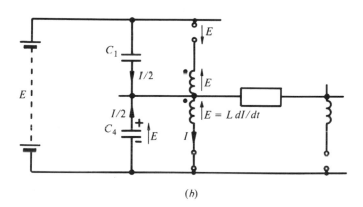

(b)

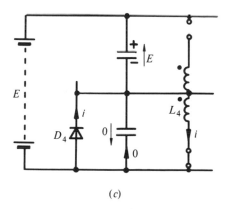

(c)

Figure 4-20 Conditions during turn-off (resistive load). (a) Before T_4 is fired. (b) Immediately after T_4 is fired. (c) After the capacitor voltage attempts to reverse.

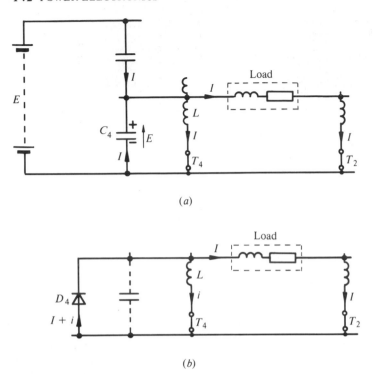

Figure 4-21 Conditions during turn-off (inductive load). (*a*) Immediately after T_4 is fired. (*b*) When diode D_4 conducts.

act in parallel to set up the start of an oscillation with L_4 of frequency $f = 1/2\pi$ $\sqrt{(2CL)}$. When the charge on capacitor C_4 has fallen to $(1/2)E$, the voltage on thyristor T_1 will become anode positive, hence giving the time available for the turn-off of thyristor T_1. If there were no diodes, then this oscillation would eventually lead to a voltage magnitude above $2E$ across capacitor C_1 settling finally back to E. However, the presence of the diode D_4 will prevent the voltage on capacitor C_4 reversing, the diode taking up the decaying current in L_4 as shown in Fig. 4-20*c*, leaving C_1 charged to E. The current in thyristor T_4 will now collapse to zero due to losses in D_4 and L_4.

In most cases, the load will be inductive, having an inductance much higher than L (in the commutation network). In this case, the load current I will continue, giving the current distribution shown in Fig. 4-21*a* immediately after thyristor T_4 is fired. Now the discharge current in capacitor C_4 is double that shown in Fig. 4-20*b*, as the capacitor must provide both the current in T_4 and the load. Assuming the load current does not change over the commutation period, then when diode D_4 starts to conduct, conditions will be as in Fig. 4-21*b* with the current in T_4 decaying, and the load current freewheeling via T_2 and D_4 with zero load voltage.

The role of the circuit of Fig. 4-19 as an inverter is fully discussed in Sec. 5-4;

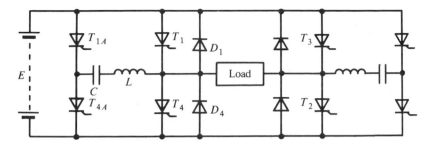

Figure 4-22 Impulse-commutation with auxiliary thyristors.

in this chapter attention is given only to methods of turn-off with a d.c. source as supply.

The circuit of Fig. 4-22 is one in which the load may be connected to the battery in either sense by either having $T_1 T_2$ on or $T_3 T_4$ on. This circuit uses a series combination of capacitance and inductance, together with auxiliary thyristors ($T_{1A} T_{4A}$) to transfer the load current.

The turn-off of thyristor T_1 in Fig. 4-22 is achieved by firing the auxiliary thyristor T_{1A}. The waveforms during commutation are shown in Fig. 4-23, the commutation occurring in three distinct stages. The load is assumed to be inductive enough for the load current I_L to remain sensibly constant over the period of commutation.

Firing thyristor T_{1A} places the CL circuit in series with thyristor T_1 as shown in Fig. 4-24a, with the capacitor current growing. At time t_1 the capacitor current equals the load current, causing thyristor T_1 to turn off as its current attempts to reverse. The capacitor current continues to grow, with the excess over the load current being transferred to diode D_1 as shown in Fig. 4-24b. After the capacitor voltage reverses, the capacitor current drops below the load current; diode D_1 then ceases to conduct, and current is transferred to diode D_4. Reference to Fig. 4-24c now shows that the CL circuit is placed across the battery, giving a boost to the capacitor current, hence taking the charge on capacitor C to a higher level than would be the case with a simple CL oscillation. The commutation is complete when the capacitor current attempts to reverse, turning off auxiliary thyristor T_{1A} at t_3, and leaving diode D_4 to freewheel the load current.

To summarize the conditions shown in Fig. 4-23, i_C must have a peak value somewhat higher than I_L. The turn-off time of thyristor T_1 is the conduction period of diode D_1, with the reverse voltage being the forward volt-drop on D_1. The rate of change of voltage across thyristor T_1 is high, so extra components are required in practice to limit this rise. A higher load current will mean an earlier transfer of current to diode D_4, with more charging current to capacitor C, giving an eventual higher charge (voltage) which makes it easier to perform the next commutation.

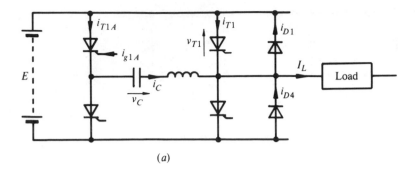

(a)

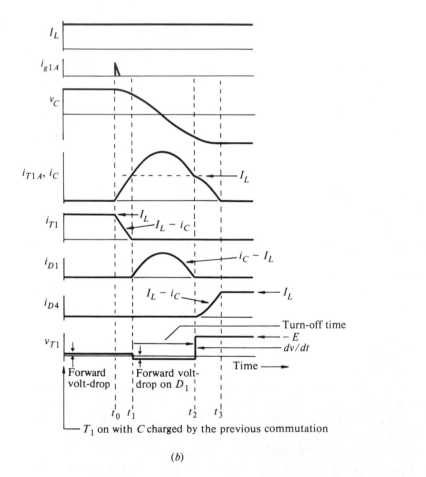

(b)

Figure 4-23 Conditions during turn-off with an inductive load. (a) Circuit reference. (b) Waveforms.

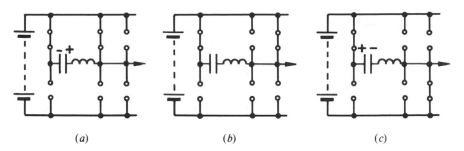

Figure 4-24 Stages in the commutation shown in Fig. 4-23b. (a) From t_0 to t_1. (b) From t_1 to t_2. (c) From t_2 to t_3.

4-5 SUMMARY WITH FORMULAE

The circuits described in this chapter were treated almost entirely from their ability to turn off thyristors linked to a d.c. source. The applications of the circuits in regard to the type of loads they control are discussed in the later chapters.

As all the circuits contained either resistor-capacitor or capacitor-inductor combinations, the formulae relating to the behaviour of these circuits is given below.

The curves relating current and voltage variations to time in a resistor-capacitor circuit are shown in Fig. 4-25. The equations to the time variations of voltage and

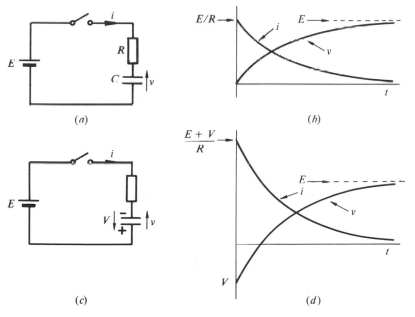

Figure 4-25 Charging conditions in RC circuit. (a), (b) with C initially uncharged. (c), (d) with C being initially charged to V.

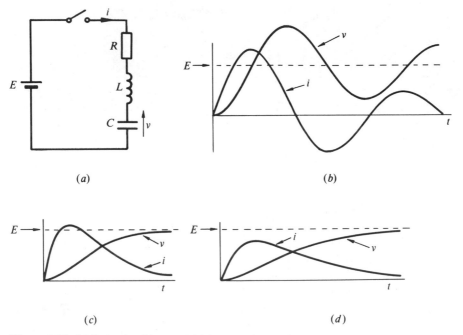

(a)   (b)

(c)   (d)

Figure 4-26 RLC circuit with zero initial conditions. (a) Circuit reference. (b) Underdamped. (c) Critically damped. (d) Overdamped.

current may be written as the steady-state value (when $t = \infty$), plus the decaying exponential component, so for Fig. 4-25:

$$i = Be^{-t/T} \tag{4-1}$$

$$v = E + Ae^{-t/T} \tag{4-2}$$

where the time constant

$$T = RC \tag{4-3}$$

The values of A and B for a particular case can be determined by substituting values for i and v at the time $t = 0$.

The resistor-inductor-capacitor series circuit will have one of the three responses shown in Fig. 4-26. Most of those met in the turn-off circuits are underdamped.

Equations for the current i and capacitor voltage v in Fig. 4-26 are:

For the underdamped condition where $R^2 < 4L/C$

$$i = Ae^{-\alpha t}\sin(\omega t + \delta) \tag{4-4}$$

with $di/dt = A[e^{-\alpha t}\omega \cos(\omega t + \delta) - \alpha e^{-\alpha t}\sin(\omega t + \delta)]$

$$v = E + Be^{-\alpha t}\sin(\omega t + \kappa) \tag{4-5}$$

with $dv/dt = B[e^{-\alpha t}\omega \cos(\omega t + \kappa) - \alpha e^{-\alpha t}\sin(\omega t + \kappa)]$

where
$$\alpha = \frac{R}{2L} \tag{4-6}$$

and
$$\omega = \left[\frac{1}{LC} - \left(\frac{R}{2L} \right)^2 \right]^{1/2} \tag{4-7}$$

ω is known as the damped frequency. Constants A, B, δ and κ may be determined from the initial conditions at $t = 0$.

If the resistance can be taken as negligible, i.e., $R = 0$, then we have the un-damped case where substituting $R = 0$ in Eqs. (4-4) to (4-7) gives

$$i = A \sin (\omega_n t + \delta) \tag{4-8}$$

$$v = E + B \sin (\omega_n t + \kappa) \tag{4-9}$$

where
$$\omega_n = \left(\frac{1}{LC} \right)^{1/2} \tag{4-10}$$

ω_n is known as the undamped natural frequency.

For the overdamped condition where $R^2 > 4L/C$

$$i = Ae^{-at} + Be^{-bt} \tag{4-11}$$

$$v = E + Ce^{-at} + De^{-bt} \tag{4-12}$$

where
$$a = \frac{R}{2L} - \left[\left(\frac{R}{2L} \right)^2 - \frac{1}{LC} \right]^{1/2} \tag{4-13}$$

$$b = \frac{R}{2L} + \left[\left(\frac{R}{2L} \right)^2 - \frac{1}{LC} \right]^{1/2} \tag{4-14}$$

Constants A, B, C and D may be determined from the initial conditions at $t = 0$.

For the critically damped case where $R^2 = 4L/C$,

$$i = e^{-at}(At + B) \tag{4-15}$$

$$v = E + e^{-at}(Ct + D) \tag{4-16}$$

where
$$\alpha = R/2L \tag{4-17}$$

Constants A, B, C and D may be determined from the initial conditions at $t = 0$.

4-6 WORKED EXAMPLES

Example 4-1

Specify the component sizes and rating requirements of the thyristors in a d.c. chopper circuit, operating at 200 Hz, as shown in Fig. 4-3 for a 5 Ω resistive load, being fed from a 24 V battery. Allow a turn-off time of 100 μs for each thyristor. Neglect all losses.

Suggest minimum on- and off-times for the chopper.

SOLUTION The waveforms are shown in Fig. 4-3. To select the capacitor value, inspection of v_{T1} shows the turn-off time available to thyristor T_1 to be midway up an exponential rise from $-E$ to $+E$. Using Eq. (4-2) with $t = 0$ at start of T_1 turn-off, $v = E + Ae^{-t/T}$, where $v = -E$ at $t = 0$, therefore $A = -2E$, giving $v = E - 2Ee^{-t/T}$. When $v = 0$, $t =$ turn-off time $= 100\ \mu s$, therefore $0 = 24 - (2 \times 24)e^{-(100 \times 10^{-6})/T}$, giving $T = 1.44 \times 10^{-4}$ s. From Eq. (4-3), $T = RC$, giving $C = (1.44 \times 10^{-4})/5 = 29\ \mu F$.

Inspection of the waveform for v_{T2} shows the turn-off time available to thyristor T_2 to be the quarter cycle time of the LC oscillation, hence

$$1/(4 \times 100 \times 10^{-6}) = 1/2\pi\sqrt{(LC)}, \text{giving } L = 0.14\ \text{mH}.$$

The magnitude of the oscillating current in LC can be found by equating $\frac{1}{2}CV^2 = \frac{1}{2}LI^2$, V and I being peak values, therefore $I = \left(\dfrac{29 \times 10^{-6} \times 24^2}{0.14 \times 10^{-3}}\right)^{1/2} = 10.9$ A

Steady load current $= 24/5 = 4.8$ A.

Specification of thyristor T_1:

Steady d.c. current $= 4.8$ A.
Peak current $= 4.8 + 10.9 = 15.7$ A.

Additional loss ($i^2R \times$ time) during LC oscillation $= R'\left\{\displaystyle\int_0^{200 \times 10^{-6}} [(4.8 + 10.9 \sin \omega t)^2 - 4.8^2]\,dt\right\}$ where $R' =$ thyristor resistance, $\omega = 1/\sqrt{(LC)}$, this giving an additional loss of $0.0252R'$ joules. Additional power loss $= 0.0252R' \times 200 = 5.04\,R'$ watts. If T_1 is on continuously, power $= 4.8^2R' = 23.04\,R'$ watts. Steady current rating of thyristor $= \left(\dfrac{23.04 + 5.04}{23.04}\right)^{1/2} \times 4.8 = 5.3$ A. Allowing a safety margin, together with the 15.7 A peak, a thyristor of (say) 7.5 A continuous rating would be suitable. The peak reverse and forward voltages are 24 V. The dv/dt value from the waveform of the rising exponential $\left(\text{from } \dfrac{dv}{dt} = \dfrac{2E}{T}e^{-t/T}\right)$ gives values of 0.33 V/μs at the start of commutation, to 0.17 V/μs when $t = 100\ \mu$s at the time the anode voltage becomes positive. The dI/dt requirement will be given by 24/(stray inductance), so if (say) the limit were 10 A/μs, then $24/(10 \times 10^6) = 2.4\ \mu$H would be needed in the supply circuit.

Specification of thyristor T_2:

T_2 carries only the capacitor charging current $i = \dfrac{2 \times 24}{5}e^{-t/T}$. Loss in thyristor in one cycle $= i^2R \times$ time $= R'\displaystyle\int_0^\infty (9.6e^{-t/T})^2 dt = 0.0067R'$ joules, giving a mean power loss of $0.0067R' \times 200 = 1.34R'$ watts, with the r.m.s. current value $= \sqrt{(1.34)} = 1.16$ A. With a peak current of 9.6 A, the 200 Hz frequency and (allowing a safety margin) a thyristor with a 1.5 A r.m.s rating would be suitable. Peak voltage = 24 V. The voltage equation is $v = -24 \cos \omega t$ with $dv/dt = 24\omega \sin \omega t$, giving a maximum dv/dt value of $24/\sqrt{(LC)} = 0.38$ V/μs. dI/dt would only be limited by the supply inductance, if (say) the limit were 10 A/μs then $(2 \times 24)/(10 \times 10^6)$ = 4.8 μH would be required (this overides the 2.4 μH requirement for T_1).

Diode specification:

Loss per cycle = $R' \times \int_0^{200 \times 10^{-6}} (10.9 \sin \omega t)^2 dt = 0.0119 R'$ joules, giving a mean power loss of $0.0119 R' \times 200 = 2.38 R'$ watts, with an r.m.s. current requirement of $\sqrt{(2.38)} = 1.55$ A, say 2 A.

Minimum on-time is that of half-cycle of LC oscillation = 200 μs.

Minimum off-time is that required for the charge on the capacitor C to reach (say) $0.8E$, giving a time of 333 μs.

Example 4-2

What effect would the circuit losses have on the performance of the d.c. chopper designed in Example 4-1, given the following characteristics:

thyristor T_1 $v = 1.2 + 0.55i$
thyristor T_2 $v = 1.3 + 0.22i$
with a holding current of 5 mA
diode D $v = 0.6 + 0.15i$
and the inductor resistance is 0.3 Ω. Symbol references are to Fig. 4-3a.

SOLUTION The capacitor initial charging current = $(24 - 1.3)e^{-t/(5+0.22)C}$ ceasing when it equals 5 mA, that is, at $t = 1275$ μs, making the maximum voltage to which the capacitor can charge = $(24 - 1.3)(1 - e^{-t/T}) = 22.7$ V. (Note that the holding current is of no consequence.)

Using Eq. (4-5) and substituting at $t = 0, v = 22.7$ V and $dv/dt = 0$ gives the capacitor voltage v when T_1 is fired as

$$v = (4.8 \times 0.05) + 1.2 + 0.6 + Be^{-\alpha t} \sin(\omega t + \kappa)$$

(4.8 × 0.05 is the iR drop in T_1 due to load current, 1.2 and 0.6 are T_1 and D constant volt-drops, with the total loop resistance = $0.15 + 0.3 + 0.05 = 0.5$ Ω), giving $v = 2.04 + 20.795e^{-\alpha t} \sin(\omega t + 1.457)$, where from Eqs. (4-6) and (4-7), $\alpha = 1785$ and $\omega = 15\,592$.

$i = C dv/dt$ will be zero when $\omega t = \pi$, giving $v = -12.38$ V, hence voltage on capacitor C to commutate thyristor T_1 equals 12.38 V.

Turn-off time for T_1 is found from the equation for the voltage across T_1 which is $24 - 5i$, where $i = (1/5.22)(24 - 1.3 + 12.38)e^{-t/(5+0.22)C} = 24 - 33.6 e^{-t/5.22C}$, which is zero at $t = 51$ μs.

The effect of the losses is to almost halve the charge on the capacitor and the turn-off time available to commutate the main thyristor.

The load voltage when on = $[(24 - 1.2)/(5 + 0.05)] \times 5 = 22.6$ V.

The losses will vary with conditions but, taking the current levels determined in

Example 4-1, the loss at 200 Hz (for maximum on-period) is:

in T_1, loss $= (5.3^2 \times 0.05) + (4.8 \times 1.2) = 7.2$ W,

in T_2, loss $= 1.16^2 \times 0.22 = 0.3 + $ say 0.2 (for $1.3i$ loss) $= 0.5$ W,

in D, loss $= 1.55^2 \times 0.15 = 0.4 + $ say 0.1 (for $0.6i$ loss) $= 0.5$ W,

in inductor, loss $= 1.55^2 \times 0.3 = 0.7$ W.

Total loss $= 7.2 + 0.5 + 0.5 + 0.7 = 8.9$ W, compared to a nominal full-load power of $4.8^2 \times 5 = 115.2$ W.

The commutation loss in $T_1 = 7.2 - (4.8 \times 1.2) - (4.8^2 \times 0.05) = 0.3$ W, hence the total switching loss $= 2$ W $= 0.01$ J per cycle.

The losses would be less, and the error in turn-off time less noticeable, if a higher battery voltage were employed, as the thyristor and diode volt-drops would be less significant.

Example 4-3

Neglecting losses, what effect will adding an inductance of 0.1 H to the load have on the operation of the chopper of Example 4-1? The load will have a commutating (freewheeling) diode across it.

SOLUTION With reference to Fig. 4-3a, the capacitor will initially be charged via the series circuit of CT_2 and RL of the load. When the voltage on C reaches the battery level, the load voltage will be clamped to zero by the diode across the load. Hence, C is charged to the battery level. When T_2 next fired, the load current will dictate the charging rate of C.

Using Eq. (4-5), with $R = 5\,\Omega$, $C = 29\,\mu$F and $L = 0.1$ H (of load), then $R^2 \ll 4L/C$, hence initially C is charged in quarter cycle time of LC undamped oscillation, which is 2.7 ms.

The maximum load current $= 24/5 = 4.8$ A, and the load inductance is high enough to keep this current at 4.8 A while the charge on C reverses after T_2 is fired, therefore $dv/dt = 4.8/(29 \times 10^{-6})$, hence the C charge is reduced to zero in $145\,\mu$s and reversed in a further $145\,\mu$s when the diode clamps the load voltage to zero. Therefore the available turn-off time for T_1 is $145\,\mu$s, with a minimum off-time at full load of $290\,\mu$s.

Example 4-4

An ideal chopper operating at a frequency of 500 Hz supplies a load of 3 Ω having an inductance of 9 mH from a 60 V battery. Assuming the load is shunted by a perfect commutating diode, and the battery to be lossless, determine the load current waveform for on/off ratios of (i) 1/1, (ii) 4/1, (iii) 1/4.

Calculate the mean values of the load voltage and current at each setting.

SOLUTION The ideal chopper will place the full battery voltage across the load during the on-period, with zero load voltage during the off-load.

During the on-period, the battery is switched to a series RL load, having an initial

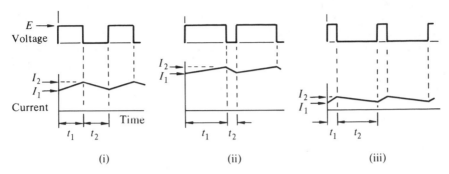

Figure 4-27

current I_1. During the off-period, the load current decays in the RL load through the diode, having an initial value of I_2. Figure 4-27 shows the general waveform for each setting.

Let t_1 = on-time, t_2 = off-time.

On-period, $i = I_1 + (\dfrac{E}{R} - I_1)(1 - e^{-t/T}) = I_2$ when $t = t_1$ \hfill (1)

Off-period, $i = I_2 e^{-t/T} = I_1$ when $t = t_2$ \hfill (2)

From Eqs. (1) and (2) above,

$$I_2 = \frac{E}{R} \times \frac{1 - e^{-t_1/T}}{1 - e^{-(t_1 + t_2)/T}} \qquad I_1 = I_2 e^{-t_2/T}$$

In the problem, $E/R = 60/3 = 20$ A, $T = L/R = 0.009/3 = 0.003$ s, at 500 Hz, $t_1 + t_2 = 1/500 = 0.002$ s.
(i) 1/1 ratio gives $t_1 = t_2 = 0.001$ s, therefore $I_2 = 11.65$ A, and $I_1 = 8.35$ A.
(ii) 4/1 ratio gives $t_1 = 0.0016$ s, $t_2 = 0.0004$ s, therefore $I_2 = 16.99$ A, and $I_2 = 14.87$ A.
(iii) 1/4 ratio gives $t_1 = 0.0004$ s, $t_2 = 0.0016$ s, therefore $I_2 = 5.13$ A, and $I_1 = 3.01$ A.

Mean load voltages and currents are:
(i) $V_{\text{mean}} = 60 \times (1/2) = 30$ V $\qquad I_{\text{mean}} = V_{\text{mean}}/R = 30/3 = 10$ A
(ii) $V_{\text{mean}} = 60 \times (4/5) = 48$ V $\qquad I_{\text{mean}} = 48/3 = 16$ A
(iii) $V_{\text{mean}} = 60 \times (1/5) = 12$ V $\qquad I_{\text{mean}} = 12/3 = 4$ A

Example 4-5

Determine the size of the capacitor required in the simple parallel-capacitor chopper shown in Fig. 4-5, when the load is 6 Ω supplied from a 36 V battery. The required turn-off time for thyristor T_1 is 80 μs. If $R_2 = 5R_1$, determine the approximate minimum on-time for thyristor T_1. Neglect thyristor losses.

SOLUTION The capacitor C is initially charged to E, hence, when thyristor T_2 is fired, the voltage across C (and T_1) = $v = E - 2Ee^{-t/T}$. At the end of turn-off

time, v will fall to zero, when $t/T = 0.693$, therefore with $T = R_1 C$ and $t = 80\ \mu s$, $C = (80 \times 10^{-6})/(0.693 \times 6) = 19.2\ \mu F$.
$R_2 = 5R_1 = 30\ \Omega$, therefore when T_1 is fired, C will change as $v = E - 2Ee^{-t/T}$, where $T = R_2 C$, and (say) T_1 must be on until $v = 0.8E$ which gives t the minimum on-time as 1.33 ms.
If T_2 were fired with the capacitor charged to only $0.8E$, then $C = 22.7\ \mu F$ to ensure the 80 μs turn-off time.

Example 4-6

Using the same load specification as in Example 4-1, determine the component sizes and ratings for the circuit shown in Fig. 4-6. Neglect losses.

SOLUTION Initially C is charged to $2E$, so when T_2 is fired, the T_1 voltage is $v = E - 3Ee^{-t/RC}$, which is zero at time $t = 100\ \mu s$, hence $C = 18.2\ \mu F$. ($R = 5\ \Omega$). With T_1 on, when T_3 is fired CL will oscillate with an initial charge voltage of E, hence the voltage across T_2 will be $v = E - 2E \cos [t/\sqrt{(LC)}]$; when $v = 0$, $t = 100\ \mu s$, giving $L = 0.5$ mH. The peak current associated with LC oscillation from $\frac{1}{2}C(2E)^2 = \frac{1}{2}LI^2$ gives $I = 9.16$ A.
Peak current in $T_2 = 3E/R = 14.4$ A (3 because C after one cycle will be charged to $2E$).
Steady load current in $T_1 = 4.8$ A.
Following similar reasoning to that given in Example 4-1, T_1 carries 4.8 A plus an oscillating current of 9.16 A, which together gives a continuous value of 5.4 A at 200 Hz operation. The r.m.s. values of the currents in the other thyristors are: T_2, 1.34 A; and T_3, 1.59 A.

Example 4-7

The series resonance turn-off circuit of Fig. 4-8 has component values of $L = 3$ mH, $C = 20\ \mu F$, and $R = 50\ \Omega$. If the d.c. source is 100 V, determine the required rating of the thyristor. Neglect all losses.
Determine the load power if the thyristor is fired at a frequency of (i) 400 Hz, (ii) 200 Hz.

SOLUTION To determine the load power, it is necessary to determine an expression for the thyristor current and load voltage. Figure 4-28 shows the equivalent circuit when the thyristor is turned on (at time $t = 0$). Assume at $t = 0$, $v =$ capacitor voltage $= V_0$, and also at $t = 0$, $i_1 = 0$.
The voltage equations to the circuit are:

$$E = L\frac{di_1}{dt} + \frac{1}{C} \int i_2 dt$$

$$0 = \frac{1}{C} \int i_2 dt - i_1 R + i_2 R$$

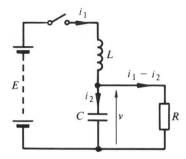

Figure 4-28

Transforming these equations into Laplace form gives

$$\frac{E}{s} = sL\bar{i}_1 + \frac{1}{C}\left[\frac{\bar{i}_2}{s} + \frac{V_0 C}{s}\right]$$

$$0 = \frac{1}{C}\left[\frac{\bar{i}_2}{s} + \frac{V_0 C}{s}\right] + \bar{i}_2 R - \bar{i}_1 R$$

Solving these equations yields

$$i_1 = \frac{E}{R} - e^{-\alpha t}\left[\frac{E}{R}\cos\omega t + \left(\frac{\alpha E}{\omega R} - \frac{E}{\omega L} + \frac{V_0}{\omega L}\right)\sin\omega t\right]$$

where

$$\alpha = \frac{1}{2CR}, \quad \omega = \left[\frac{1}{LC} - \left(\frac{1}{2CR}\right)^2\right]^{1/2}$$

$$v = E - L\frac{di_1}{dt} = E - e^{-\alpha t}\left[(E - V_0)\cos\omega t + \frac{\alpha}{\omega}(E + V_0)\sin\omega t\right]$$

i_1 is the thyristor current, v is the load voltage while the thyristor is on.
Using the values given in the problem,

$$i_1 = 2 - e^{-500t}[2\cos 4051t + (0.082\,V_0 - 7.89)\sin 4051t]$$

$$v = 100 - e^{-500t}[(100 - V_0)\cos 4051t + (12.34 + 0.1234\,V_0)\sin 4051t]$$

The thyristor current will cease when $i_1 = 0$, which is when $t = 0.94$ ms, for the initial turn-on of the thyristor (when $V_0 = 0$). The maximum value of i_1 will be at $t = 0.47$ ms, where the current is 8.41 A. At turn-off ($t = 0.94$ ms) the load voltage $v = 153.9$ V. The reverse voltage across the thyristor at turn-off $= 153.9 - 100 = 53.9$ V, decaying to zero in 431 μs when $100 = 153.9 e^{-t/RC}$, giving the available turn-off time.
The peak thyristor current $= 8.41$ A and, based on (say) an off-period equal in length to the on-period, the r.m.s. current will be of the order of $8.41/2 = 4.2$ A.
(i) With a cyclic rate of 400 Hz the thyristor will be fired-on every 2.5 ms. Using the earlier data, the load voltage will have decayed to $v = 153.9 e^{-(2.5-0.94)10^{-3}/RC} =$

32.3 V at the end of the first cycle. Using this value of V_0 and making further corrections, V_0 finally settles to approximately 30 V at the start and end of each cycle, with thyristor on-time of 1.035 ms. Voltage calculations give $v = 100 - e^{-500t}[70 \cos 4051t + 16 \sin 4051t]$ during the thyristor on-time, and $v = 130 e^{-t/RC}$ during the capacitor discharge period of $2.5 - 1.035 = 1.465$ ms.

The load energy per cycle will equal the battery output energy per cycle $= \int E i_1 \, dt$

$$= \int_0^{1.035 \times 10^{-3}} 100[2 - e^{-500t}(2 \cos 4051t - 5.43 \sin 4051t)] dt$$

$$= [200t - e^{-500t}(0.126 \cos 4051t + 0.065 \sin 4051t)]_0^{0.001035}$$

$$= 0.207 + 0.071 + 0.126 = 0.404 \text{ J}$$

Power $= 0.404 \times 400 = 161.6$ W

(ii) If the thyristor is fired at 200 Hz, giving a cycle duration of 0.005 s, then after the first input cycle the load voltage will have decayed to $v = 153.9 e^{-(5-0.94) \times 10^{-3}/RC} = 2.6$ V, which in practical terms can be considered zero. The energy drawn from the battery during each cycle with $V_0 = 0$ is 0.484 J, giving the power $= 0.484 \times 200 = 96.8$ W.

Example 4-8

Determine suitable component values for the parallel resonance turn-off circuit of Fig. 4-9 if the d.c. source is 100 V, the load 20 Ω, and a turn-off time of 80 μs is required for the thyristor. Assume the circuit is critically damped and neglect losses. Determine the time for which the load is on before the thyristor starts to turn off.

SOLUTION Referring to the equation for the *CLR* current at turn-off and using Eq. (4-15), $i = e^{-\alpha t}(At + B)$, where $\alpha = R/2L$. At the start of turn-off, $t = 0$, $i = 100/20 = 5$ A. After 80 μs, the current will fall back to 5 A (load voltage = 100 V). Reference to the waveforms in Fig. 4-9b shows that the capacitor voltage at the start of turn-off is less than 100 V, say 87 V, therefore assume $Ldi/dt = 87$ at $t = 0$. Using these three conditions yields $A = 302\,000$, $B = 5$, and $\alpha = 22\,040$, giving $L = R/2\alpha = 0.454$ mH. For critical damping $R^2 = 4L/C$, hence $C = 4.54$ μF. When the thyristor is turned on, the *LC* circuit oscillates such that $i = -(V/\omega L) \sin \omega t$, where $\omega - 1/\sqrt{(LC)}$, the time to when the thyristor turns off being given when $i = 100/20 = 5$ A, that is, when $t = 166$ μs. Peak thyristor current $= 5 + (V/\omega L) = 15$ A.

Example 4-9

Determine the on-time of the thyristor in Example 4-8 if the inductor is replaced by an ideal saturable reactor, and the capacitor has a value of 3.82 μF. The reactor has a saturated value of 0.34 mH rising to an unsaturated value of 50 mH below a current of 0.5 A.

SOLUTION The waveforms shown in Fig. 4-11 refer to this example. When the thyristor is turned on, the LC circuit oscillates at first such that $i = (V/\omega L) \sin \omega t$, where $\omega = 1/\sqrt{(3.82 \times 10^{-6} \times 50 \times 10^{-3})} = 2288$, and $V = 100$, hence $i = 0.874 \sin \omega t$ A. The reactor must saturate before the current reaches 0.874 A, in fact saturating when i reaches 0.5 A, that is, at $t = 266 \, \mu s$. At this time the capacitor voltage will have fallen to $100 \cos \omega t = 82$ V.

As the reactor changes from 50 mH to 0.34 mH, the current must change from 0.5 A to 6.06 A in order to maintain an unchanged stored magnetic energy $(\frac{1}{2} LI^2)$. The oscillating current is now $i = 10.6 \sin (\omega t + 0.61)$, where $\omega = 27\,748$ and $t = 0$ at the start of the oscillation, this current falling to 0.5 A at $t = 90 \, \mu s$.

To maintain the energy balance, the current now falls from 0.5 A to 0.04 A, and the time to when it comes out of saturation in the opposite sense is $21 + 266 = 287 \, \mu s$. The current now changes to the saturated state of 6.06 A, hence turning off the thyristor, as this is above the load-current value of 5 A.

The total time for the thyristor on-period is $266 + 90 + 287 = 643 \, \mu s$.

Example 4-10

Determine suitable values for L and C in the circuit of Fig. 4-12, where the d.c. source is 72 V feeding a load of 10 Ω having an inductance of 0.05 H. Ignore losses and allow a turn-off time of 70 μs. Describe the sequence of events during each cycle.

SOLUTION The maximum load current $= 72/10 = 7.2$ A. To turn off the thyristor, allow a peak LC current of (say) 11 A. The capacitor C will charge to 72 V, hence $11 = 72/\omega L$. When turning off, assume the load inductance holds the load current to 7.2 A, then the turn-off time is given by time in the oscillation $11 \sin \omega t$ between when it reaches 7.2 A to when it falls to 7.2 A, that is, $\omega t = 2 \arccos (7.2/11) = 1.71$, giving $\omega = 1.71/(70 \times 10^{-6}) = 24\,500$.

Using $\omega L = 72/11$ then $L = 0.267$ mH, and from $\omega = 1/\sqrt{(LC)}, C = 6.24 \, \mu F$.

Note that the load time constant $= 0.05/10 = 5$ ms, which is much longer than the LC oscillation cycle time of 0.25 ms.

The load current will remain sensibly constant due to its inductance, and say it is 4 A at an intermediate setting. During the off-period after the LC oscillation, the load current will freewheel in the commutating diode with capacitor C charged to 72 V. When the thyristor is fired, the load current of 4 A will flow from the d.c. source; the LC oscillation commences building up current in the thyristor. When the LC current reverses, the thyristor current will reduce until the oscillating current reaches 4 A; the thyristor then turns off. The oscillating current build-up now switches to the diode D, less the 4 A load component. When the oscillating current falls back to 4 A, the diode turns off, and the 4 A continues falling at a rate which reverses the load voltage, bringing the commutating diode in, while the LC current dies away when C is again charged to 72 V.

Example 4-11

Taking the component and load values given in Example 4-10, derive, neglecting losses, expressions for each current in the various arms of the resonant controller given in Fig. 4-13. Make the assumption that the load current is 4 A when the thyristor is turned on.

SOLUTION Figure 4-29 shows the circuit to be analysed with the given initial conditions.

Using the Laplace transform method

$$\frac{E}{s} = (L_1 + L_2)\, s\bar{i}_1 - (L_1 + L_2)I_0 - (L_1 s\bar{i}_2 - L_1 I_0) + R\bar{i}_1$$

$$\frac{E}{s} = \frac{\bar{i}_2}{sC} + \frac{V_0}{s} + L_2 s\bar{i}_1 - L_2 I_0 + R\bar{i}_1$$

At turn-on $t = 0$, $i_1 = 4$, $i_3 = 0$, $i_2 = 4$.

Substituting values of $E = 72$ V, $V_0 = 72$ V, $I_0 = 4$ A, $C = 6.24\ \mu F$, $L_1 = 0.267$ mH, $L_2 = 0.05$ H, and $R = 10\ \Omega$, and solving the equations, the expressions for the currents are

$$i_1 = 7.2 - 3.2\, e^{-\alpha t}$$

$$i_2 = 4 \cos \omega t - 11 \sin \omega t$$

$$i_3 = 7.2 - 3.2\, e^{-\alpha t} - 4 \cos \omega t + 11 \sin \omega t$$

where $\alpha = 10/0.05 = 200$, $\omega = 1/\sqrt{(0.267 \times 10^{-3} \times 6.24 \times 10^{-6})} = 24\,500$.

These solutions are approximate, omitting terms of small magnitude. As $L_1 \ll L_2$

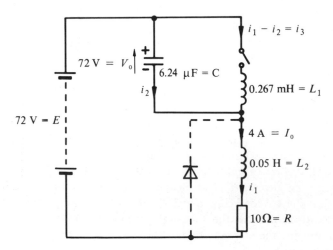

Figure 4-29

then the load current is the solution ignoring the LC oscillation, and the thyristor current is the LC oscillation current ignoring the load except for the 4 A.

The thyristor will turn off when i_3 attempts to reverse, by which time there will be little change in the load current.

Example 4-12

A load of 10 A is to be controlled by the external pulse commutation circuit of Fig. 4-14. If the supply is 200 V, suggest suitable values for the capacitor and inductors, assuming they are lossless and allowing a turn-off time of 100 μs.

SOLUTION Initially when the d.c. source is connected, capacitor C will charge to $2 \times 200 = 400$ V. When T_2 is fired, the equivalent circuit is as shown in Fig. 4-30a. At $t = 0$, i_1 the current in L_1 will be the load current of 10 A. Using Eq. (4-8) or other means, $i_1 = 10 \cos \omega t + (400/\omega L_1) \sin \omega t$, where $\omega = 1/\sqrt{(L_1 C)}$. Make L_1 such that the peak value of i_1 is approximately 20 A, that is, double the load value. Then (say) $20 = 400/\omega L_1$, that is, $\omega L_1 = 20$. When $v = L_1 di_1/dt$ falls to 200 V, the thyristor T_1 voltage will start to go positive, so this defines the turn-off time, therefore

$$- 10 \, \omega L_1 \, \sin \omega t + 400 \cos \omega t - 200$$

where $t = 100$ μs. From these equations, $L_1 = 3.11$ mH, $\omega = 6435$, and $C = 7.76$ μF. When the thyristor T_1 voltage goes positive, the diode D_2 will start to conduct, giving the equivalent circuit shown in Fig. 4-30b, which will apply until i_1 falls to zero, when T_2 turns off. A new set of equations apply starting at $t = 0$, with $i_1 = 20$ A, $i_2 = 0$ A, and $v = 200$ V. Inductor L_2 must be much greater than L_1, say $L_2 = 0.02$ H. A higher value would make the $L_2 C$ oscillation rather slow. Solving the equations for these conditions,

$$i_1 = 8655t + 8.04 \sin \omega t + 17.31 \cos \omega t + 2.69$$

$$i_2 = 8655t + 2.69 - 2.69 \cos \omega t$$

$$v = 27 + 173 \cos \omega t - 372 \sin \omega t$$

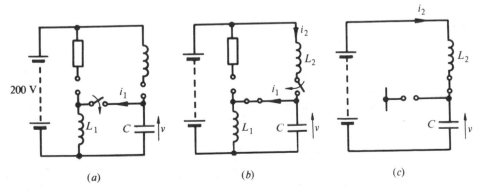

(a)   (b)   (c)

Figure 4-30

where $\omega = 6919$. When i_1 falls to zero, T_2 turns off giving $t = 332.7$ μs, when at this time $i_2 = 7.36$ A, and $v = -365$ V.

The circuit conditions will now change to those shown in Fig. 4-30c where, starting again with $t = 0$, $v = -365$ V, and $i_2 = 7.36$ A. Solving the equations for i_2 and v gives

$$i_2 = 11.1 \sin \omega t + 7.36 \cos \omega t$$

$$v = 200 - 565 \cos \omega t + 374 \sin \omega t$$

where $\omega = 2538$. Diode D_2 turns off when $i_2 = 0$, at $t = 1007$ μs, leaving $v = 878$ V. The turn-off time for T_2 is given by the time to when $v = 0$, that is, $t = 270$ μs, which is above the 100 μs specified.

Note that after commutation the voltage on the capacitor is 878 V, whereas it was only 400 V at the start. By ignoring losses, further calculations would show greater rises in this voltage at the end of each commutation. In practice, non-linear losses would limit the rise.

Example 4-13

Using the data of Example 4-12, determine the capacitor voltage after three cycles if the diode D_2 is replaced by a thyristor T_3 fired after thyristor T_2 (Fig. 4-14a) is turned off. Take the losses in the $L_1 T_2 C$ circuit to be represented by 3 Ω and those in the $L_2 T_3 C$ circuit by 9 Ω.

SOLUTION Equations (4-4) and (4-5) relate to the conditions, as by replacing the diode, only Figs. 4-30a and c apply.

For the $L_2 T_3 C$ response, $\alpha = 225$, $\omega = 2528$, and for the $L_1 T_2 C$ response, $\alpha = 482$, $\omega = 6419$.

When the source is connected, T_3 fired, with $i_2 = 0$ and $v = 0$ at $t = 0$. $i_2 = 3.96 e^{-\alpha t} \sin \omega t$, which is zero when $t = 1243$ μs, $v = 200 + 200.8 e^{-\alpha t} \sin (\omega t + 1.482) = 351$ V, when i_2 falls to zero.

T_2 fired, $i_1 = 10$ A, and $v = 351$ V at $t = 0$.
$i_1 = 20.88 e^{-\alpha t} \sin (\omega t + 0.5)$, which is zero when $t = 412$ μs. $v = -789 e^{-\alpha t} \sin (\omega t - 0.461) = -622$ V when i_1 falls to zero.

T_3 fired, $i_2 = 0$ A, and $v = -622$ V at $t = 0$.
$i_2 = 16.26 e^{-\alpha t} \sin \omega t$, which is zero when $t = 1243$ μs. $v = 200 - 825 e^{-\alpha t} \sin (\omega t + 1.482) = 824$ V when i_2 falls to zero.

T_2 fired, $i_1 = 10$ A, and $v = 824$ V at $t = 0$.
$i_1 = 43.2 e^{-\alpha t} \sin (\omega t + 0.234)$, which is zero when $t = 453$ μs. $v = -836 e^{-\alpha t} \sin (\omega t - 1.404) = -670$ V when i_1 falls to zero.

T_3 fired, $i_2 = 0$ A, and $v = -670$ V at $t = 0$.
$i_2 = 17.2 e^{-\alpha t} \sin \omega t$, which is zero when $t = 1243$ μs. $v = 200 - 873 e^{-\alpha t} \sin (\omega t + 1.482) = 857$ V when i_2 falls to zero.

T_2 fired, $i_1 = 10$ A, and $v = 857$ V at $t = 0$.

$i_1 = 44.8\,e^{-\alpha t} \sin(\omega t + 0.225)$, which is zero when $t = 454\,\mu$s. $v = -868\,e^{-\alpha t}$ $\sin(\omega t - 1.413) = -696$ V when i_1 falls to zero.

It can be seen that the final charge on the capacitor will be of the order of 700 V.

Example 4-14

The coupled-pulse commutation circuit of Fig. 4-16 is to control a 60 A load from a 48 V d.c. source. Using thyristors with a turn-off time of 30 μs, suggest suitable values for the circuit components. The load is to be switched at a maximum rate of 1 kHz.

SOLUTION In common with many other turn-off circuits, design calculations can only be approximate, modifications being made in practice after experimental testing. The components can be assumed lossless in the first instance.

When thyristor T_1 is initially fired, assume capacitor C will charge to 48 V. When thyristor T_2 is fired, the capacitor will discharge via the d.c. source and load at 60 A, assuming level load current. Therefore $dv/dt = 60/C$, where $dv/dt = 48/(30 \times 10^{-6})$, the capacitor (and T_1) voltage reversing after 30 μs, giving $C = 37.5\,\mu$F. The capacitor will take a further 30 μs to be charged to 48 V when T_2 turns off.

The half-cycle time of the $L_1 C$ oscillation determines the minimum on-time of T_1 before T_2 can be fired. Load cycle time $= 1/10^3 = 1000\,\mu$s. Say minimum on-time $= 200\,\mu$s (hence allowing control of load voltage down to approximately 20%), this allowing at least 100 μs to turn-off T_2. Therefore, $1/2\pi\sqrt{(L_1 C)} = 10^6/(200 \times 2)$, giving $L_1 = 108\,\mu$H.

At first switching of T_1 the circuit is as shown in Fig. 4-31, with zero initial conditions. Circuit analysis yields that the maximum value of v is approximately $(E/R)\sqrt{(L_2/C)}$, taking $M = \sqrt{(L_1 L_2)}$. If C is to be charged to the d.c. source level of 48 V, then $L_2 = 24\,\mu$H, taking $E/R = 60$ A.

Turns ratio of coupled coils $= \sqrt{(L_1/L_2)} = 2.12$.

At subsequent firing, the capacitor will be initially charged to 48 V, so that an additional voltage of approximately $(E/R)\sqrt{(L_2/C)}$ is given to the capacitor. Maximum capacitor voltage $48 + 48 = 96$ V.

Peak value of $T_1 DC$ oscillating current $= 96/\omega L_1 = 96\sqrt{(C/L_1)} = 56$ A for half-cycle of duration 200 μs.

Thyristor T_2 will carry 60 A for the time taken for the capacitor voltage to reverse from 96 V to 48 V, that is for time $= [37.5 \times 10^{-6} \times (96 + 48)]/60 = 90\,\mu$s.

Rating of thyristor T_1: r.m.s. current is (say) 70 A from load current and short-time oscillating current, $dv/dt = 60/(37.5 \times 10^{-6}) = 1.6$ V/μs; $di/dt = 48/(24 \times 10^{-6}) = 2$ A/μs, P.F.V. = 96 V.

Rating of thyristor T_2: r.m.s. current $= 60/\sqrt{(1000/90)} = 18$ A, that is, 60 A for 90/1000 part of each load cycle, $dv/dt = 56/(37.5 \times 10^{-6}) = 1.5$ V/μs; $di/dt = 96/$ (stray inductance in CT_2 loop), P.F.V. = 96 V.

Rating of diode D: r.m.s. current is 18 A calculated from the intermittent oscillating peak current of 56 A. In the $L_2 C$ oscillation following turn-off of T_2, the P.R.V. may reach 96 × turns ratio of L_1/L_2, that is, 204 V.

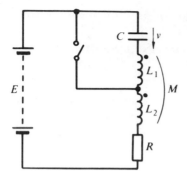

Figure 4-31

It must be emphasized that the above calculations are very approximate, giving only guidance to component values, experimental testing being required to finalize component requirements.

Example 4-15

Determine the size of capacitor required in the circuit of Fig. 4-17; the supply is 60 V, each load is 3 Ω, and the turn-off time of the thyristors is 25 μs. If the loads are switched at 1500 Hz, determine the percentage increase in load power due to the capacitor charging requirements.

SOLUTION Inspection of the waveforms in Fig. 4-17 shows the turn-off time is t in the equation, $0 = -60 + (2 \times 60)(1 - e^{-t/3C})$. Taking t as 25 μs gives $C = 12\ \mu$F.

Capacitor charging current $= [(2 \times 60)/3]\ e^{-t/RC}$.

$$\text{Loss in load} = \int i^2 R\, dt = \int_{t=0}^{t \to \infty} (40\ e^{-t/RC})^2\ 3\, dt = 0.0864 \text{ J}$$

Basic energy in each load in one cycle $= \dfrac{1}{2 \times 1500} \times \dfrac{60}{3} \times 60 = 0.4$ J. % increase in load power $= 100(0.0864/0.4) = 21.6\%$.

Example 4-16

The complementary impulse-commutated bridge of Fig. 4-19 is used to supply a resistive load of 5 Ω from a 200 V d.c. source. Suggest suitable component values allowing a thyristor turn-off time of 40 μs. Assume components are ideal and lossless.

SOLUTION Figure 4-20 defines conditions during turn-off. The oscillation involves a total stored energy of $\frac{1}{2}LI^2 + \frac{1}{2}CV^2$ at a frequency of $1/2\pi\sqrt{(2CL)}$. The turn-off time of 40 μs represents the time for the capacitor voltage to fall to half its initial value.

Initially the capacitor will discharge into the inductor, increasing its current such that $i = I_{max}\sin(\omega t + \phi)$, where $i = 200/5$ at $t = 0$. The capacitor voltage = $L di/dt$, hence the time to when di/dt has fallen to half its initial value is the turn-off time available to the thyristor. In choosing a value for I_{max} one must compromise between ensuring commutation while avoiding excessive commutation losses. Choose (say) I_{max} in this case as $1.5 \times (200/5) = 60$ A, then $i = 60 \sin(\omega t + \phi)$, where $i = 40$ A at $t = 0$, giving $\phi = 0.73$. $di/dt = 60 \omega \cos(\omega t + \phi)$, which is half its initial value at $\omega t = 0.459$, therefore $\omega = 0.459/(40 \times 10^{-6}) = 11\,478$.

When $(\omega t + \phi) = \pi/2$, the capacitor voltage is zero, and the inductor current is at its peak value of 60 A. The charge lost on C_4 equals that gained on C_1, hence the additional stored energy in inductor L comes from the battery and equals

$$\int_{\omega t=0}^{\omega t+\phi=\pi/2} 200(60/2)\sin(\omega t + \phi)dt = 4472/\omega = \tfrac{1}{2}L(60^2 - 40^2).$$

Also $\omega = 11\,478 = 1/\sqrt{(2CL)}$, from which $L = 390\,\mu$H and $C = 9.74\,\mu$F.

Referring to Fig. 4-20c, the current in diode D_4 will start at 60 A, decaying to zero when the stored energy $\tfrac{1}{2}LI^2 = 0.7$ J is lost at each commutation.

dv/dt at thyristor $T_1 = i/C = 30/9.74 = 3$ V/μs.

Example 4-17

Repeat the calculation for Example 4-16 if an inductance is added to the load, making the load current sensibly level over the commutation period.

SOLUTION Conditions are as shown in Fig. 4-21 where, by considering the load inductance to be very much larger than L, circuit analysis will yield that the current i in L is approximately $I_{max} \sin(\omega t + \phi) - I$.

Taking the same criterion as in Example 4-16 (that i will have a peak value of 60 A), then $i = 100 \sin(\omega t + \phi) - 40$, with $i = 40$ A at $t = 0$, giving $\phi = 0.927$. $di/dt = 100 \omega \cos(\omega t + \phi)$, which is half its initial value at $\omega t = 0.3388$, therefore $\omega = 0.3388/(40 \times 10^{-6}) = 8470$.

The battery current = $(100/2) \sin(\omega t + \phi)$. At $(\omega t + \phi) = \pi/2$, the capacitor voltage is zero and the inductor current is at its peak of 60 A. The energy drawn from the battery is

$$\int_{\omega t=0}^{\omega t+\phi=\pi/2} 200 \times 50 \sin(\omega t + \phi)dt = 0.7084 \text{ J}.$$

The load voltage equals that across L, hence the energy to the load is

$$L\frac{di}{dt} \times I \times \text{time} = \int_{\omega t=0}^{\omega t+\psi=\pi/2} L \times 40 \times 100 \omega \cos(\omega t + \phi)dt = 800L.$$

Energy lost in C_4 is transferred to C_1. Additional energy stored in L is $\frac{1}{2}L(60^2 - 40^2) = 1000L$. Therefore $0.7084 = 800L + 1000L$, giving $L = 394\ \mu H$. From $\omega = 1/\sqrt{(2CL)}$ we get $C = 17.7\ \mu F$.

Example 4-18

Determine suitable component values for the impulse-commutation circuit of Fig. 4-22, given the d.c. source to be 200 V and the load (inductive) current to be 40 A. Allow a thyristor turn-off time of 40 μs.

SOLUTION Conditions during turn-off are shown in Figs. 4-23 and 4-24.
The turn-off time exists from t_1 to t_2, which from Fig. 4-23b shows a simple LC oscillation starting at 40 A and ending at 40 A, being the top of a sinewave as shown by i_{D1}. The equation to this current is $i = I_{max} \sin(\omega t + \phi)$. I_{max} must be significantly greater than 40 A, but not so high as to lead to excessive component rating and loss. Select (say) $I_{max} = 60$ A, then $i = 60 \sin(\omega t + \phi)$, being 40 at $t = 0$, hence $\phi = 0.73$. Midway between t_1 and t_2, the current is at its peak when $(\omega t + \phi) = \pi/2$, where $t = $ half T_1 turn-off time $= (40/2)\mu s$, hence $\omega = 42\,051 = 1/\sqrt{(LC)}$.
The energy available for commutation $= \frac{1}{2}L \times 60^2 = \frac{1}{2}CV^2$, where V is the capacitor voltage at t_0 in Fig. 4-23. Selecting $V = 2 \times 200 = 400$ V, we have two equations relating C and L which yield $C = 3.57\ \mu F$ and $L = 159\ \mu H$.
The time from t_2 to t_3 is a period when the CL circuit is connected to the d.c. source as shown in Fig. 4-24c, hence an additional boost is given to the capacitor which, in the steady state, is sufficient to replace the losses.

Example 4-19

Using the data of Example 4-18, demonstrate how different losses will affect the capacitor voltage. Also demonstrate the effect of different load currents.

SOLUTION Referring to Fig. 4-23, assume that the capacitor voltage v is 400 V at t_0. Ignoring losses at t_1 and t_2, $\frac{1}{2}Cv^2 = \frac{1}{2}C\,400^2 - \frac{1}{2}L\,40^2$, giving $v = 298$ V. If losses reduce v by 10% at t_2, then $v = 268$ V. Equations (4-8) and (4-9) will give approximately conditions between t_2 and t_3, where if t is the time starting at t_2 then $i = 40$ A, $v = 268$ V, $di/dt = (200 - 268)/L$, and $dv/dt = 40/C$, all at $t = 0$; hence $v = 200 + 275 \sin(\omega t + 0.25)$ and $i = 41.3 \sin(\omega t + 1.82)$. When $i = 0$ at $(\omega t + 1.82) = \pi$, then v is 475 V at t_3, hence the capacitor has received a further charge.
If the losses reduce v by 20% at t_2, calculations give $v = 238.5$ V at t_2, with $v = 469$ V at t_3.
These calculations show that the capacitor voltage would settle to a value somewhat above 400 V.
Taking a voltage loss of 10%, starting with $v = 400$ V at t_0, but a load current of 30 A, then conditions at t_2 will be $i = 30$ A, $v = 312$ V, $di/dt = (200 - 312)/L$,

and $dv/dt = 30/C$; hence $v = 200 + 229 \sin(\omega t + 0.51)$, and $i = 34.3 \sin(\omega t + 2.08)$. When $i = 0$ at $(\omega t + 2.08) = \pi$, then v is 429 V at t_3.

Recalculating at $i = 20$ A gives 393 V at t_3.

These calculations show that a reduction in the load current reduces the capacitor charge.

In practice, conditions in this circuit are so complex that the final values of the components need to be determined under experimental test conditions.

Example 4-20

Using the data of Example 4-18, but with short thyristor turn-off times, recalculate the required capacitor and inductor sizes.

SOLUTION Following the same reasoning as in Example 4-18 the values are: for a thyristor turn-off time of 35 μs, $\omega = 48\,059$, $C = 3.12\ \mu F$, $L = 138.7\ \mu H$,

at 30 μs, $\omega = 56\,069$, $C = 2.68\ \mu F$, $L = 118.9\ \mu H$

at 25 μs, $\omega = 67\,283$, $C = 2.23\ \mu F$, $L = 99.1\ \mu H$

at 20 μs, $\omega = 84\,104$, $C = 1.78\ \mu F$, $L = 79.3\ \mu H$.

It can be seen that by using higher-grade thyristors with faster turn-off times, the component values (and consequently the commutation losses) can be reduced.

FIVE

FREQUENCY CONVERSION

Frequency conversion techniques can be considered to cover those methods by which it is possible to take a fixed frequency or d.c. source, and convert this energy to provide a load with a different or variable frequency supply.

Cycloconversion is concerned mostly with direct conversion of energy to a different frequency by synthesizing a low-frequency wave from appropriate sections of a higher-frequency source.

Inverters broadly cover those circuits which have a d.c. source, and by appropriate switching of rectifying devices enable an alternating voltage to be synthesized, for feeding to an a.c. load.

5-1 CYCLOCONVERTER

A cycloconverter can be considered to be composed of two converters connected back to back as shown in Fig. 5-1a. The load waveforms of Fig. 5-1b show that in the general case, the instantaneous power flows in the load fall into one of four periods. The two periods, when the product of load voltage and current is positive, require power flow into the load, dictating a situation where the converter groups rectify, the positive and negative groups conducting respectively during the appropriate positive and negative load-current periods.

The other two periods represent times when the product of load voltage and current is negative, hence the power flow is out of the load, demanding that the converters operate in the inverting mode.

5-1-1 Principle

The principle of the cycloconverter can be demonstrated by using the simplest possible single-phase input to single-phase output with a pure resistance as load, as shown in Fig. 5-2a. Each converter is a bi-phase half-wave connection, the positive group labelled P and the negative group for reverse current labelled N.

The load-voltage waveform of Fig. 5-2b is constructed on the basis that P group only is conducting for five half cycles, the thyristors being fired without delay, that is, P is acting as if it were a diode rectifier. For the next five half cycles, N group only conducts to synthesize the negative half cycle to the load. Inspection of the voltage waveform in Fig. 5-2b clearly shows an output frequency of one fifth

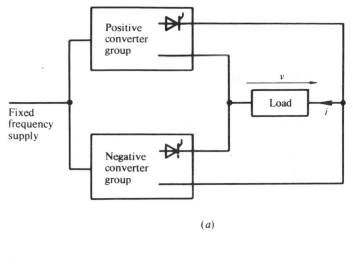

(a)

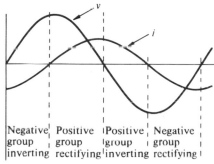

(b)

Figure 5-1 General cycloconverter layout. (a) Block diagram representation. (b) Ideal load waveforms.

of the input, the shape tending to a square wave having a large low-order harmonic content. The individual thyristor conducting periods are shown in the current waveforms of Fig. 5-2c, the current from the supply being a continuous sinewave.

A closer approximation to a sinewave can be synthesized by phase delaying the firing of the thyristors as shown in Fig. 5-2d. The load receives a full half cycle of the source at its peak period, with an increasing degree of firing delay as the load voltage zero is approached. This phase-controlled operation reduces the harmonic content of the load-voltage waveform, as compared to the earlier waveform. The plotted current waveforms of Fig. 5-2e show that the supply current is very distorted, with a component at the output frequency, that is, a sub-harmonic of the supply frequency.

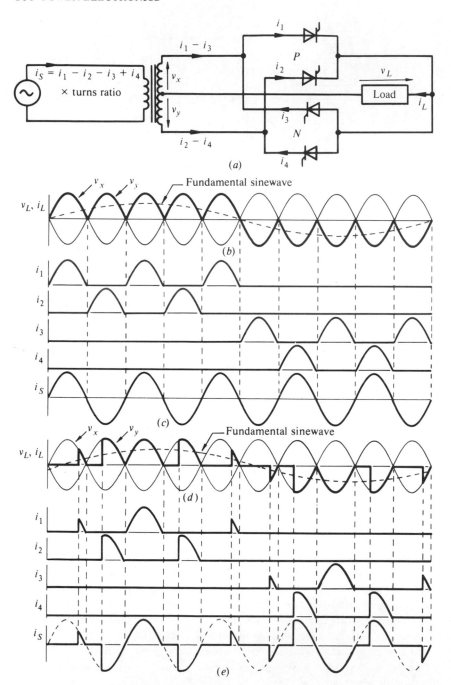

Figure 5-2 Single-phase bi-phase converters to load of pure resistance. (*a*) Circuit. (*b*) Load voltage with full conduction of each thyristor. (*c*) Current waveforms associated with (*b*). (*d*) Load voltage with phase control of each thyristor. (*e*) Current waveforms associated with (*d*).

5-1-2 Blocked group operation

Inspection of Figs. 5-1*a* and 5-2*a* reveals that if at any instant thyristors are conducting in both the positive and negative groups, then a short-circuit exists on the supply via the thyristors. To avoid this eventuality, a reactor can be inserted between the groups to limit the circulating current, or the firing control circuitry can be arranged so that neither group is fired while current is flowing in the other group. This circulating current-free (or blocked group) operation involves blocking the firing of one group, until the current in the other group has ceased.

The operation of the blocked group cycloconverter with various loads can be readily explained by reference to the three-pulse connection shown in Fig. 5-3, with the associated waveforms in Figs. 5-4 to 5-6.

Taking the cycloconverter to be feeding a load of pure resistance, set to give a maximum output load voltage, the waveforms are as shown in Fig. 5-4. The desired sinusoidal output voltage is shown at a frequency such that one output cycle occupies just less than five input cycles. The thyristors are fired at such angles as to follow as closely as possible the fundamental sinewave. As the load is resistance, the voltage waveforms will contain zero periods. The output of the negative group differs in waveform to that of the positive group because, as the output wave is not composed of a whole number of input cycles, successive output half cycles start at different instants relative to the input cycle. The current waveforms show the considerable unbalance imposed on the supply.

When the load is inductive, the waveforms are as shown in Fig. 5-5, these being at a condition of maximum voltage. The load current will lag the voltage and, as the load-current direction determines which group is conducting, the group on-periods are delayed relative to the desired output voltage. The group thyristors are fired at such angles to acheive an output as close as possible to a sinewave, but

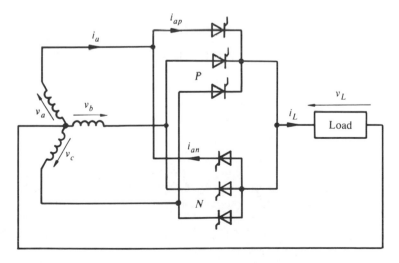

Figure 5-3 Single-phase load fed from a three-pulse cycloconverter.

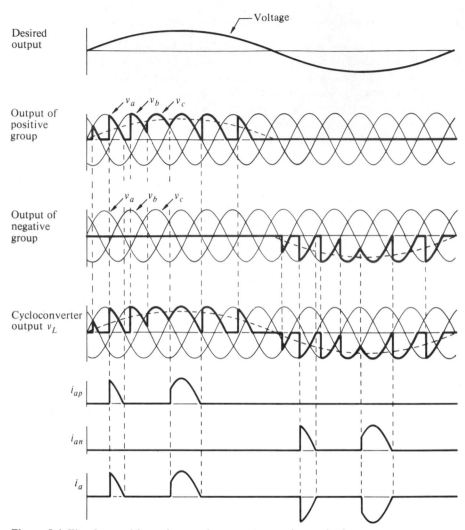

Figure 5-4 Waveforms with maximum voltage to a pure resistance load.

now the lagging load current takes each group into the inverting mode. The group will cease conducting when the load current reverses. The load-voltage waveform is shown as a smooth transfer between groups, but in practice a short gap would be present to ensure cessation of current in, and the regaining of the blocking state in, the outgoing group before the incoming group is fired. The waveforms drawn assume the current is continuous within each load half-cycle. The effects of overlap as decribed in Sec. 3-1 will in practice be present in the waveforms.

The load-current waveform in Fig. 5-5 is assumed to be sinusoidal, although in practice it would contain a ripple somewhat smaller than that in the voltage waveform. A light load inductance would result in discontinuous current, giving short

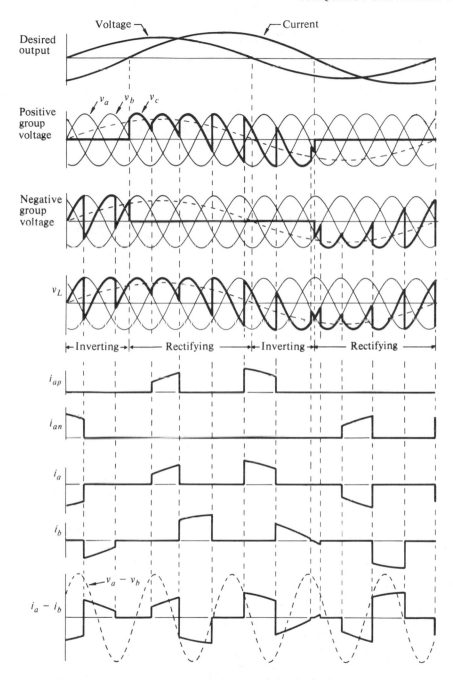

Figure 5-5 Waveforms with maximum voltage to an inductive load.

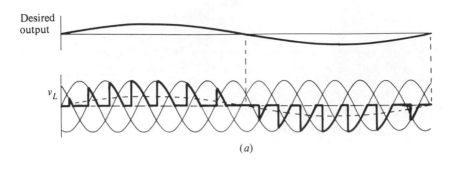

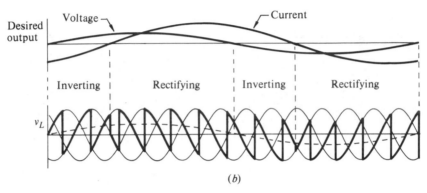

Figure 5-6 Waveforms when the load voltage is at half maximum. (*a*) Pure resistance load. (*b*) Inductive load, current continuous.

zero-voltage periods. Each thyristor will conduct the appropriate block of load current, having the branch currents shown. If one assumed the supply to be via a delta primary transformer, the current $i_a - i_b$ would represent the transformer input line current. The input-current waveform shows changes in shape from cycle to cycle but, where the input and output frequencies are an exact multiple, the waveform will repeat over each period of output frequency.

A reduction in the output voltage can be obtained by firing angle delay as shown in Fig. 5-6. Here firing is delayed, even at the peak of the output voltage, so that control is possible over the magnitude of the output voltage. Comparison of Fig. 5-6 with Fig. 5-5 indicates a higher ripple content when the output voltage is reduced.

The three-pulse cycloconverter when feeding a three-phase load can be connected as shown in Fig. 5-7a, with a total of 18 thyristors. A six-pulse cycloconverter can be based on either six-phase half-wave blocks or the bridge connection as shown in Fig. 5-7b, when 36 thyristors are required.

Examples of the cycloconverter output waveforms for the higher-pulse connections are shown in Fig. 5-8, with an output frequency of one-third of the input frequency. It is clear from these waveforms that the higher the pulse-number, the closer is the output waveform to the desired sinusoidal waveform. The output

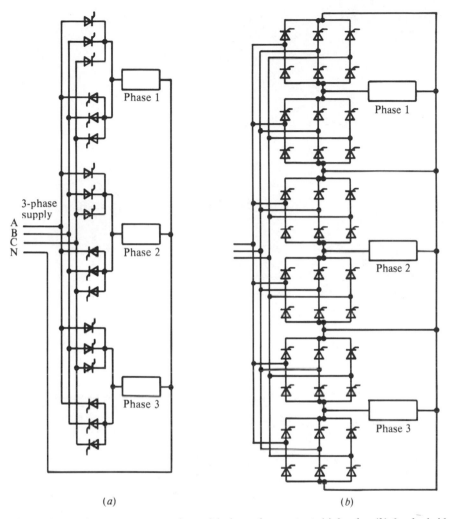

Figure 5-7 Cycloconverter connections with three-phase output. (*a*) 3-pulse. (*b*) 6-pulse bridge.

frequency is in general limited to about one-half to one-third of the input frequency, the higher-pulse connections permitting a higher limit.

The peak value of the output voltage is the mean level of the direct voltage which each group can supply. This statement can be justified by inspection of the waveforms, when at the peak of the output voltage the group conducting is acting as if it were a diode rectifier.

Hence, from Example 2-22 for a *p*-pulse cycloconverter, the peak value of the output voltage is

$$V_{0(\text{max})} = \frac{p}{\pi} \sin \frac{\pi}{p} V_{s(\text{max})} \tag{5-1}$$

where $V_{s(\text{max})}$ is the maximum value of the supply voltage.

When the output voltage is reduced in magnitude by firing delay α, then

$$V_{0(max)} = \frac{p}{\pi} \sin \frac{\pi}{p} V_{s(max)} \cos \alpha \tag{5-2}$$

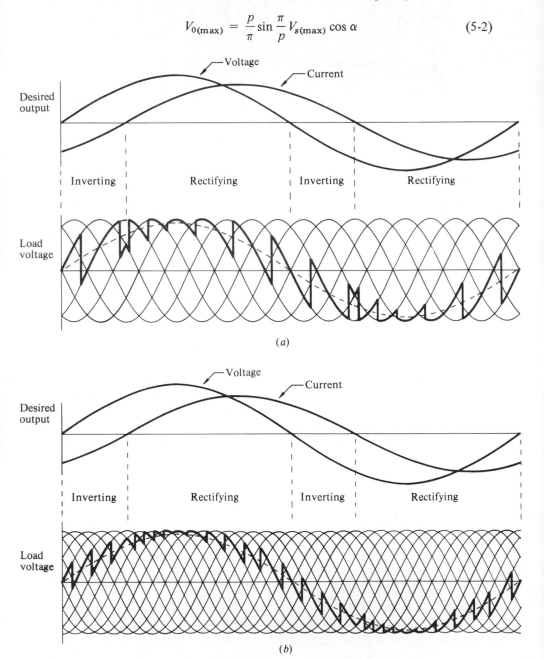

Figure 5-8 Cycloconverter load-voltage waveforms with a lagging power factor load. (*a*) 6-pulse connection. (*b*) 12-pulse connection.

When the three-pulse cycloconverter feeds a three-phase balanced load as in Fig. 5-7, the current loading on the supply is much more evenly balanced. The waveforms to illustrate this are shown in Fig. 5-9 for a frequency ratio of 4/1 with a load of 0.707 power factor lagging. It has been assumed that the load current is

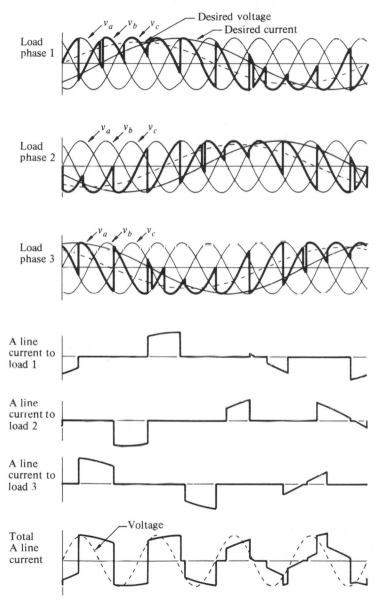

Figure 5-9 Development of total input current to 3-pulse cycloconverter with three-phase lagging power factor load.

sinusoidal, although in practice it must contain ripple. The total load current is not identical from one cycle to the next, does obviously contain harmonics, and its fundamental component lags the supply voltage by a larger amount than the load power factor angle.

The thyristors of a cycloconverter are commutated naturally, and whether the load be resistive, inductive, or capacitive, the firing of the thyristors must be delayed to shape the output. The net result is that the a.c. supply input current will always lag its associated voltage.

5-1-3 Circulating current mode

The previous section specified cycloconverter operation where either the positive or negative groups were conducting, but never together. If a centre-tapped reactor is connected between the positive group P and negative group N as shown in Fig. 5-10, then both groups can be permitted to conduct. The reactor will limit the circulating current i_c, the value of its inductance to the flow of load current from either group being one quarter of its value to the circulating current, because inductance is proportional to the square of the number of turns.

Typical waveforms are shown in Fig. 5-11 for the three-pulse cycloconverter shown in Fig. 5-10. Each group conducts continuously, with rectifying and inverting modes as shown. The mean between the two groups will be fed to the load, some of the ripple being cancelled in the combination of the two groups. Both groups synthesize the same fundamental sinewave. The reactor voltage is the instantaneous difference between the two group voltages.

The circulating current shown in Fig. 5-11 can only flow in one direction, the thyristors preventing reverse flow. Hence, the current will build up during the reactor voltage positive periods until in the steady state it is continuous, rising and falling as shown.

When load current flows, the group currents will be as shown in Fig. 5-12. Ignoring the ripple voltage components, the two group voltages are identical sine-

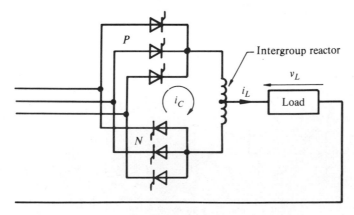

Figure 5-10 3-pulse cycloconverter with intergroup reactor.

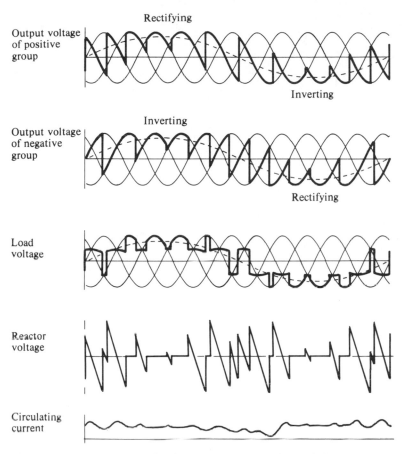

Figure 5-11 Waveforms of a 3-pulse cycloconverter with circulating current but without load current.

waves at the output frequency, in opposition as shown in Fig. 5-12a. Hence, no circulating current arises from the fundamental voltage. At the start of conduction (say, from the positive group), the growth of i_p will induce a voltage across the reactor which is reflected into the negative group circuit, the reverse bias of its thyristors preventing any current i_n flow. After i_p reaches its peak and starts to fall, a reverse voltage is induced across the reactor, permitting current flow from the negative group. The action of the reactor is such as to maintain its stored magnetic energy level, hence as i_p falls, i_n rises at an equal rate. In the reverse half load cycle, i_p falls to zero, i_n rising to its peak. As the reactor stored magnetic energy does not change, the reactor voltage is zero. The net outcome is that the waveforms of current are as shown in Fig. 5-12b where $i_L = i_p - i_n$, and each group current has a mean value equal to half the maximum value of the load current. The circulating current due to the ripple voltage will be superimposed on to these fundamental current components.

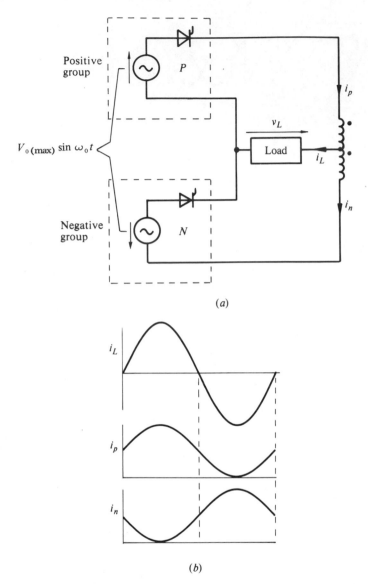

(a)

(b)

Figure 5-12 Idealized cycloconverter in circulating current mode. (*a*) Equivalent circuit. (*b*) Waveforms.

The continuous current of each group in the circulating current mode imposes a higher loading on each group compared to the blocked group operation. In practice, the circulating current mode would only be used when the load current is low, so that continuous load current with a better waveform can be maintained. At the higher levels of load current, the groups would be blocked to prevent circulating current. Control circuits would be used to sense the level of the load current, allowing firing pulses to each group at low current levels, but blocking firing to one or other group at the higher current levels. The reactor would be designed to saturate at the higher current levels when the cycloconverter is operating in the blocked group mode, thus permitting a smaller core.

5-1-4 Control

The mean output voltage of a converter is proportional to the cosine of the firing delay angle. The firing delay required of an individual thyristor is that angle where the required instantaneous value of the desired output (relative to its maximum possible value) is equal to $\cos \alpha$, α being the firing delay angle.

Referring to Fig. 5-13, waveforms are drawn giving the maximum possible output. Taking the positive group at its positive peak, the output waveform shows zero firing delay. The next switching requires a firing delay to satisfy the drop in the desired output. The intersection or crossing points of a reference sinewave (the desired output) to cosine waves drawn with their peaks at the instant of zero firing delay determines the instant of firing to each cycloconverter thyristor.

Figure 5-13 shows how the output of the positive group is constructed, noting how, when rectifying, the firing angle is less than $90°$ as shown by α_{p1}; but when inverting, in the negative half cycle firing delay extends beyond $90°$ as shown

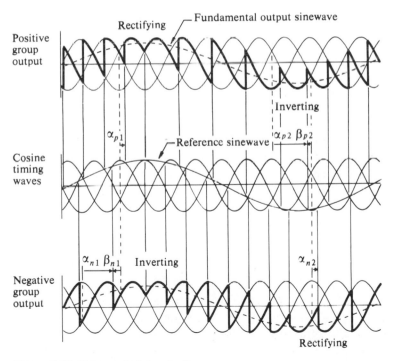

Figure 5-13 Firing angle determination.

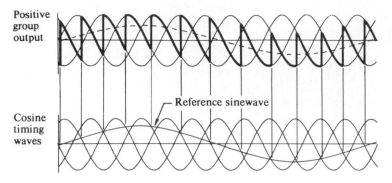

Figure 5-14 Firing angles at reduced output voltage.

by α_{p2}, with β_{p2} the associated firing advance. The output of the negative group is determined in a similar manner.

One aspect of the firing angle determination which is highlighted in Fig. 5-13 is with reference to the negative group when inverting at the instant of maximum output voltage. As drawn, the angle of firing advance is zero, which, as described in Sec. 3-1, must be avoided due to overlap. In practice, a firing pulse must be delivered to the thyristor at an early enough instant to allow overlap to be completed.

To reduce the output voltage, the reference sinewave is reduced to the desired level as shown in Fig. 5-14. One effect of the reduced output is to increase the ripple content of the output wave.

The control requirements of a cycloconverter are necessarily complex, but the very basic layout is shown in block diagram form in Fig. 5-15. A signal, detecting current flow in one converter, will be fed into the firing circuit of the other group, so as to inhibit its firing pulses when blocked group operation is required.

5-2 ENVELOPE CYCLOCONVERTER

The cycloconverter can be controlled so that conduction in either group is fully on, acting as if the devices were diodes. In this manner, the firing control to the thyristors would be on continuously during one half of the load cycle, and completely off in the other half cycle. A multi-pulse converter could be composed of diodes with thyristors (or triacs) in the a.c. input lines acting as on/off switches for each half load cycle. The output waveforms of such a cycloconverter are shown in Fig. 5-16, being the envelopes of the input waves, hence the name *envelope cyclo-converter.*

Clearly the control circuits required to synthesize the output from an envelope cycloconverter are much simpler than the phase-controlled cycloconverter, but this mode of control has limitations. The output wave tends to be rectangular, hence has more harmonic content. The ratio of output frequency to input frequency is not continuously variable, but is restricted to ratios based on whole or part cycles

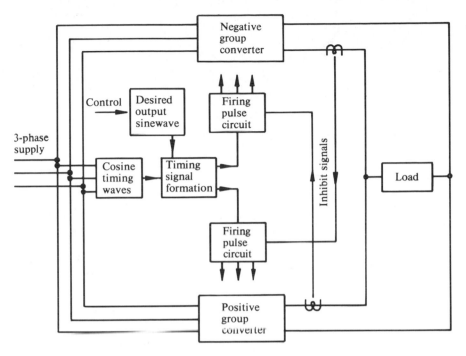

Figure 5-15 Simple diagram of control requirements.

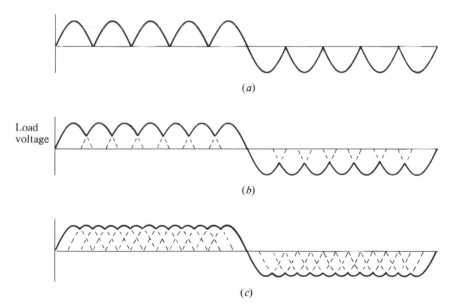

Figure 5-16 Envelope cycloconverter rectangular waveforms. (*a*) 2-pulse. (*b*) 3-pulse. (*c*) 6-pulse.

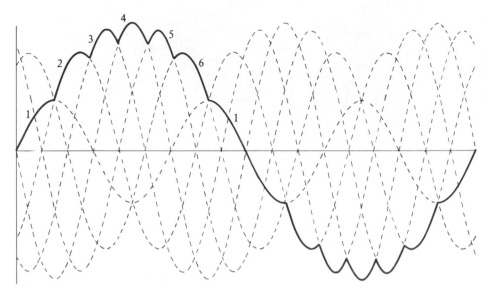

Figure 5-17 An envelope cycloconverter with small ripple.

of the input. A load which is at a lagging (or leading) power factor demands that each group has periods of operation in the inverting mode. The envelope cycloconverter basically can only rectify, so its use is restricted to loads at or near to unity power factor.

If each group of the cycloconverter is six-phase half-wave as in Fig. 2-15a, but with different magnitudes of phase voltages, it is possible to commutate the phases naturally to give the 3/1 ratio as shown in Fig. 5-17. This waveform can be seen to be a close approximation to a sinewave.

5-3 SINGLE-PHASE CENTRE-TAPPED INVERTER

As described in Sec. 4-4 in relation to Figs. 4-17 and 4-18, an alternating voltage to a load can be generated from a d.c. source by the use of a centre-tapped transformer as shown in Fig. 5-18a. Basically, by alternately switching the two thyristors, the d.c. source is connected in alternative senses to the two halves of the transformer primary, so inducing a square-wave voltage across the load in the transformer secondary. The capacitor shown in Fig. 5-18a is required for commutation as described for Fig. 4-17a, but as the capacitor is effectively in parallel with the load via the transformer an inductor L is required in series with the d.c. source to prevent the instant discharge of capacitor C via the source when thyristor switching occurs.

When one thyristor is on, the d.c. source voltage E appears across half the transformer primary, which means the total primary voltage is $2E$, hence the

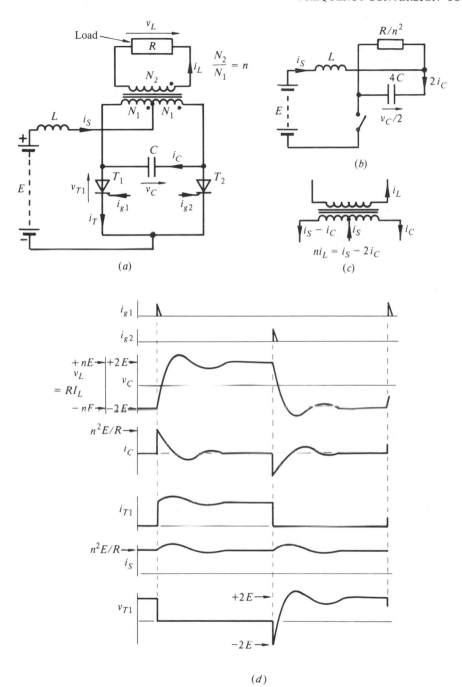

Figure 5-18 Single-phase centre-tapped inverter. (*a*) Connection. (*b*) Equivalent circuit when thyristor T_1 is fired. (*c*) Current distribution in transformer when T_1 is on. (*d*) Waveforms.

capacitor is charged to $2E$. The firing of the other thyristor now turns off the first thyristor, by the principle of parallel capacitor commutation.

If the transformer is assumed perfect, then there must be an ampere-turn balance between windings at all times. In practice, the d.c. source voltage E across the winding can only be maintained by a changing flux which in turn demands a magnetizing current, but in the simple analysis this magnetizing current is ignored.

To analyse the circuit of Fig. 5-18a when supplying a load of pure resistance, it must be reduced to the equivalent circuit of Fig. 5-18b where the capacitor is equivalent to $4C$ to take account of the 2/1 ratio in total to half winding turns of the transformer primary. The ampere-turn balance of the transformer is shown in Fig. 5-18c. Typical waveforms for the circuit are shown in Fig. 5-18d.

An improvement in the load waveform so that it approximates more closely to a sinewave can be brought about if the component values are chosen such that the level section of the load-voltage waveform is avoided, that is, the thyristor is fired shortly after the load voltage reaches its peak from the previous thyristor switching.

For loads other than pure resistance, the load current will be out of phase with the voltage. For these conditions, two diodes are added as shown in Fig. 5-19a to feedback the stored load energy during those periods when the load current reverses relative to the voltage.

When the load is inductive, the load current rises and decays as indicated in Fig. 5-19b. When thyristor T_1 is on, current flows from c to a, with c being positive with respect to a, power being delivered to the load. When thyristor T_2 is fired to reverse the load voltage, thyristor T_1 is turned off, but the load current cannot suddenly reverse, so its direction of flow in the primary winding is unchanged. With thyristor T_1 off, the only path for this current in the winding is from d to c via diode D_2 and the d.c. source. While diode D_2 conducts, thyristor T_2 will turn off once the commutation sequence is over, the voltage of d being negative relative to c, meaning that power is being removed from the load and fed back to the d.c. source.

Referring to Fig. 5-19b, at time t_2 the load current falls to zero, diode D_2 ceases to conduct, and thyristor T_2 can take up conduction, reversing the load current, and power now flows into the load. To ensure thyristor T_2 does take up the current flow at t_2, a train of firing pulses to the thyristor gate is required. A similar sequence of events occurs in the first half cycle, where the initial firing of thyristor T_1 turns off thyristor T_2, current changes to diode D_1, and thyristor T_1 eventually takes up the load current at t_1.

The feedback diodes could be connected to the ends of the winding, but this would result in the loss of the commutation energy in inductor L. By connecting the diodes a short distance in from the winding ends, the stored energy in L can be recovered after commutation, thus reducing commutation losses.

When the load is at a leading power factor, the simplified waveforms of Fig. 5-19c demonstrate that current will transfer to the diodes at t_3 and t_4 respectively, before the thyristors are fired to reverse the load voltage. In practice, the waveforms will not be sinewaves, but Fig. 5-19c serves to show the principle of operation at a leading power factor.

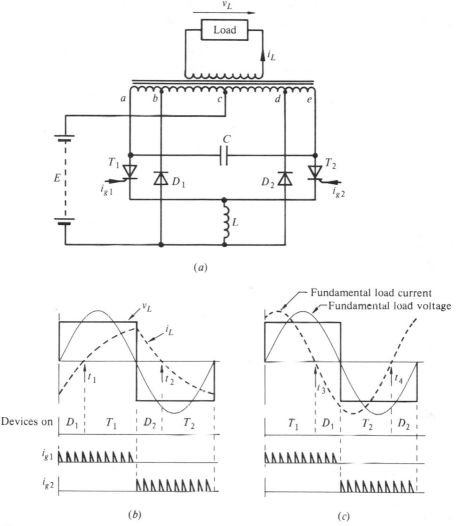

Figure 5-19 Operation with reactive loads. (*a*) Centre-tapped inverter with feedback diodes. (*b*) Lagging power factor load. (*c*) Leading power factor load.

5-4 SINGLE-PHASE BRIDGE INVERTER

The basic circuit for the single-phase bridge inverter without the commutating elements is shown in Fig. 5-20*a*. The commutation circuit and its action was given in Sec. 4-4, where with (say) reference to Fig. 4-19 the turning-off of thyristor T_1 is initiated by the firing of the complementary thyristor T_4. If, as is the case with an inductive load, the load current does not immediately reverse, then once commutation is complete, thyristor T_4 will cease to conduct with the load current being transferred to diode D_4. Typically, the commutation period is very short relative to the inverter load frequency period, and in this section commutation is assumed to be ideal, and hence neglected in developing the waveforms of Figs. 5-20b and c.

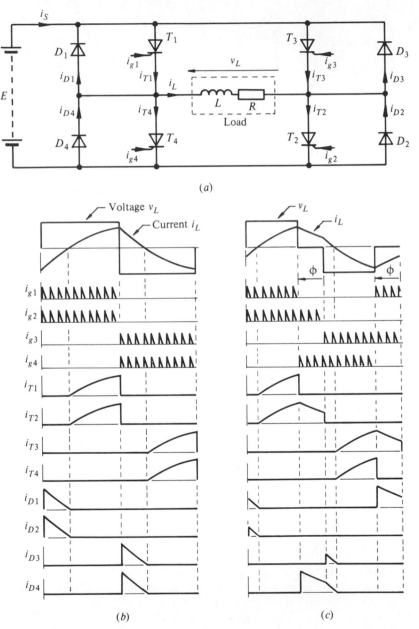

(a)

(b)   (c)

Figure 5-20 Basic single-phase bridge inverter. (a) Circuit. (b) Square-wave output. (c) Quasi-square-wave output.

If the load of Fig. 5-20a were pure resistance, then the alternate firing of thyristors $T_1 T_2$ and $T_3 T_4$ will place the d.c. source in alternate senses across the load, giving a square wave. However, with an inductive load, the current waveform is delayed, although the voltage wave is still square.

The generation of a square-wave load voltage with an inductive load is shown in the waveforms of Fig. 5-20b. The thyristors are fired by a continuous train of gate pulses for 180° of the inverter output voltage. Looking at the latter end of the positive half cycle, the load current is positive and growing exponentially; however, when thyristors T_3 and T_4 are gated to turn off thyristors T_1 and T_2, the load voltage reverses but not the load current. The only path for the load current is via diodes D_3 and D_4, which connects the d.c. source to the load, giving a reverse voltage with the stored inductive energy of the load being returned to the d.c. source until the load current falls to zero. Once the load current ceases, thyristors T_3 and T_4 can conduct so as to feed power into the load, the load current now growing exponentially. Because the thyristors require re-firing at the instant of the load current zero, a train of firing pulses is required at the gates, as this instant can be at any time in the half-cycle.

Control of the voltage can be obtained from a fixed-level d.c. source by introducing zero periods into the square wave, giving a shape known as a *quasi-square wave* as shown in Fig. 5-20c. The quasi-square wave can be generated by phase-advancing the firing of the complementary pair of thyristors $T_1 T_4$ compared to thyristors $T_2 T_3$. In Fig. 5-20c this advance is shown as an angle ϕ, that is, the firing pulse train start of thyristor T_1 (and T_4) is advanced ϕ degrees before that of thyristor T_2 (and T_3).

Taking the instant in the load-voltage waveform of Fig. 5-20c where thyristor T_4 is fired to turn off thyristor T_1, the load current transfers to diode D_4 but, as thyristor T_2 is still on, the load current follows a path via D_4 and T_2, effectively short-circuiting the load, giving zero load voltage. Now when thyristor T_3 is fired to turn off thyristor T_2, the only path for the load current is via diode D_3, connecting the d.c. source to the load in the negative sense, with thyristors T_3 and T_4

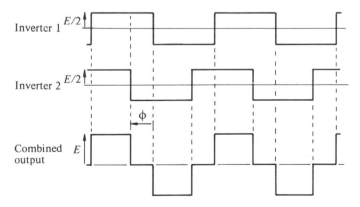

Figure 5-21 Combined output of phased inverters giving quasi-square wave.

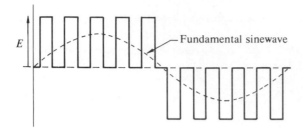

Figure 5-22 Inverter controlled to give notched waveform.

taking up conduction immediately after zero load current. The current waveforms in the thyristors (and diodes) are not now identical.

An alternative way of producing a quasi-square wave of controllable width is to combine (add) the outputs of two phase-shifted square-wave inverters as shown in Fig. 5-21. By phase-shifting inverter 2 an angle ϕ compared to inverter 1, the combined output has zero periods of length ϕ.

The voltage level of the quasi-square wave of fixed width can be controlled by a reduction of the d.c. source voltage.

Another form of voltage control is by notching, as shown by the waveform of Fig. 5-22, where the thyristors in the inverter circuit are turned on and off so as to produce zero periods of equal length, the d.c. source being a fixed level of voltage E.

An improvement to the notched waveform is to vary the on/off periods such that the on-periods are longest at the peak of the wave as shown in Fig. 5-23. This form of control is known as *pulse-width modulation*, and it can be observed that the area of each pulse corresponds approximately to the area under the sinewave between the adjacent mid-points of the off-periods. The pulse-width modulated wave has a much lower low-order harmonic content than the other waveforms.

To determine the firing instants required to synthesize correctly the pulse-width modulated wave, a method which can be used is to generate a reference sinewave of the desired frequency within the control circuits, and to then compare this sinewave to an offset triangular wave as shown in Fig. 5-24. The crossover points of the two waves determine the firing instants. Figure 5-24a shows a maximum output, a reduction to half this value being made by simply reducing the reference sinewave to half value as shown in Fig. 5-24b. Figure 5-24c shows how

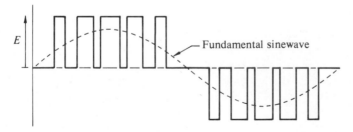

Figure 5-23 Inverter controlled to give pulse-width modulated waveform.

a reduction in frequency of the reference sinewave increases the number of pulses within each half cycle.

The justification of the use of the triangular wave can be given by reference to Fig. 5-25. The section of the upper reference sinewave within the period shown

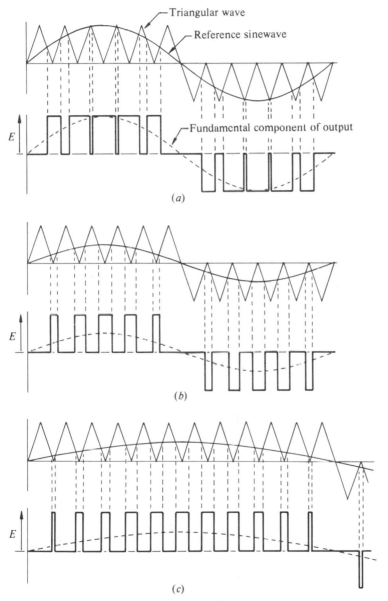

Figure 5-24 Formation of firing instants for pulse-width modulated wave. (*a*) At maximum output voltage. (*b*) At half maximum output voltage. (*c*) At half voltage and half frequency.

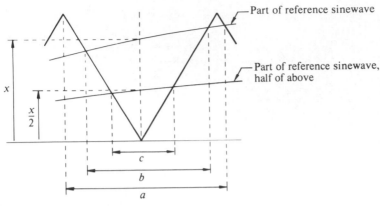

Figure 5-25 Justifying the use of a triangular wave.

will give a pulse of width b. Reduction of the reference sinewave height to half will give a pulse of width c. Assuming the section of sinewave to be a straight line, c is half the width of b, the pulse height being unaltered, therefore the pulse area is then halved, corresponding to the reduction in height of the desired sinewave. Similar arguments apply if the two sections of sinewave shown are at different times within the same reference sinewave, width a corresponding to the maximum of the sinewave.

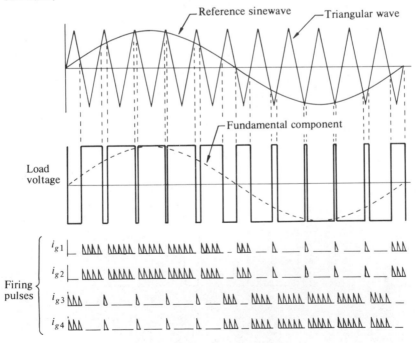

Figure 5-26 Pulse-width modulation by source alternation.

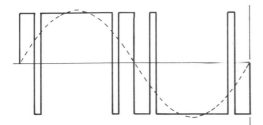

Figure 5-27 Low-order harmonic reduction.

The high number of pulses within the output cycle does lead to a large increase in the number of high-order harmonics, but these are more easily filtered than the low-order harmonics, an inductive load severely attenuating these harmonics in the current waveform.

As an alternative to the pulse-width modulation control described above, the inverter (Fig. 5-20) can be controlled so that the d.c. source is always connected to the load, by firing thyristors T_1 and T_2 as a pair, with thyristors T_3 and T_4 as the other pair, hence avoiding the zero periods. In this manner, the pulse-width modulated wave shown in Fig. 5-26 is generated, having short reverse periods during the output half cycle. To determine the thyristor firing instants, the high-frequency triangular wave is modulated by the reference sinewave, this time the triangular wave not having any offset as in Fig. 5-24.

The high number of commutations taking place in each cycle with the notched and pulse-width modulated waveforms does result in high commutation losses in the thyristors within the inverter. In choosing between the quasi-square-wave inverter and the pulse-width modulated inverter, consideration must be given to the additional cost of control circuitry and switching losses in the one and to the higher low-order harmonic content of the output in the other.

A method which avoids excessive commutations within the output cycle, but reduces the low-order harmonics, is shown in Fig. 5-27. By reversing the output voltage for a short interval in each half cycle at particular angles, it is possible to eliminate two harmonic components, say the third and fifth components. With a fixed d.c. source, it is possible to control the output level of this voltage by combining two such waveforms of Fig. 5-27, phase-displaced to the principle demonstrated in Fig. 5-21.

5-5 THREE-PHASE BRIDGE INVERTER

The basic circuit of the three-phase bridge inverter is shown in Fig. 5-28. The commutation circuits can be one of the types shown in Sec. 4-4, but these are omitted from Fig. 5-28 so as not to confuse the explanation of its operation.

Similar to the three-phase rectifier bridge, the inverter can be controlled so that each thyristor conducts for $120°$ of the output cycle. The waveforms shown in Fig. 5-29 to the circuit reference of Fig. 5-28 relate to a pure resistance load. It is assumed that at the end of the $120°$ period a commutation circuit is initiated to turn off the appropriate thyristor.

The waveforms in Fig. 5-29*b* show that the load currents are quasi-square wave, with each thyristor conducting the load current for one-third cycle. By reference to the circuit in Fig. 5-29*a*, considering the thyristors as switches, the d.c. source is switched in six steps to synthesize the three-phase output. The rate at which the thyristors are switched determines the load frequency. The stepped waveform shown for the line voltage will be modified if any inductance is present in the load, as the transfer of the load current into the diodes will effectively maintain the switches (shown in Fig. 5-29*a*) closed for a longer period than $120°$.

It is more usual to operate the inverter so that each thyristor can conduct over $180°$. In this manner, the d.c. source is connected to the load via one thyristor on one side with two in parallel on the other side.

The waveforms of Fig. 5-30 show $180°$ conduction, the line voltage being quasi-square wave. The load currents are stepped and each thyristor conducts over $180°$. An advantage of this form of control is that commutation circuits like those of Figs. 4-19 and 4-23 can be used, as when (say) thyristor T_4 is turned on it will commutate-off the complementary thyristor T_1.

If the load being supplied by the inverter contains inductance, then the current in each arm of the load will be delayed to its voltage as shown in Fig. 5-31. When thyristor T_1 is fired, thyristor T_4 is turned off but, because the load current cannot reverse, the only path for this current is through diode D_1 (see Fig. 5-28). Hence the load phase is connected to the positive end of the d.c. source but, until the load current reverses at t_1, thyristor T_1 will not take up conduction. Similar arguments apply in the reverse half cycle at t_2.

Voltage control of the three-phase inverter can be implemented by adding two

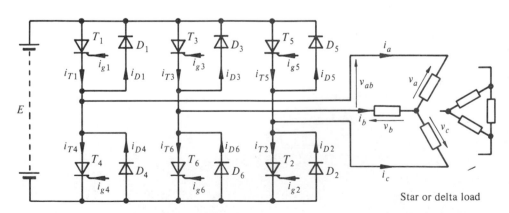

Figure 5-28 Basic three-phase bridge inverter.

single-phase inverter outputs, one phase delayed to the other, the combined outputs of three such circuits being linked via transformers to give a three-phase output. In this manner, outputs would be as shown in Fig. 5-32.

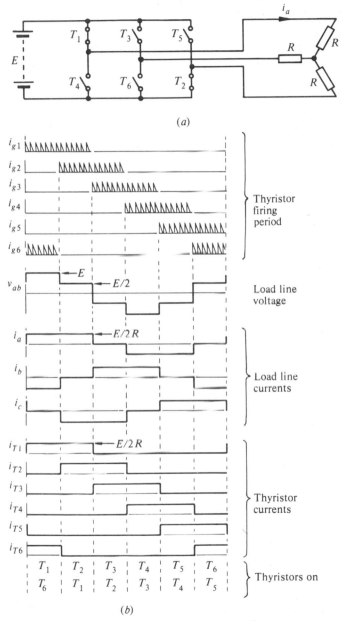

(a)

(b)

Figure 5-29 Three-phase bridge inverter with 120° firing and resistive load. (a) To illustrate switching sequence, thyristors T_1 and T_2 on. (b) Waveforms.

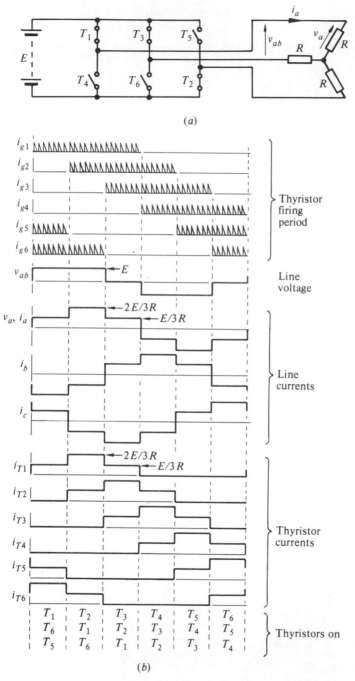

Figure 5-30 Three-phase bridge inverter with 180° firing and resistive load. (*a*) To illustrate switching sequence, with thyristors T_1, T_2 and T_3 on. (*b*) Waveforms.

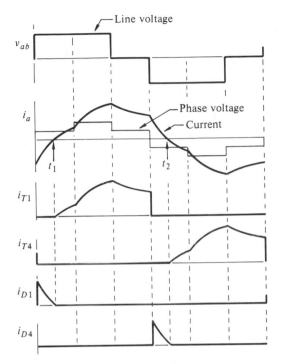

Figure 5-31 Waveforms for 180° firing with an inductive load.

The pulse-width modulation technique can be used as illustrated by Fig. 5-33 (to circuit reference Fig. 5-28), where the three reference sinewaves modulate the high-frequency triangular wave to determine the firing instants of each thyristor. The explanation to these waveforms is similar to that developed for the single-phase inverters.

With the control as in Fig. 5-33, one or other of the devices in each arm is conducting at all times, connecting the load line to either the positive or negative side of the d.c. source. Taking, for example, the arm to phase A with devices numbered 1 and 4 in Fig. 5-28, if i_a is positive, thyristor T_1 conducting, then when thyristor T_4 is fired, T_1 turns off and the load current transfers to diode D_4. If, however, i_a were negative, diode D_1 would have been conducting and, when thyristor T_4 fired, it would have taken up the load current immediately; and in this condition, thyristor T_1 did not require turning off, as it was in any case not on.

Referring to Fig. 5-33, the firing pulses must be continuous into the gate of the

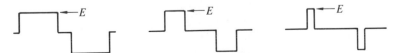

Figure 5-32 Voltage control by combining phased inverters.

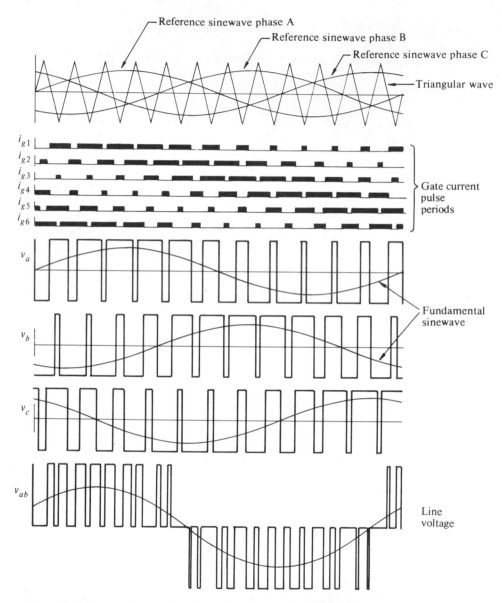

Figure 5-33 Pulse-width modulated waveform for three-phase bridge inverter.

thyristor when current is required from that arm, so that the thyristor can take up the load current when it reverses under inductive loading. If the instantaneous load current is reversed to the voltage, then the diode on is that in parallel with the thyristor receiving firing pulses. Hence, in the period shown when (say) i_{g1} is present either thyristor T_1 or diode D_1 is on.

5-6 CONSTANT-CURRENT SOURCE INVERTER

In the previous sections, inverters were described which were fed from a constant-voltage source in which the load was supplied with a stepped-voltage waveform. The constant-current inverter is one in which the current from the d.c. source is effectively constant, irrespective of events in the inverter, over a period of a few cycles. In practice, the constant-current inverter is one supplied from a d.c. source via a large inductance as shown in Fig. 5-34, so that changes in the inverter voltage can occur, balanced by $L \, di/dt$, but with only a small di/dt, hence effectively maintaining a constant level of supply current over short periods.

With the constant-current inverter, it is possible to use a simpler commutation circuit employing only capacitors. In Fig. 5-34a, with the thyristors T_1 and T_2 on, both capacitors are charged with their left-hand plates positive. When thyristors T_3 and T_4 are fired, the capacitors are placed across thyristors T_1 and T_2 respectively, turning them off, current now flowing via $T_3 C_1 D_1$ load, $D_2 C_2 T_4$. The capacitor voltages will be reversed and in time, depending on the load voltage, diodes D_3 and

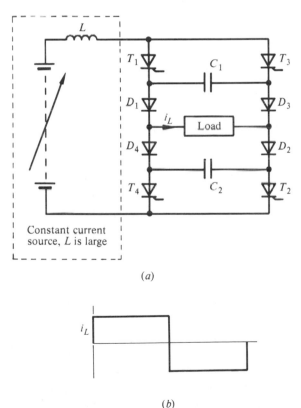

(a)

(b)

Figure 5-34 Single-phase constant-current source inverter. (a) Circuit. (b) Load-current waveform.

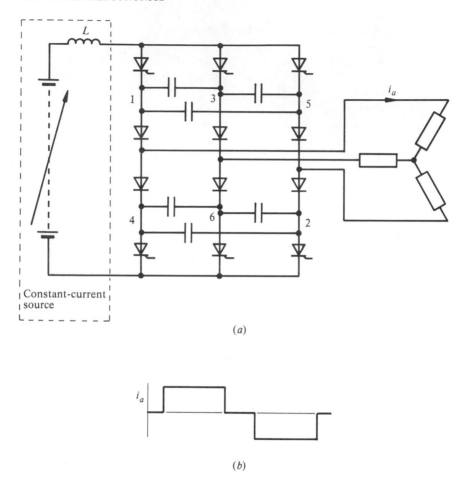

(a)

(b)

Figure 5-35 Three-phase constant-current source inverter. (a) Circuit. (b) Line-load current.

D_4 will turn on, the supply current then being transferred over a time interval from D_1 to D_3 and D_4 to D_2, diodes D_1 and D_2 finally ceasing to conduct when the load current is completely reversed. The capacitor voltages will reverse ready for the next half cycle.

In Fig. 5-34, the diodes serve to isolate the capacitors from the load voltage. The load current is square wave, ignoring the commutation period, with the load voltage typically being sinewave in shape but containing voltage spikes at the commutating instants. A typical application would be the supply to an induction motor. Over a long period, the current level would change according to the load demands, final steady-state conditions being dependent on the d.c. source voltage level.

The three-phase constant-current inverter is shown in Fig. 5-35. At any one time, only two thyristors are on. When thyristor T_3 is fired, thyristor T_1 is turned off by the common capacitor. Likewise, when thyristor T_4 is fired, thyristor T_2 is turned off. The load current is quasi-square wave, with each thyristor conducting for $120°$.

5-7 INVERTER DEVICES

The inverter circuits shown in the earlier sections are illustrated using conventional thyristors as the switching elements. As described, the circuits require extensive commutation networks using capacitors and inductors to bring about turn-off, but the earliest inverters developed used the conventional thyristor because that was the only device available. More recent developments in device technology have seen the introduction of a number of new devices appropriate for inverter application. The reverse blocking capability of the conventional thyristor is not utilized in the inverter circuits because of the need to include inverse parallel connected feedback diodes.

The asymmetrical thyristor has an improved turn-off performance, so its use in inverters reduces the size of the commutation networks and associated losses. The commutation networks can be eliminated by using gate turn-off thyristors, because the anode current can be stopped by removal of gate current. The gating circuits are clearly more complex than with the other thyristors, but this must be set off against the saving in commutation components. The overall losses and total equipment size are less, as is the acoustic noise generated within the components.

The transistor family of devices is now very widely used, Fig. 5-36 illustrating the use of bipolar transistors, the feedback diodes still being essential to the inverter operation. The bipolar transistor can be switched much faster than the thyristor but does require high levels of base current during the on-period, the base current having to be carefully controlled in magnitude to keep the transistor just into saturation. Too high a level of base current will extend the turn-off time and possibly delay the removal of base current which is necessary when a load fault

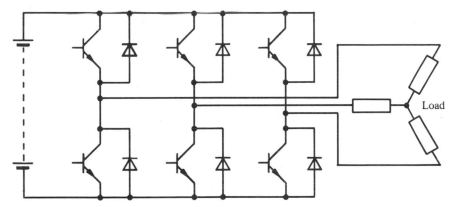

Figure 5-36 Basic three-phase power transistor inverter.

occurs. The Darlington arrangement of the transistor (see Fig. 1-14) considerably reduces the base drive requirements at the expense of slightly increasing the switching times. Care must be taken to ensure an elapse of a few microseconds after turn-off of the transistor before its complementary transistor is turned on, otherwise there is a risk of reconduction in the outgoing transistor, giving a short-circuit across the d.c. source via the two transistors.

By far the fastest switching device is the power MOSFET, used in inverters at the lower power ratings commeasurable with its more limited voltage and current rating. Being a voltage-controlled device the gate firing controls are less onerous and it is possible to implement the gating controls directly from microelectronic circuits.

Using commutation networks with the conventional thyristors limits the inverter frequency to approximately 100 Hz. The gate turn-off thyristors extend the frequency up to about 2 kHz, the bipolar transistor to 10 kHz, and the power MOSFET to 25 kHz. These figures for frequencies are only an approximate guide, because it is necessary to carefully balance the conduction and switching losses together with the load losses, cost, and cooling of the equipment to arrive at an optimum design. A limiting feature can be the reverse recovery time of the feedback diodes.

The object of using the faster switching devices is not only to enable the inverter to give a higher range of frequencies but primarily to enable the inverter to be controlled in the pulse-width modulation mode. The more switchings that can be made within each cycle the more low-order harmonics can be eliminated. By eliminating the low-order harmonics a much smaller filter can be employed to convert the stepped waveform into a sinewave. A load such as an induction motor which is highly inductive will attenuate the high-order harmonics naturally, resulting in the motor current being almost sinusoidal.

At low power levels it is possible to shape the output voltage of the inverter by operating the transistors in the controlled mode, thereby deliberately introducing a voltage drop across the transistor. This involves control in the safe operating area as shown in Fig. 1-13b. This type of control is prohibited at medium and higher power levels because of the transistor loss causing overheating of the transistor and because of the reduction in efficiency.

The physical layout of the inverter circuit is very important when switching at very high frequencies. Stray inductance of even a low value can, when associated with very high rates of change of current, produce very high voltage transients. Care has to be taken to keep leads short and to locate the gate circuitry close to the power devices.

Protection is covered in detail in Chapter 10 but it is worth noting here that protection of the inverter devices is very important in the event of a load short-circuit or other fault condition. The thyristors with their internal regenerative action remain fully conducting under fault conditions, enabling fuses to be used as the protecting agent. With the transistor, extra collector current will take it out of saturation, a high collector-emitter voltage being developed, leading to high losses internal to the transistor and destruction without the current being raised to a level sufficient to operate a fuse. Likewise the current in a power MOSFET is self-limiting with a drain-source voltage developing which will overheat the device. For both the transistor and the power MOSFET it is necessary to detect the fault and immediately turn the device off by removing the base current or gate voltage respectively, or alternatively by removing the d.c. input voltage to the inverter.

5-8 INVERTER REVERSE POWER FLOW

To reverse the power flow through an inverter, one could consider the a.c. side to be a generating source, feeding a d.c. load via a rectifier. If the thyristors were omitted, then the inverter becomes a simple rectifier. However, the constraints in regard to how the circuit functions differ from those of the rectifiers described in Chapter 2. Referring to Fig. 5-37a the d.c. voltage is a fixed value, the capacitor emphasizing the fixed-voltage concept. In the rectifier circuits, the d.c. load voltage contained large ripple components, but here the ripple variations will be in the current waveforms.

In practice, the a.c. load most likely to generate is an induction motor being accelerated by a mechanical torque to above synchronous speed. The current from such a generator is at a leading power factor. Figure 5-37b–d illustrates the stages of current into the load drifting until the lag is over 90°, when the current reference can be reversed for the generating mode.

The waveforms of Fig. 5-37e show the need to fire the thyristors so as to maintain continuous current in each phase of the load, the diodes now conducting for a longer period than the thyristors, so that power is being fed into the d.c. element. For simplicity, the a.c. waveforms of current are shown as sinewaves but, in practice, they must contain harmonic components.

The direction of power flow through the inverter is reversed automatically when the a.c. element changes to a generator without any change needed in the sequence or duration of the 180° train of firing pulses.

5-9 WORKED EXAMPLES

Example 5-1

A 3-pulse cycloconverter feeds a single-phase load of 200 V, 50 A at a power factor of 0.8 lagging. Estimate the required supply voltage, thyristor rating, and power factor of the supply current. Neglect device and supply impedance volt-drops.

SOLUTION Figure 5-3 refers to this circuit. Using Eq. (5-1),

$$200\sqrt{2} = \frac{3}{\pi} \sin \frac{\pi}{3} \times \sqrt{2}V_s,$$ giving V_s the r.m.s. supply voltage as 242 V/ph.

To determine the thyristor ratings, the worst case will be when the output frequency is so low that a converter group is acting as a rectifier feeding a d.c. load

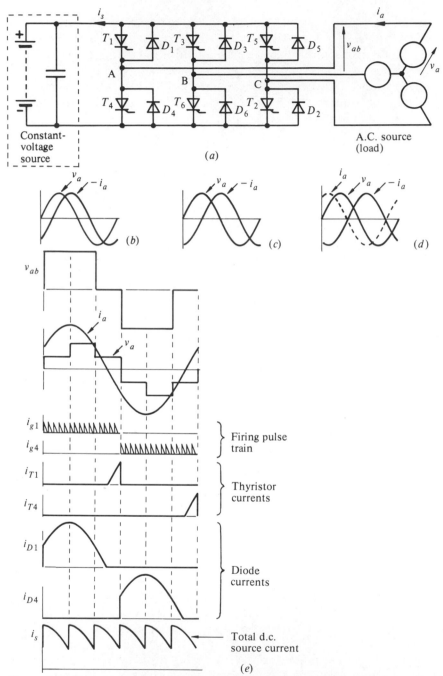

Figure 5-37 Illustrating reverse power flow through three-phase inverter. (*a*) Circuit reference. (*b*) Power flow into an a.c. load, lagging power factor. (*c*) A.C. load current lagging 90°, zero power factor. (*d*) Load current lagging over 90°, that is, generating with leading power factor. (*e*) Waveforms assuming sinewave currents.

for a long period, with a current equal to the maximum value of the cycloconverter load. Maximum value of the cycloconverter load = $50\sqrt{2}$ A, giving from Eq. (2-10) for the thyristor, $I_{\text{rms}} = \dfrac{50\sqrt{2}}{\sqrt{3}} = 41$ A. P.R.V. $= \sqrt{3} V_{s(\text{max})} = \sqrt{3} \times 242\sqrt{2} = 593$ V.

For each cycle of load current, each input line will conduct short blocks of current for one-third of the time. The r.m.s. value of a complete sinewave = 50 A, hence the r.m.s. value for one-third period conduction equals $\sqrt{(50^2/3)} = 28.9$ A.

Input power per phase = $1/3 \times$ load power = $\frac{1}{3} \times 200 \times 50 \times 0.8 = 2667$ W.

$$\text{Power factor} = \frac{\text{power}}{V_{\text{rms}} I_{\text{rms}}} = \frac{2667}{242 \times 28.9} = 0.38.$$

Note that if firing delay (angle α) is used to reduce the load voltage, the power factor is further reduced by $\cos \alpha$.

Example 5-2

Construct the input supply-current waveform of a three-pulse cycloconverter feeding a balanced three-phase load taking current at a power factor of 0.8 lagging. Take the ratio of input to output frequency as 3. Hence estimate the power factor of the input current using the values given in Example 5-1.

SOLUTION The constructed waveforms are shown in Fig. 5-38, drawn to the circuit reference of Fig. 5-7a. Taking the load current to be sinusoidal, the A line current into each load phase group is drawn. The total A line current is the sum of the three group currents.

Given a very high pulse-number circuit with a high-frequency ratio, the total current would be sinusoidal with the same maximum value as the load current. However, with three-pulse, 3/1 ratio, the current i_A contains steps, although it can be roughly considered sinusoidal.

Taking the supply current to be sinusoidal, it will have an r.m.s. value of 50 A, that is, equal to the load. Taking the values of Example 5-1,

$$\text{power factor} = \frac{3 \times 200 \times 50 \times 0.8}{3 \times 242 \times 50} = 0.66 = \cos 49°.$$

The fundamental sinewave at a lagging angle of $49°$ is shown on the waveform.

Example 5-3

A six-pulse cycloconverter is fed from a 415 V three-phase supply having a reactance of 0.3 Ω/phase. Determine the output load voltage for firing delay angles of $0°$ and $45°$, given a load current of 40 A.

SOLUTION The maximum value of the load voltage is the mean value of a group acting as a rectifier. Using Eq. (3-26), with zero firing delay $\alpha = 0°$, then

$$0.3 \times 40 = 415\sqrt{2} \sin \frac{\pi}{6} (1 - \cos \gamma), \text{ giving } \gamma = 16.44°.$$

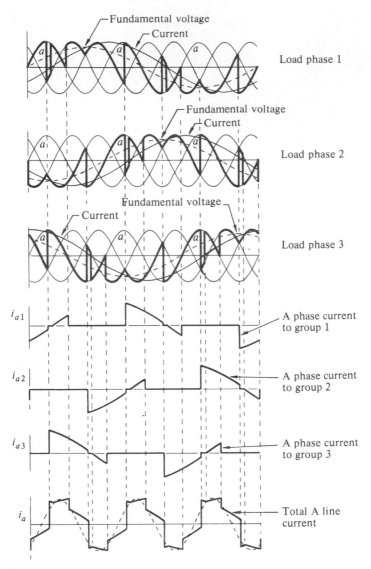

Figure 5-38

Using Eq. (3-23), $V_{\mathrm{mean}} = \dfrac{6 \times 415\sqrt{2}}{2\pi} \sin \dfrac{\pi}{6}(1 + \cos 16.44°) = 549$ V,

this being the maximum value of the load voltage. Load voltage $= 549/\sqrt{2} = 388$ V. With $\alpha = 45°$, calculations yield: $\gamma = 3.225°$, $V_{\mathrm{mean}} = 385$ V, load voltage $= 272$ V. In practice, the load voltage would be reduced by device volt-drops and the fact that the maximum load voltage may not quite reach the mean value of the group operating as a rectifier.

Example 5-4

A three-pulse cycloconverter operating in a circulating-current mode is fed at 240 V/phase, 50 Hz. Determine the specification of the intergroup reactor, taking the r.m.s. load current to be 10 A when the control is changed to blocked group operation.

SOLUTION The waveforms drawn in Fig. 5-11 show that the possible peak-voltage excursion is $1.5V_{max}$, decaying to zero after a time of $\pi/3\omega$, that is, the reactor voltage is the last $60°$ in the half cycle of $\sqrt{3}V_{max} \sin \omega t$.

From flux change $= \int v \, dt$,

$$\text{peak flux linkage} = \int_{2\pi/3\omega}^{\pi/\omega} \sqrt{3} \times 240\sqrt{2} \sin \omega t \, dt = 0.93 \text{ Wb-turns.}$$

Reference to Fig. 5-12 shows the peak reactor current is equal to the peak load current, that is, $10\sqrt{2}$ A.

Taking the total reactor turns as N, the maximum core flux as Φ, and the maximum winding current below saturation as $I (= 10\sqrt{2})$, we have $N\Phi = 0.93$ and m.m.f. $= NI/2 = 5\sqrt{2}N$.

Using a reactor with a non-magnetic gap in the core of length l and cross-sectional area a, and taking saturation in the iron to commence around 1.8 T, then $\Phi = 1.8a$

and $H = \dfrac{B}{\mu_0} = \dfrac{1.8}{4\pi \times 10^{-7}} = \dfrac{\text{m.m.f.}}{l}$.

Hence gap volume $= al = \dfrac{1}{N \times 0.93 \times 1.8} \times \dfrac{5\sqrt{2}N}{1.8} \times 4\pi \times 10^{-7} = 2.95 \times 10^{-6}\,\text{m}^3.$

To avoid fringing, take a gap with a circular cross-section of diameter ten times gap length. We get a gap of length 3.3 mm, with cross-sectional diameter (say) 33 mm, which would need, from $N = 0.93/\Phi$, approximately 600 turns.

As the coil conductor must be rated to the full-load current, it would probably be economical to increase the core cross-sectional area, so reducing the number of turns. Typically, for transformer cores, an excitation of 70 AT/m will start taking the core into saturation, so if the core length were (say) 0.3 m, only 3 turns would be required at $5\sqrt{2}$ A, but the saturation level would be ill-defined without a gap in the core. On test, the gap could be adjusted to give the required characteristic.

Example 5-5

Determine the ratios of the magnitudes of the phase voltages required to synthesize the low-frequency output voltage of an envelope cycloconverter as shown in Fig. 5-17.

SOLUTION The most convenient way to solve this problem is to plot a sinewave of peak value 100, and sketch the phase voltages to it, so that the envelope waveform has similar areas above and below the fundamental sinewave. Such a sketch will give peak values of:

$$V_1 \triangleq 42, V_2 = V_6 \triangleq 81, V_3 = V_5 \triangleq 100, \text{ and } V_4 \triangleq 106.$$

A more accurate determination is hardly justified when overlap and device volt-drops are in practice present in the waveform.

Example 5-6

A 50 Hz, single-phase, centre-tapped inverter is required to feed an a.c. load from a 72 V d.c. source. Taking the transformer to step-up 3/1 turns ratio into a load of 18 Ω pure resistance, determine the size of the inductance and capacitance required, neglecting all losses, thyristor volt-drops, and assuming the transformer to be perfect. Design to allow a turn-off time of 60 μs. Sketch the load-voltage waveform shape. Determine the transformer core size and thyristor ratings.

SOLUTION Figure 5-18 refers to the circuit. The equivalent circuit Fig. 5-18b has a resistance of $18/3^2 = 2\,\Omega$, with an initial source current of $72/2 = 36$ A, and an initial voltage across $4C$ of 72 V (i.e. 144/2), assuming waveform shapes as shown. Using symbols i_1 and i_2 for the currents in the resistor and capacitor respectively in Fig. 5-18b, we can write Laplace transform equations:

$$L(s\bar{i}_1 - 36 + s\bar{i}_2) + 2\bar{i}_1 = \frac{72}{s}$$

and
$$-2\bar{i}_1 + \frac{\bar{i}_2}{s4C} - \frac{72}{s} = 0$$

Solution of these yields
$$\bar{i}_1 = \frac{36}{s} - \frac{72s}{(s+\alpha)^2 + \omega^2}$$

where
$$\alpha = \frac{1}{16C} \quad \text{and} \quad \omega = \left(\frac{1}{4LC} - \frac{1}{256C^2}\right)^{1/2}$$

Transforming this for the underdamped case,

$$i_1 = 36 - 72\, e^{-\alpha t}(\cos \omega t - \frac{\alpha}{\omega} \sin \omega t)\, \text{A}$$

this being the current in the resistor.

Looking at the waveforms in Fig. 5-18d, the load current (voltage) oscillation after one cycle must have decayed to under 50% of its initial value (72 A), so as to ensure the thyristor current does not attempt to reverse. Say the oscillation has died to 10% after one cycle. Approximately, the time of one-sixth cycle (i.e. when $\cos \omega t \triangleq 0.5$ at $i_1 = 0$) of oscillation is 60 μs, that required to turn off the thyristor. Therefore $\omega = (2\pi/6)/(60 \times 10^{-6}) = 17\,453$. Now $e^{-\alpha t} = 0.1$ after one cycle when $t = 2\pi/\omega$, giving $\alpha = 6396$.

To check the turn-off time to when the thyristor is forward-biased, using above values makes $i_1 = 0$ at $t = 34\,\mu s$, a time considerably less than the original 60 μs. Recalculating values using an initial time of (say) $60 \times (60/34) = 105\,\mu s$ yields

$\omega = 9974$, $\alpha = 3654$, with a turn-off time of $60\,\mu s$. Hence $C = 17.1\,\mu F$ and $L = 130\,\mu H$.

If the circuit parameters are chosen to give a critically damped response, then $\omega = 0$, that is, $1/4LC = 1/256C^2$, and $i_1 = 36 - 72\,e^{-\alpha t}(1 - \alpha t)\,A$. Putting in the turn-off time $t = 60\,\mu s$ when $i_1 = 0$, then $\alpha = 5250$, giving $C = 11.9\,\mu F$ and $L = 762\,\mu H$.

If the circuit parameters are chosen to give an overdamped response, then

$$i_1 = 36 - \frac{72}{\beta - \gamma}[\beta\,e^{-\beta t} - \gamma\,e^{-\gamma t}]\,A$$

where

$$\beta = \frac{1}{16C} + \left(\frac{1}{256C^2} - \frac{1}{4LC}\right)^{1/2}$$

and

$$\gamma = \frac{1}{16C} - \left(\frac{1}{256C^2} - \frac{1}{4LC}\right)^{1/2}$$

If taking $i_1 = 0$ at $t = 60\,\mu s$, with (say) $\beta = 3\gamma$, then $\beta = 8044$, $\gamma = 2681$, giving $C = 11.66\,\mu F$ and $L = 995\,\mu H$.

The above three load waveforms are shown in Fig. 5-39. The scale of the load voltage $= i_1 \times (18/3^2) \times 3 = 6i_1$.

To find the core size, the following procedure can be used:

From $v = N\dfrac{d\theta}{dt}$, the flux change $= \dfrac{1}{N}\displaystyle\int v\,dt$. In each half cycle the flux changes from $-\Phi_{max}$ to $+\Phi_{max}$, therefore $2\Phi_{max} = $ area under the rectangular waveform in Fig. 5-39.

Therefore $2\Phi_{max} = (1/N) \times 216 \times 0.01$, giving $\Phi_{max} = 1.08/N$.

Given a secondary of N turns, the core area $= \Phi_{max}/B_{max}$, where B_{max} is the maximum level of the core flux density. To avoid core saturation at the initial switching, B_{max} as above must be somewhat below saturation level.

To find the equations for the thyristor current, the capacitor current is required. The capacitor voltage $=$ load voltage $\times 2/(\text{turns ratio})$, and the capacitor current $i_c = C(dv/dt)$.

$i_c = (36 \cos 9974t + 42.5 \sin 9974t)\,e^{-3654t}\,A$ for the underdamped case.

$i_c = 18\,e^{-5250t}(2 - 5250t)\,A$ for the critically damped case.

$i_c = 40.5\,e^{-8044t} - 4.5\,e^{-2681t}\,A$ for the overdamped case.

The thyristor current $i_T = $ source current, which from Fig. 5-18c is $i_T = i_1 + 2i_c$, giving

$i_T = 36 + (111.4 \sin 9974t)\,e^{-3654t}\,A$ for the underdamped case.

$i_T = 36 + 189\,000t\,e^{-5250t}\,A$ for the critically damped case.

$i_T = 36 - 27\,e^{-8044t} + 27\,e^{-2681t}\,A$ for the overdamped case.

The peak values of the thyristor currents are:

103 A at $t = 122\,\mu s$ for the underdamped case.

49.25 A at $t = 190\,\mu s$ for the critically damped case.

46.4 A at $t = 205\,\mu s$ for the overdamped case.

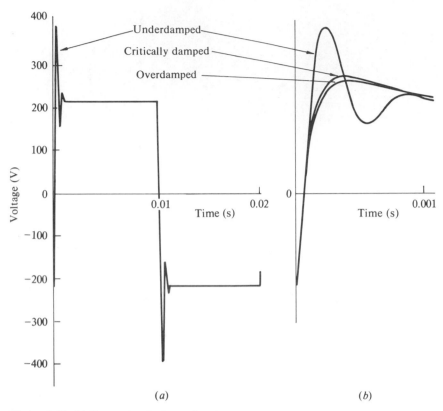

Figure 5-39 (*a*) One cycle. (*b*) Start of cycle to an expanded scale.

Ignoring the transient component, each thyristor carries 36 A for one half cycle, giving a required r.m.s. value of $36/\sqrt{2} = 26$ A, (say) 30 A to allow for additional losses due to the transient.

The P.F.V. will be the peak load voltage $\times (1/3) \times 2$, giving 262 V, 183 V, and 176 V respectively for the three cases.

Because components have been assumed perfect, it appears that the thyristor current reaches 36 A in zero time when switched, giving an infinite rate of rise of current. In practice, the transformer leakage reactance would limit this rise, or extra inductance could be added in series with the thyristor.

The thyristor voltage goes positive after 60 μs when from $dv/dt = i/C$ we get values of 2.5 V/μs, 1.86 V/μs, and 1.82 V/μs respectively for the three cases.

Example 5-7

A 400 Hz, single-phase, centre-tapped inverter is required to generate an output close to sinewave shape from a 72 V d.c. source into a 10 Ω load via a 1/1 ratio transformer. Determine the required values of capacitance and inductance. Neglect all losses.

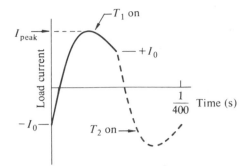

Figure 5-40

SOLUTION Figure 5-18 refers to this problem, the output voltage being the start of the underdamped transient as shown in Fig. 5-40.

Let (say) I_0 be the same as $E/R = 72/10 = 7.2$ A and, using similar reasoning to that used in Example 5-6, using the different resistance value, the load current is

$$i = 7.2 - 14.4\, e^{-\alpha t}(\cos \omega t - \frac{\alpha}{\omega} \sin \omega t)\, A$$

where $\alpha = \dfrac{1}{80C}, \omega = \left(\dfrac{1}{4LC} - \dfrac{1}{6400C^2}\right)^{1/2}$

At $t = 0, i = -7.2$ A; and at $t = 1/(2 \times 400), i - 7.2$ A.

No unique solution for the values of L and C exists, but if the design allows for (say) 10% attenuation in the half cycle, then $e^{-\alpha t} = 0.9$, at $t = 1/800$, giving $C = 148.3\,\mu$F, and $L = 120\,\mu$H, with $I_{\text{peak}} = 20.6$ A.

If $e^{-\alpha t} = 0.8$, then $C = 70.0\,\mu$F, $L = 256\,\mu$H, and $I_{\text{peak}} = 19.6$ A.

If $e^{-\alpha t} - 0.7$, then $C = 43.8\,\mu$F, $L = 415\,\mu$H, and $I_{\text{peak}} = 18.6$ A.

If $e^{-\alpha t} = 0.5$, then $C = 22.5\,\mu$F, $L = 833\,\mu$H, and $I_{\text{peak}} = 16.6$ A.

If $e^{-\alpha t} = 0.3$, then $C = 12.98\,\mu$F, $L = 1.521$ mH, and $I_{\text{peak}} = 14.3$ A.

If $e^{-\alpha t} = 0.1$, then $C = 6.79\,\mu$F, $L = 3.244$ mH, and $I_{\text{peak}} = 11.7$ A.

The most practical solution would probably be the last one of $C = 6.79\,\mu$F and $L = 3.244$ mH, as the earlier ones demand too large a capacitor to be economical.

Example 5-8

In the circuit of Example 5-6, if the inductance in the d.c. source is very high, calculate the required size of the capacitor.

SOLUTION If (say) the source inductance were 1 H, then a voltage of 72 V across it would only change the current by $(72/1) \times (1/100) = 0.72$ A in one half cycle of the 50 Hz output. Hence, to all practical purposes, the source can be considered as a constant current source at 36 A.

Referring to Fig. 5-18b, the current into the parallel combination of the equivalent load (R/n^2) with the capacitor $4C$ will be constant at 36 A. At commutation, the capacitor has a reversed charge and, letting the current in R/n^2 be i, then $(36 - i) = 4C(dv/dt)$, where $v = iR/n^2$, the solution of which is

$$i = 36 - 72\,e^{-t/8C} \qquad (n = 3, \ R = 18)$$

The turn-off time of 60 μs is reached when i is zero, giving $C = 10.82\,\mu$F.
With the large source inductance, several cycles will elapse from the connection of the d.c. source, to the time when the load reaches its steady-state condition.

Example 5-9

Recalculate the value of capacitor determined for Example 5-8 if an inductance of 0.045 H is added to the load.

SOLUTION Now in the equivalent circuit of Fig. 5-18b there is an inductance of $0.045/n^2 = 0.005$ H, $n = 3$ from Example 5-6. The equivalent circuit is as shown in Fig. 5-41, where the source current is constant at 36 A, and the initial values of i and v are -36 A and -72 V respectively at time $t = 0$, the time at the start of the commutation of the thyristor.
The equation for the capacitor voltage will be for an underdamped oscillation giving $i = 36 - A\,e^{-\alpha t}\cos(\omega t + \phi)$
where $\alpha = 2/(2 \times 0.005) = 200$, and $\omega = [(50/C) - 40\,000]^{1/2}$.
An approximate value for C can be estimated by assuming the capacitor current to remain at its initial value of 72 A with v collapsing to zero in the turn-off time of 60 μs. Hence $72 = 4C \times (72/60 \times 10^{-6})$, giving $C = 15\,\mu$F.
Correcting the value of C by substitution into the exact equation so that $v = 0$ at $t = 60\,\mu$s gives the value of C as between 14.9 and 15 μF, so showing the approximate solution to be accurate.

Example 5-10

Using the load data of Example 5-9, redesign for values of capacitor and source inductance, this time assuming a small source inductance.

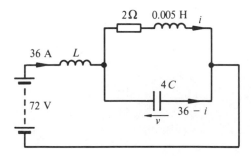

Figure 5-41

SOLUTION This is the case of a constant-voltage source, the source inductance being only sufficiently large to allow turn-off to take place.

Referring to Fig. 5-41 we can now assume the load inductance is large enough to prevent load current change over the commutation period.

Neglecting source losses, $v = 72 - V \cos(\omega t + \phi)$, where initially at $t = 0$,

$v = -72\,\text{V}$ and $dv/dt = 72/4C$, with $\omega = 1/\sqrt{(4LC)}$.

At the turn-off time of $t = 60\,\mu s$, $v = 0$.

A solution to these equations is $C = 18.6\,\mu F$, $L = 211\,\mu H$ (by taking $\phi = 0.7$). Other solutions are possible.

Note that the load voltage change of $72\,\text{V}$ in $60\,\mu s$ is taken across the inductance of $0.005\,\text{H}$, giving a current change of $0.86\,\text{A}$, hence assuming a constant load current of $36\,\text{A}$ is justified.

Example 5-11

A load of $18\,\Omega$ with an inductance of $0.045\,\text{H}$ is fed from a $50\,\text{Hz}$ centre-tapped inverter with a d.c. source of $72\,\text{V}$ via a transformer of $3/1$ turns ratio. The circuit is as shown in Fig. 5-19 with the diodes tapped into the primary winding at 10% from the end. Sketch the load-voltage and current waveforms of the inverter.

Determine the periods of conduction of the thyristors and diodes neglecting commutation conditions.

How would the circuit performance differ if the diodes were connected to the winding ends with the thyristors?

SOLUTION The data in this example correspond to Example 5-10, the equivalent circuit being that shown in Fig. 5-42 with the load values referred to the primary. In fact, the load voltage will be $3v_L$ and the load current $i_L/3$.

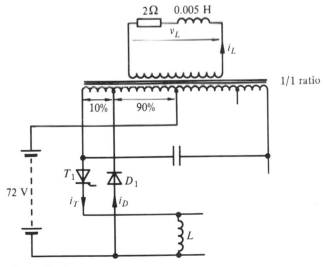

Figure 5-42

When thyristor T_1 is fired, the load current i_L is virtually unchanged over the commutation period, being maintained by the capacitor current, the load voltage following the capacitor voltage ($\div 2$).

Assuming perfect devices, when the voltage across the 90% tapping reaches 72 V, diode D_1 will conduct, the load voltage being maintained at $72/0.9 = 80$ V. Ignoring the commutation period and taking $t = 0$ when D_1 commences conduction, then at $t = 0$, $v_L = 80$ V, $i_L = -36$ A; hence $i_L = \dfrac{80}{2} - \left(\dfrac{80}{2} + 36\right) e^{-t/T}$ A, where $T = 0.005/2 = 0.0025$ s. i_L will fall to zero at $t = 1.6$ ms. D_1 conducts current for 1.6 ms of value $i_L/0.9$. When D_1 ceases conduction, T_1 now turns on, $v_L = 72$ V, $i_L = \dfrac{72}{2} - \dfrac{72}{2} e^{-t/T}$ A, where $t = 0$ at the instant T_1 takes over from D_1. During the period of drop in secondary voltage from 80 to 72 V, the capacitor will discharge, starting positive current flow.

To summarize, during each half cycle period of $1/(50 \times 2) = 10$ ms, $i_L = 40 - 76\, e^{-400t}$ A for the first 1.6 ms, then $i_L = 36 - 36\, e^{-400t}$ A for the last 8.4 ms, finishing at 34.8 A, which is close enough to be assumed 36 A considering the other simplifying assumptions. The load waveforms are shown in Fig. 5-43. When D_1 starts to take current from T_1, a path is formed for the stored magnetic energy in L to be transferred into the transformer primary via the tapped section. When all of the stored energy is dissipated, T_1 will turn off. Without the tapped section, D_1 and T_1 would form a short-circuit across L, with the stored magnetic energy being dissipated as losses within D_1 and T_1.

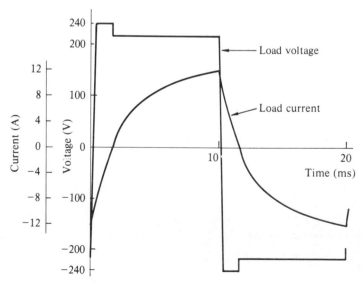

Figure 5-43

Example 5-12

A single-phase inverter (as shown in Fig. 5-20) supplies from a 200 V d.c. source, a load of 8 Ω resistance with inductance 0.02 H. If the inverter is operating at 50 Hz, determine the load-voltage and current waveforms for the first two cycles with (i) square-wave output, (ii) quasi-square-wave output with an on-period of 0.5. Neglect losses and commutation effects. Determine the steady-state load power for each case.

SOLUTION Time constant of the load $= T = 0.02/8 = 0.0025$ s.
(i) At the start the load current is zero.
First half cycle, $v_L = 200$ V,
$$i_L = 25 - 25\,e^{-400t}\,\text{A} = 24.5\ \text{A at end of half cycle } (t = 1/100).$$
Second half cycle, $v_L = -200$ V,
$$i_L = -25 + (25 + 24.5)\,e^{-400t}\,\text{A} = -24.1\ \text{A at end of half cycle}$$
with i_L zero at $t = 1.71$ ms.
Third half cycle, $v_L = 200$ V,
$$i_L = 25 - (25 + 24.1)\,e^{-400t}\,\text{A} = 24.1\ \text{A at end of half cycle}$$
with i_L zero at $t = 1.69$ ms.
Fourth half cycle, $v_L = -200$ V,
$$i_L = -25 + (25 + 24.1)\,e^{-400t}\,\text{A} = -24.1\ \text{A at end of half}$$
cycle.
Steady-state conditions have been reached with the waveform shapes as in Fig. 5-20b.
(ii) This particular quasi-square wave will have an on-period of 5 ms and a zero period of 5 ms in each half cycle of 10 ms. Starting with zero current the following apply:
First on-period, $v_L = 200$ V,
$$i_L = 25 - 25\,e^{-400t}\,\text{A, ending at } t = 5\text{ ms at } 21.6\ \text{A}.$$
First zero-period, $v_L = 0$,
$$i_L = 21.6\,e^{-400t}\,\text{A, ending at } t = 5\text{ ms at } 2.9\ \text{A}.$$
Second on-period, $v_L = -200$ V,
$$i_L = -25 + (25 + 2.9)\,e^{-400t}\,\text{A, ending at } -21.2\ \text{A}.$$
Second zero-period, $v_L = 0$,
$$i_L = -21.2\,e^{-400t}\,\text{A, ending at } -2.9\ \text{A}.$$
Third on-period, $v_L = 200$ V,
$$i_L = 25 - (25 + 2.9)\,e^{-400t}\,\text{A, ending at } 21.2\ \text{A}.$$
Steady-state conditions have been reached, the waveforms being as shown in Fig. 5-20c.
To determine the mean load power, the energy input to the load during each half cycle can be determined, and then divided by the half cycle time.
For the square wave,

$$\text{mean power} = \frac{1}{0.01}\int_0^{0.01} 200(25 - 49.1\,e^{-400t})\,dt = 2590\,\text{W}.$$

For the quasi-square wave,

$$\text{mean power} = \frac{1}{0.01}\int_0^{0.005} 200(25 - 27.9\, e^{-400t})\, dt = 1294\,\text{W}.$$

Example 5-13

The inverter (and load) of Example 5-12 is controlled to give a notched waveform with five on-periods in each half cycle. Determine the load-current waveform over the first two cycles, when control is such as to give half maximum output.

SOLUTION The principle of the notched waveform is shown in Fig. 5-22. With five on-periods and equal off-periods, each period is 1 ms long.
For each on-period, $i_L = 25 - (25 - I_1)\, e^{-400t}$ A, where I_1 is the load current at the start of the on-period. For each off-period $i_L = I_2\, e^{-400t}$ A, where I_2 is the current at the start of that period, that is, the current at the end of the previous period. The waveform plot is shown in Fig. 5-44.

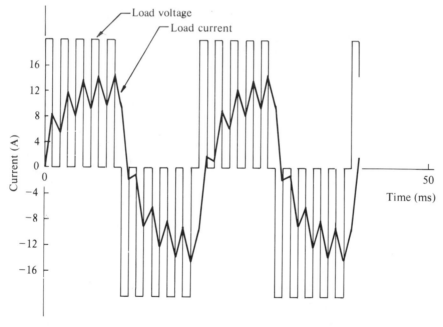

Figure 5-44

Example 5-14

The inverter (and load) of Example 5-12 is controlled to give a pulse-width modulated waveform from a triangular wave of frequency 500 Hz, with the reference set to one half maximum output. Plot the load-current waveform over the first two cycles.

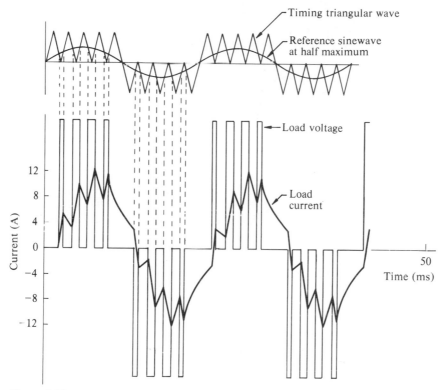

Figure 5-45

SOLUTION The principle of the pulse-width modulated waveform at half maximum is shown in Fig. 5-24b.

Figure 5 45 shows the timing waveforms and the load waveforms, the exponential change in load current being determined as in the previous example except that the on- and off-periods differ.

Example 5-15

A three-phase bridge inverter is fed from a d.c. source of 200 V. If the load is star-connected of 10 Ω/phase pure resistance, determine the r.m.s. load current, the required r.m.s. current rating of the thyristors, and the load power for (i) 120° firing, (ii) 180° firing.

SOLUTION (i) Figure 5-29 relates to the waveforms.
Peak load and thyristor current = 200/(2 × 10) = 10 A,

$$\text{r.m.s. load current} = \left(\frac{10^2 + 10^2 + 0^2}{3}\right)^{1/2} = 8.16 \text{ A,}$$

thyristor r.m.s. current = $10/\sqrt{3} = 5.8$ A,
load power = $8.16^2 \times 10 \times 3 = 2000$ W.

(ii) Figure 5-30 relates to the waveforms.

Peak load current $= (2 \times 200)/(3 \times 10) = 13.33$ A,

r.m.s. load current $= \left(\dfrac{6.66^2 + 13.33^2 + 6.66^2}{3} \right)^{1/2} = 9.43$ A,

thyristor r.m.s. current $= \left(\dfrac{6.66^2 + 13.33^2 + 6.66^2}{6} \right)^{1/2} = 6.67$ A,

load power $= 9.34^2 \times 10 \times 3 = 2667$ W.

Example 5-16

A three-phase bridge inverter is fed from a d.c. source of 240 V. If the control is by 180° firing, quasi-square-wave output, determine the load line-current waveform at 50 Hz output given the load is (i) delta-connected, with each phase 6 Ω resistance and 0.012 H inductance, (ii) star-connected, with each phase 2 Ω resistance and 0.004 H inductance.

SOLUTION (i) Referring to the waveforms of Fig. 5-31, each phase of the delta will have a quasi-square wave of 120° duration applied to it. The current i_{ab} in the load phase across lines A to B, starting with zero initial current, is as follows:

For first 120°, $v_{ab} = 240$ V, giving $i_{ab} = 40 - 40\,e^{-t/T}$, where $t = 0$ at the start of this period, and $T = L/R = 0.002$ s. At the end of this period $t = 1/150$, with $i_{ab} = 38.57$ A.

For the next 60°, $v_{ab} = 0$, giving $i_{ab} = 38.57\,e^{-t/T}$, ending at $t = 1/300$, with $i_{ab} = 7.29$ A.

For the next 120°, $v_{ab} = -240$ V, giving $i_{ab} = -40 + (40 + 7.29)\,e^{-t/T}$, ending at $t = 1/150$, with $i_{ab} = -38.31$ A.

For the next 60°, $v_{ab} = 0$, giving $i_{ab} = -38.31\,e^{-t/T}$, ending at $t = 1/300$, with $i_{ab} = -7.24$ A.

Likewise for the next cycle, the expression for the four periods will be $40 - (40 + 7.24)\,e^{-t/T}$, $38.31\,e^{-t/T}$, $-40 + (40 + 7.24)\,e^{-t/T}$, $-38.31\,e^{-t/T}$, ending with $i_{ab} = -7.24$ A, that is, steady-state conditions have effectively been reached in the second cycle.

The current in line $A = i_a = i_{ab} - i_{ca}$, where the current i_{ca} is identical to above but displaced by 240° in the cycle as shown in Fig. 5-46.

For first 60°, $i_a = 40 - 47.24\,e^{-t/T} - 38.31\,e^{-t/T} = 40 - 85.55\,e^{-t/T}$ A

next 60°, $i_a = 40 - 8.92\,e^{-t/T} + 40 - 47.24\,e^{-t/T} = 80 - 56.16\,e^{-t/T}$ A

next 60°, $i_a = 38.31\,e^{-t/T} + 40 - 8.92\,e^{-t/T} = 40 + 29.39\,e^{-t/T}$ A

next 60°, $i_a = -40 + 85.55\,e^{-t/T}$ A

next 60°, $i_a = -80 + 56.16\,e^{-t/T}$ A

last 60°, $i_a = -40 - 29.39\,e^{-t/T}$ A

(ii) The star-connected load values are equivalent to the delta-connection above, and ought to yield the same current expressions.

From Fig. 5-31, the phase voltage is stepped with a peak value of $240 \times (2/3) = 160$ V and half step value 80 V.

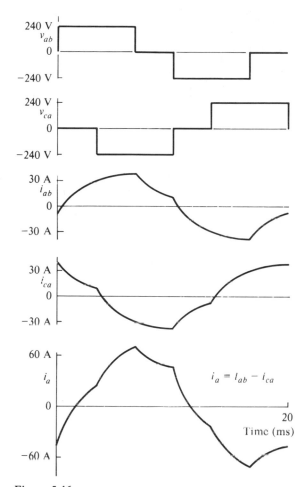

Figure 5-46

Using the initial value from (i) for $i_a = -45.55$ A, then:

in first 60°, $v_a = 80$ V, $i_a = 40 - (40 + 45.55) e^{-t/T}$ A

next 60°, $v_a = 160$ V, $i_a = 80 - (80 - 23.84) e^{-t/T}$ A

next 60°, $v_a = 80$ V, $i_a = 40 - (40 - 69.39) e^{-t/T}$ A

next 60°, $v_a = -80$ V, $i_a = -40 + 85.55 e^{-t/T}$ A

next 60°, $v_a = -160$ V, $i_a = -80 + 56.16 e^{-t/T}$ A

last 60°, $v_a = -80$ V, $i_a = -40 - 29.39 e^{-t/T}$ A

These calculations give identical expressions for (i) and (ii).

Example 5-17

A single-phase load is to be supplied from a constant-current source inverter. If the load is 12Ω resistance, the d.c. source voltage 120 V, the operation frequency

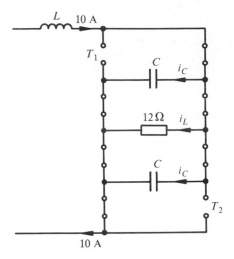

Figure 5-47

20 Hz, then determine suitable values for the source inductance and turn-off capacitors. Allow the thyristors a turn-off time of 50 μs and neglect losses.

SOLUTION Figure 5-34 refers to this circuit. The selection of the inductance size hinges on how constant the source current is when changes occur in the load impedance. The steady load current will be $120/12 = 10$ A. Suppose a load short-circuit increases the load current by 1 A in one cycle, then $di/dt = 1 \times 20$, and with $L(di/dt) = 120$, we have $L = 120/20 = 6$ H.

Immediately after turn-off of thyristors T_1 and T_2, all the diodes are conducting as shown in Fig. 5-47, with the capacitor currents being identical. The voltage difference between source and load is taken across L by a very small rate of current change, small enough so that the 10 A can be assumed constant.

The equations to the circuit are:

$$2i_C + i_L = 10$$

and $$12i_L = \frac{1}{C} \int i_C \, dt \quad \text{with } C \text{ initially charged to} -120 \text{ V.}$$

Using Laplace transforms, these equations become:

$$2\bar{i}_C + \bar{i}_L = \frac{10}{s}$$

and $$12\bar{i}_L = \frac{\bar{i}_C}{sC} - \frac{120}{s}$$

The solution to these equations yields:

$$i_L(t) = 10 - 20 \, e^{-t/24C} \text{ A}$$

and
$$i_C(t) = 10\,e^{-t/24C}\,\text{A}$$

The voltage across thyristor T_1 (or T_2) will become forward biased after the load voltage, that is, i_L is zero. Using the turn-off time of $50\,\mu s$ yields a value for C of $3\,\mu F$.

Example 5-18

If to the load given in Example 5-17 is added an inductance of 0.06 H, analyse the conditions during commutation.

SOLUTION Adding the inductance component to the equation derived in Example 5-17 we have:

$$2\bar{i}_C + \bar{i}_L = \frac{10}{s}$$

$$12\bar{i}_L + 0.06(s\bar{i}_L + 10) = \frac{\bar{i}_C}{sC} + \frac{V_0}{s}$$

Taking $C = 3\,\mu F$ and the initial capacitor voltage as $-120\,\text{V}\,(= V_0)$, the solution to these equations is:

$$i_L = 10 - e^{-100t}(20\cos 1664t + 1.2\sin 1664t)\,\text{A}$$

$$i_C = e^{-100t}(10\cos 1664t + 0.6\sin 1664t)\,\text{A}$$

$$v_C = \text{load voltage} = 120 + e^{-100t}(1990\sin 1664t - 240\cos 1664t)\,\text{V}$$

Examination of these expressions reveals that the capacitor current will fall to zero after $980\,\mu s$, when the diodes D_1 and D_2 (see Fig. 5-34) will turn off, the load current now being fully reversed to 10 A and the load voltage steady at 120 V. However, at $980\,\mu s$ the capacitor will be charged to 1933 V, which is excessive and could break down the thyristor in the off-state. Also this value means that the initial assumption of the capacitor being charged to 120 V was incorrect.

To reduce the voltage imposed on the thyristor, the capacitor size must be increased. If (say) C were made $20\,\mu F$, with an initial capacitor voltage of $-1100\,\text{V}$, then the expressions become:

$$i_L = 10 - e^{-100t}(20\cos 638t + 28.75\sin 638t)\,\text{A}$$

$$i_C = e^{-100t}(10\cos 638t + 14.37\sin 638t)\,\text{A}$$

$$v_C = \text{load voltage} = 120 + e^{-100t}(593\sin 638t - 1220\cos 638t)\,\text{V}$$

Examination of these expressions reveals that i_C falls to zero in 4 ms, leaving the capacitor charged to just over 1000 V. The turn-off time available to the thyristors is that given when v_C is zero, that is, after 1.59 ms, which is very considerably above the minimum required. Hence, in this circuit, it is the voltage limitation which determines the capacitor size.

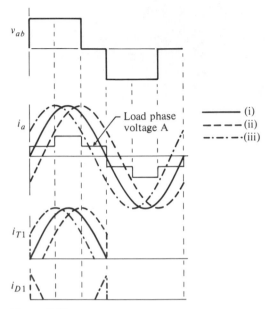

Figure 5-48

Example 5-19

A three-phase inverter is connected to a constant direct-voltage source. If the load takes a current of sinusoidal waveform, sketch the thyristor and diode current waveforms when the load current is (i) in phase with the voltage, (ii) lagging by 30°, and (iii) leading by 30°. Neglect all losses, assume perfect conditions, and 180° firing.

SOLUTION Taking as circuit reference Fig. 5-28, the ideal load phase voltage will be stepped as in Fig. 5-30. The required waveforms are shown plotted in Fig. 5-48. With the in-phase condition of (i), only the thyristors conduct.

With lagging power factor, the diode D_1 conducts after thyristor T_1 is turned off. With a leading power factor, thyristor T_1 takes the load current directly it is fired, but thyristor T_1 current ceases naturally before the end of 180°, when the reverse current transfers to diode D_1.

Example 5-20

A three-phase inverter is connected to a direct-voltage source of 240 V. If the a.c. side acts as a generator of sinusoidal current of peak value 20 A, plot the thyristor, diode, and d.c. source current when the generator current is leading the voltage by 60°. Determine the power flow through the inverter. Neglect all losses, assume perfect conditions, and 180° firing.

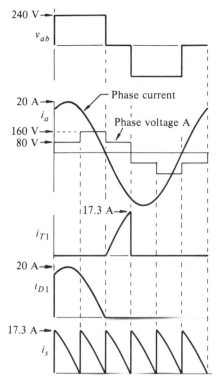

240 V — v_{ab}

20 A — Phase current

i_a

Phase voltage A

160 V —
80 V —

17.3 A —

i_{T1}

20 A —

i_{D1}

17.3 A —

i_s

Figure 5-49

SOLUTION Using Fig. 5-37 as reference, the required waveforms are shown in Fig. 5-49.

The power flow can be calculated from the d.c. source current i_S multiplied by 240 V.

$$\text{Mean power} = \frac{1}{2\pi/6} \int_{2\pi/3}^{\pi} 240 \times 20 \sin \theta \, d\theta = 2292 \, \text{W}.$$

SIX

SOME APPLICATIONS

The applications of semiconductor power devices fall into one of two major areas — one is those which involve motor control, and the other involves non-motor applications. In this chapter, attention will be given to those applications not directly involving motor control.

6-1 CONTACTOR

The contactor is a switching device which connects (on) or isolates (off) a load from the supply. The mechanical contactor requires an operating movement, and in the device shown in Fig. 6-1a is actuated by energization of the magnetic plunger carrying the contacts. The semiconductor alternatives are shown respectively in Fig. 6-1b and c, as a triac for low power, and as two thyristors for heavier power applications.

The mechanical contactor will take a time to operate which may extend to a few cycles. The semiconductor devices will turn on immediately a firing pulse is received at the gate and, by suitable controls in the firing circuit, the turn-on can be timed to any desired angle within the cycle. To maintain the semiconductor devices in the on-state, continuous firing pulses are required at the gate, as turn-on is required after each current zero. Turn-off will be at a current zero, the semiconductor devices turning off at the first zero after removal of the firing pulses. The mechanical contactor will draw an arc as the contacts part, turning off at a current zero when the gap is sufficient to maintain the restriking voltage.

The thyristor (or triac) on-state voltage-drop will result in heat generation requiring cooling, however; it is possible to short-circuit the thyristors as shown in Fig. 6-1d. Switching is carried out by the thyristors, but in between times the mechanical switch carries the load current, thus avoiding excessive losses.

Voltage transients can occur at switching which will require suppression (see Chapter 10) in the form of an RC network across the contactor. The presence of this network can mean that with an open-circuit at the load, the supply voltage could be present across this open-circuit, a possibly dangerous situation. To avoid this problem, an isolating switch can be included as shown in Fig. 6-1d.

The advantage of the semiconductor contactor is the speed of response, accuracy of switching time, freedom from routine maintenance, and lack of audio-noise.

When used for motor control, the semiconductor contactor must be rated to handle starting and any transient-load currents. When used in spot welding appli-

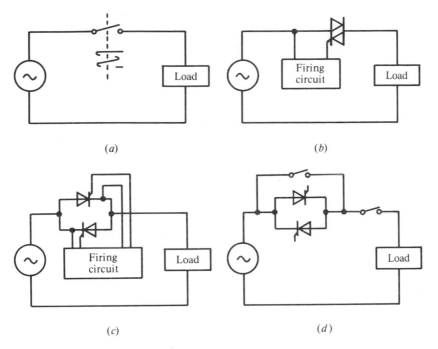

(a)

(b)

(c)

(d)

Figure 6-1 Single-phase load contactor arrangements. (*a*) Electromechanical. (*b*) Triac. (*c*) Two thyristors in inverse-parallel connection. (*d*) With short-circuiting and isolating switches.

cations, the load seen by the supply is very inductive, and here the initial switching instant will be arranged to be near to the supply voltage maximum, so as to avoid any off-set transient-current component.

A semiconductor contactor for use with d.c. supplies could be one of the circuits described in Chapter 4, say, the chopper circuit of Fig. 4-3.

6-2 HEATING

A heating load of the resistance type can be controlled to different power levels by the use of the triac connection shown in Fig. 6-2*a*. Where high power levels preclude the use of the triac, two thyristors can be used in the inverse parallel connection (see Fig. 6-1*c*). To obtain common cathodes, and hence simplify the firing circuit, two diodes can be added as shown in Fig. 6-2*b*. A further alternative is shown in Fig. 6-2*c*.

Phase angle control can be used as shown in Fig. 6-2*d*, where the start of each half cycle is delayed by an angle α. Taking the load to be a pure resistance R and the supply to have a peak value $V_{\max}$, then

$$\text{power} = \frac{V_{\text{rms}}^2}{R} = \frac{1}{\pi R} \int_{\alpha}^{\pi} (V_{\max} \sin \theta)^2 d\theta = \frac{V_{\max}^2}{2\pi R} \left(\pi - \alpha + \frac{1}{2} \sin 2\alpha \right) \qquad (6\text{-}1)$$

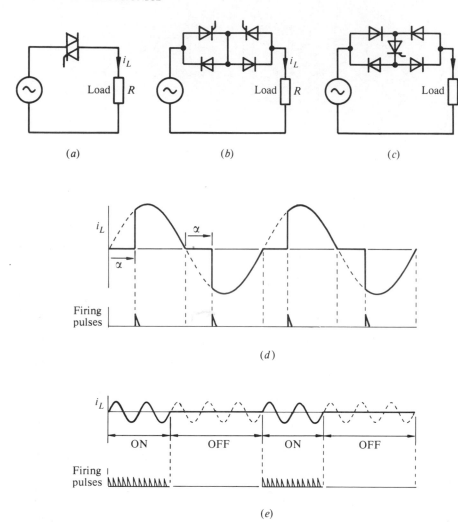

Figure 6-2 Power control in single-phase heating load. (*a*) Triac or inverse-parallel connection. (*b*) An alternative connection with common cathodes. (*c*) A single thyristor with diode bridge. (*d*) Phase angle control. (*e*) Integral cycle control.

The majority of heating loads have thermal time constants of several seconds or longer. In this case, little variation of the heater temperature will occur if control is achieved by allowing a number of cycles on, with a number of cycles off, as shown in Fig. 6-2*e*. This form of control is variously called *integral cycle control, burst firing,* or *cycle syncopation.* The time of the on-period consists of whole half cycles. The power will be given by

$$\text{power} = \frac{V_{\text{max}}^2}{2R} \times \frac{\text{on time}}{(\text{on} + \text{off}) \text{ time}} \tag{6-2}$$

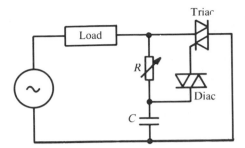

Figure 6-3 Firing circuit for phase angle control.

In respect to the loading on the supply system, phase angle control will be similar to the controlled rectifier, the load taking its fundamental component at a lagging power factor, together with harmonic currents. Integral cycle control does not suffer from these drawbacks, hence electricity supply authorities will normally insist that heating loads be controlled by the integral cycle method.

The illumination level of incandescent lamps can be controlled in the above manner, provided the phase angle control method is used, as the thermal time constant of a lamp is too short to allow the use of the integral cycle control.

A simple triac control circuit for lamp dimmers is shown in Fig. 6-3, where use is made of a diac. A diac is a gateless triac designed to break down at a low voltage. When the charge on capacitor C has reached the diac breakdown level, it will be discharged into the triac gate. Adjustment of the series resistance R determines the charging rate of capacitor C and hence the phase angle delay. Additional circuitry can be introduced into the firing circuit so that firing is automatically retarded at initial switch-on, so as to avoid the current surge experienced with a cold incandescent lamp.

Spot (resistance) welding, where high current is passed for (say) one second, can be controlled by using water-cooled inverse-parallel-connected thyristors with integral cycle control. As the load is inductive, the initial switching-on angle can be selected so as to avoid any asymmetrical component of current.

Three-phase, three-wire loads, either delta or star-connected, may be controlled as shown in Fig. 6-4. The fully-controlled circuit requires six thyristors (or three triacs), whereas the half-controlled circuit uses diodes for the return current path.

For current to flow in the fully-controlled circuit of Fig. 6-4a, at least two thyristors must conduct. If all the devices were diodes, they would take up conduction in the numbered order at 60° intervals. Hence, each thyristor must be given a second firing pulse 60° after its initial firing pulse for starting, and also in order to permit conduction through two thyristors when current conduction is discontinuous. Waveforms for a firing delay angle of 100° are shown in Fig. 6-5, where the need for the second firing pulse occurs following the zero current period in each half cycle of line current. Except for starting, the second firing pulse will not be required below firing delay angles of 90°.

To construct the waveforms, consider that, if a thyristor in each line is conduct-

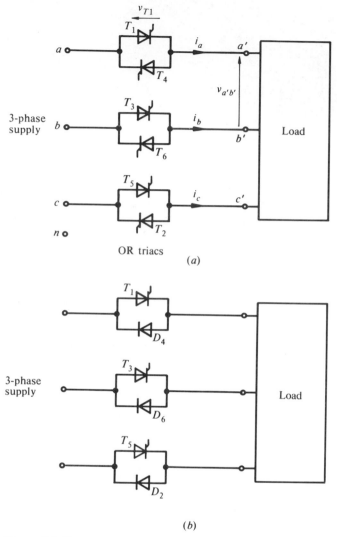

Figure 6-4 Three-phase loads. (*a*) Fully-controlled. (*b*) Half-controlled.

ing, then normal three-phase theory applies. If only two thyristors are conducting at a particular time, then two lines are acting as a single-phase supply to the load, with the load terminal of the floating line at the midpoint between the two conducting phases.

The line current waveforms for different fully-controlled loads are shown in Fig. 6-6. Up to a firing delay of 60°, conditions alternate between three thyristors on (immediately after firing) to only two thyristors on, one of the thyristors turning off when its current attempts to reverse. Above a firing delay of 60° and up to 90°, at any given time two thyristors only are conducting. Above 90° delay, there

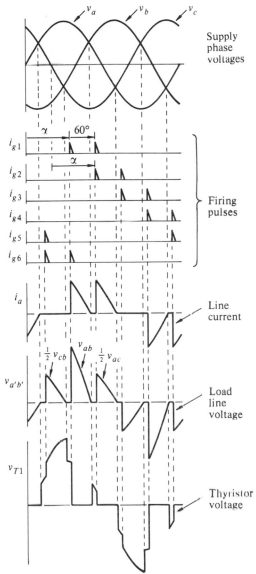

Figure 6-5 Waveforms for fully-controlled resistance load at a firing delay angle of 100°.

are periods when no thyristors are on. Zero r.m.s. current is reached when the firing delay is 150°.

Using the star-connected load of Fig. 6-6 with a resistance of R Ω/phase, when three thyristors are on,

$$i_a = \frac{V_{\text{line(max)}}}{\sqrt{3}R} \sin (\omega t + \phi),$$

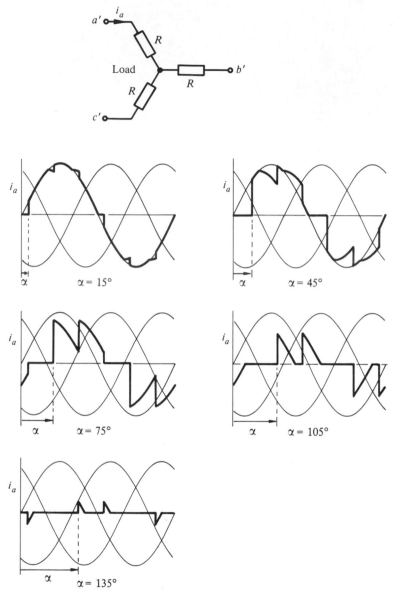

Figure 6-6 Line current at various firing angles, fully-controlled load.

but when only two thyristors are on

$$i_a = \frac{V_{\text{line(max)}}}{2R} \sin (\omega t + \phi).$$

Between firing delay angles $0°$ and $60°$,

$$\text{power} = 3I_{\text{rms}}^2 R = \frac{3V_{\text{line(max)}}^2}{R\pi} \left[\int_{\alpha}^{\pi/3} \frac{\sin^2\theta}{3} d\theta + \int_{\pi/2}^{\frac{\pi}{2}+\alpha} \frac{\sin^2\theta}{4} d\theta \right.$$

$$\left. + \int_{\frac{\pi}{3}+\alpha}^{2\pi/3} \frac{\sin^2\theta}{3} d\theta + \int_{\pi/2}^{\frac{\pi}{2}+\alpha} \frac{\sin^2\theta}{4} d\theta + \int_{\frac{2\pi}{3}+\alpha}^{\pi} \frac{\sin^2\theta}{3} d\theta \right]$$

$$= \frac{3V_{\text{line(max)}}^2}{R\pi} \left[\frac{\pi}{6} - \frac{\alpha}{4} + \frac{\sin 2\alpha}{8} \right] \tag{6-3}$$

Between firing delay angles 60° and 90°,

$$\text{power} = \frac{3V_{\text{line(max)}}^2}{R\pi} \left[\int_{\frac{\pi}{2}-\frac{\pi}{3}+\alpha}^{\frac{5\pi}{6}-\frac{\pi}{3}+\alpha} \frac{\sin^2\theta}{4} d\theta + \int_{\frac{\pi}{2}-\frac{\pi}{3}+\alpha}^{\frac{5\pi}{6}-\frac{\pi}{3}+\alpha} \frac{\sin^2\theta}{4} d\theta \right]$$

$$= \frac{3V_{\text{line(max)}}^2}{R\pi} \left[\frac{\pi}{12} + \frac{3}{16} \sin 2\alpha + \frac{\sqrt{3}}{16} \cos 2\alpha \right] \tag{6-4}$$

Between firing delay angles 90° and 150°,

$$\text{power} = \frac{3V_{\text{line(max)}}^2}{R\pi} \left[\int_{\frac{\pi}{2}-\frac{\pi}{3}+\alpha}^{\pi} \frac{\sin^2\theta}{4} d\theta + \int_{\frac{\pi}{2}-\frac{\pi}{3}+\alpha}^{\pi} \frac{\sin^2\theta}{4} d\theta \right]$$

$$= \frac{3V_{\text{line(max)}}^2}{R\pi} \left[\frac{5\pi}{24} - \frac{\alpha}{4} + \frac{\sqrt{3}}{16} \cos 2\alpha + \frac{1}{16} \sin 2\alpha \right] \tag{6-5}$$

The above equations neglect the volt-drops across the thyristors, and any other losses.

The half-controlled arrangement of Fig. 6-4b is simpler in as much as only a single firing pulse is required with the return current path being a diode. The line currents for various firing delay angles are shown in Fig. 6-7. Up to firing delays of 60°, current immediately before firing has been flowing in one thyristor and one diode respectively in each of the other two lines, so, after firing, normal three-phase operation occurs until the diode is reversed biased, when current flows in two lines only. With firing delay angles above 60°, at any one time only one thyristor is conducting, the return current being shared at different intervals by two or one diodes. Above firing delay of 120°, current flows only in one diode. The zero power condition is reached at a firing delay angle of 210°.

By similar reasoning to that used for deriving Eqs. (6-3) to (6-5), the power equations for the half-controlled connection are
Between firing delay angles 0° and 90°,

$$\text{power} = \frac{3V_{\text{line(max)}}^2}{2R\pi} \left[\int_{\alpha}^{2\pi/3} \frac{\sin^2\theta}{3} d\theta + \int_{\pi/2}^{\frac{\pi}{2}+\alpha} \frac{\sin^2\theta}{4} d\theta \right.$$

$$\left. + \int_{\frac{2\pi}{3}+\alpha}^{4\pi/3} \frac{\sin^2\theta}{3} d\theta + \int_{3\pi/2}^{\frac{3\pi}{2}+\alpha} \frac{\sin^2\theta}{4} d\theta + \int_{\frac{4\pi}{3}+\alpha}^{2\pi} \frac{\sin^2\theta}{3} d\theta \right]$$

$$= \frac{3V_{\text{line(max)}}^2}{2R\pi} \left[\frac{\pi}{3} - \frac{\alpha}{4} + \frac{\sin 2\alpha}{8} \right] \tag{6-6}$$

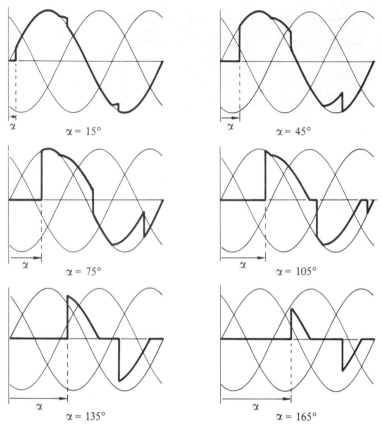

Figure 6-7 Line current at various firing angles, half-controlled load.

Between firing delay angles 90° and 120°,

$$\text{power} = \frac{3V^2_{\text{line(max)}}}{2R\pi}\left[\int_\alpha^{2\pi/3}\frac{\sin^2\theta}{3}d\theta + \int_{\pi/2}^\pi\frac{\sin^2\theta}{4}d\theta\right.$$

$$\left. + \int_{\frac{2\pi}{3}+\alpha}^{4\pi/3}\frac{\sin^2\theta}{3}d\theta + \int_{3\pi/2}^{2\pi}\frac{\sin^2\theta}{4}d\theta + \int_{\frac{4\pi}{3}+\alpha}^{2\pi}\frac{\sin^2\theta}{3}d\theta\right]$$

$$= \frac{3V^2_{\text{line(max)}}}{2R\pi}\left[\frac{11\pi}{24} - \frac{\alpha}{2}\right] \tag{6-7}$$

Between firing delay angles 120° and 210°,

$$\text{power} = \frac{3V^2_{\text{line(max)}}}{2R\pi}\left[\int_{\frac{\pi}{2}-\frac{2\pi}{3}+\alpha}^\pi\frac{\sin^2\theta}{4}d\theta + \int_{\frac{3\pi}{2}-\frac{2\pi}{3}+\alpha}^{2\pi}\frac{\sin^2\theta}{4}d\theta\right]$$

$$= \frac{3V^2_{\text{line(max)}}}{2R\pi}\left[\frac{7\pi}{24} - \frac{\alpha}{4} + \frac{\sin 2\alpha}{16} - \frac{\sqrt{3}\cos 2\alpha}{16}\right] \tag{6-8}$$

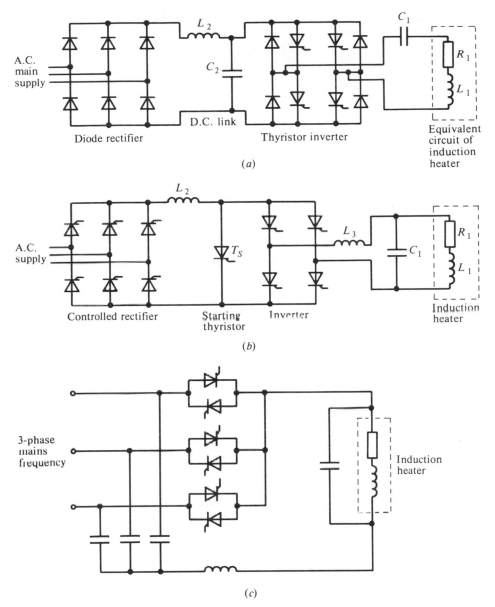

Figure 6-8 Basic power supply circuit layout for a medium-frequency induction heater. (*a*) Constant-voltage inverter. (*b*) Constant-current inverter. (*c*) Cycloinverter.

Induction heaters (melters) in the frequency range (say) 250 Hz to 10 000 Hz can be supplied from rectifier/inverter circuits as shown in Fig. 6-8. The induction heater will have a high inductance L_1 and, by suitable selection of added capacitor C_1, the load can be made a resonance circuit at the desired frequency.

The characteristic of the series resonant load in Fig. 6-8*a* means the inverter thyristors are turned off naturally without the need for commutation circuits. The firing instants of the thyristors would be controlled so as to maintain the load power constant, the heater frequency being allowed to drift. The input d.c. source to the inverter is a diode rectifier, the d.c. link voltage being maintained level by capacitor C_2, with inductor L_2 protecting against inverter-frequency current pulses being directly reflected into the diode rectifier. The function of the inverter diodes are to limit the capacitor C_1 voltage.

Placing the capacitor C_1 in parallel with the induction heater demands that the inverter is fed from a constant-current source. In Fig. 6-8*b* the inductor L_2 is very large, making the input to the inverter sensibly constant over the period of one cycle of the load frequency. Thyristor T_S is used at the start initially to build up the current in L_2. The frequency of the load is measured, phase advanced, then fed into the firing control circuits for the inverters, so that the inverter thyristors are switched at the load frequency and in phase advance of the load voltage. The small inductance L_3 dictates the overlap period when alternate arms of the inverter commutate. Control of the rectifier voltage controls the load power.

A circuit is shown in Fig. 6-8*c* where the d.c. link is avoided, directly converting power at the mains frequency to a higher frequency through what may be called a *cycloinverter arrangement.* The load self-resonates to turn off the thyristors, the thyristors being fired so as to provide the positive half cycles of load power from the most positive pair of incoming lines, and likewise the negative half cycle of load power from the two most negative incoming lines.

6-3 VOLTAGE REGULATION

Thyristors and triacs can be used in a number of configurations to control the voltage level at an a.c. load. In the previous section, connections were given in Fig. 6-2 for the single-phase and Fig. 6-4 for the three-phase loads respectively. These circuit configurations may be used to control voltage levels in loads which are either resistance or contain in addition some inductance.

Figure 6-9 shows the waveforms obtained when phase angle delay is used with an inductive load. The current continues beyond the voltage zero due to the inductance, hence giving the chopped voltage waveform shown. The rapid rise of voltage on the opposite thyristor at turn-off could lead to high dv/dt and possible turn-on. The high dv/dt is more of a problem with the triac than with the thyristor.

With inductive loads, the continuation of conduction beyond the voltage zero means that there is no control below a certain firing angle delay. Firing control is only possible after cessation of current.

One application of the above single- or three-phase regulators is to control the input voltage to a transformer, the secondary of which is connected to a diode rectifier feeding a d.c. load as shown in Fig. 6-10. Circumstances where this technique would be used are: low voltage, high current, d.c. loads, where it is more

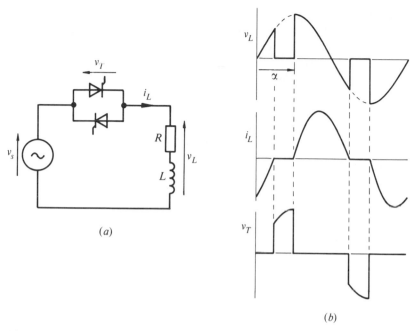

(a)

(b)

Figure 6-9 Single-phase regulator with inductive load. (*a*) Circuit, (*b*) Waveforms.

economic to use a lower-current-rated thyristor in the transformer primary. The d.c. load characteristics are similar to those where the normal controlled thyristor rectifier is employed.

Where a small degree of regulation is required, say 80–100%, then the tapped transformer as shown in Fig. 6-11 is best used. If thyristors T_1 and T_2 only are fired, then the output is v_1, that is, the output is a maximum. If thyristors T_3 and T_4 only are fired, then the output voltage is v_2, the minimum.

Any voltage between maximum and minimum is possible by firing thyristor T_3 (T_4), then later in the half cycle firing thyristor T_1 (T_2), as shown by the wave-forms in Fig. 6-11*b*. When thyristor T_1 is fired, thyristor T_3 is reverse-biased, hence it is turned off. Control of the firing angle of thyristor T_1 (T_2) determines the r.m.s. load voltage. The peak reverse and forward voltage experienced by the thyristors relate to the tapped section voltage only, provided one thyristor is conducting. Compared to the simple regulator of Fig. 6-2, the supply current waveform of

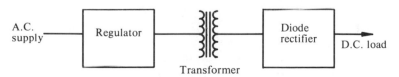

Figure 6-10 Use of the regulator to control d.c. load voltage.

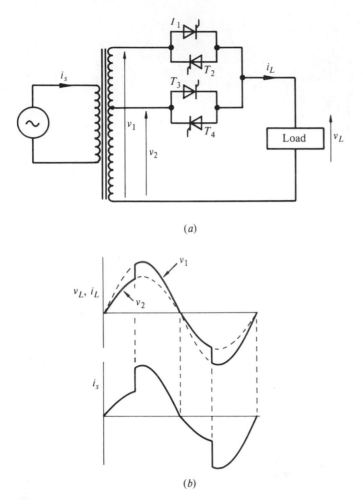

(a)

(b)

Figure 6-11 Regulation by tapped transformer. (a) Circuit. (b) Waveforms for resistance load.

the tapped transformer is less distorted, having no zero periods. The case of the inductive load is more complex, as illustrated by Example 6-9.

6-4 VOLTAGE MULTIPLIERS

A means of obtaining a direct voltage at a level above the maximum of the a.c. supply voltage is given by the circuits of Fig. 6-12. In Fig. 6-12a, capacitor C_1 is charged in the direction shown in the negative half cycle to the maximum value of the a.c. supply via diode D_1. During the positive half cycle, the charge on capacitor C_1 aids the a.c. supply, so that the capacitor C_2 is charged to twice the maximum supply voltage. In practice, with load and volt-drops, the value will be reduced.

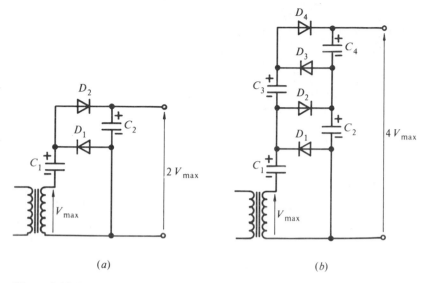

Figure 6-12 Voltage multiplier circuits. (*a*) Voltage doubler. (*b*) Cascade connection.

The cascade connection of Fig. 6-12*b* is an extension of the voltage doubler, capacitor C_1 is charged to V_{max}, while capacitors C_2, C_3 and C_4 are eventually charged to $2V_{max}$, hence giving an output voltage of $4V_{max}$ if losses are ignored.

6-5 STANDBY INVERTERS

Standby inverters are used mostly to provide an emergency supply at mains frequency (50/60 Hz), in the event of a mains failure. The form of standby may demand an uninterruptible supply, that is, in the event of mains failure the inverter immediately operates without loss of waveform. The less-demanding form of standby is where an interruption in supply can be tolerated between mains failure and emergency supply start.

A typical uninterruptible system is shown in Fig. 6-13. Here the load can be permanently fed from the inverter, the inverter d.c. source being obtained by rectifying the a.c. mains. In the event of a mains failure, the inverter will take its power supply from the battery, thus avoiding any interruption to the load. Loads which typically demand an uninterruptible supply are computers, communication links, and essential instrumentation in certain processes. The battery will have a limited capacity, and will be constantly trickle-charged to maintain its float voltage, with a higher charge rate after any discharge period.

Figure 6-13 includes an alternative direct link to the load from the a.c. supply. A change-over from mains to inverter supply will demand that the inverter be synchronized with the mains to avoid waveform distortion.

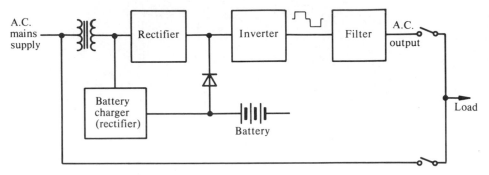

Figure 6-13 Uninterruptible inverter system.

In those situations where a short interruption in supply can be tolerated, the inverter would only be brought into operation and connected to the load after mains failure. Such circumstances might include emergency lighting where a loss of supply for (say) 0.2 second could be tolerated. The alternative of running up a diesel engine generation set as an emergency source involves a much longer gap in supply than with the static inverter systems.

The basic designs of the inverter systems were given in Secs. 5-3, 5-4 and 5-5. The specifications for standby use are for a fixed frequency with some control of output voltage levels and a near sinusoidal output voltage waveform. The raw output voltage can be quasi-square wave as in Fig. 5-20c, with control of the zero period determining the r.m.s. output voltage, but large filters will be required to remove the low-order harmonics. Alternatively, pulse-width modulation techniques can be used as illustrated in Fig. 5-23, this very much reducing the filtering requirements as low-order harmonics are not present in such waveforms. Filtering networks are described in Sec. 7-4-2.

Standby inverters for three-phase four-wire supplies are constructed from three separate single-phase inverters, feeding into the primaries of three single-phase transformers, with one end of each secondary joined to form the neutral of the three-phase supply. In this connection of the three-phase inverter, it is possible to supply unbalanced loads, while maintaining balance between the voltages.

The normal mains supply is a low-impedance source which can cope with the overloads associated with motor starting and protection demands (say) due to ensuring fuse blowing under short-circuit conditions. Care must be taken to ensure a static inverter can cope with these demands, if required, but, due to the low thermal overload capacity of semiconductor devices, severe overload specification will inevitably raise the inverter cost.

6-6 PARALLEL CONNECTIONS

In applications where the current requirement of a diode exceeds the rating of the largest devices, it is necessary to connect a number of diodes in parallel. Figure

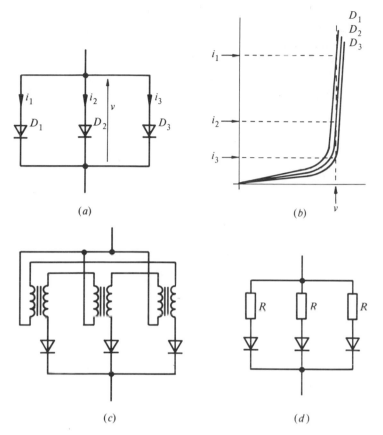

Figure 6-14 Current sharing in parallel diodes. (*a*) Plain connection. (*b*) Current distribution with unmatched diodes. (*c*) Use of current sharing reactors. (*d*) Addition of resistance.

6-14*a* and *b* illustrates the danger of using diodes with differing characteristics. Small differences in the on-state voltages can lead to large differences in the current distribution between the diodes.

One method of ensuring near equable distribution of current is to use double-wound reactors as shown in Fig. 6-14*c*, where any current unbalance induces a reactor voltage such as to bring about an m.m.f. balance. This solution is expensive and bulky. A simple solution is to add resistors as in Fig. 6-14*d*, so that the *IR* drop predominates over diode characteristic differences, but this solution is costly in losses.

The most usual means of ensuring equable current distribution without extra components is to use diodes matched for identical characteristics, both at low and overload ratings. In this manner (typically in a string of diodes) the current differences will not exceed 20%. The diodes must each be maintained at the same temperature.

Inevitably, the need for parallel diodes means total current of possibly several thousand amperes, so other considerations arise which compound the problems. The diodes can be mounted on a double-ducted oil-cooled busbar. The oil will be directed along one duct and return along the other parallel duct within the busbar, so that each diode is at the same mean temperature. The high current can set up a.c. magnetizing fields which may be minimized by careful layout of the busbars and conductors. The conductor layout must be so arranged that the inductance and resistance in each diode path are identical.

Chapter 10 deals in detail with protection, but it is worth noting here the need to fuse each diode separately, maintain all fuses at the same temperature, and to build in spare capacity in the event of the loss of one diode.

Paralleling thyristors presents more problems than with diodes. The above comments relating to diodes apply equally to thyristors, and forced equality of current can be obtained by use of the arrangements given in Fig. 6-14c and d. In general, matched thyristors are used as the best economical method. The additional problems relate to firing and holding levels.

In a parallel thyristor arrangement, once one thyristor has turned on, the anode-cathode voltage on the other thyristors falls. To ensure the turn-on of all thyristors, the firing pulse must be from the same source, have as fast a rise time as possible, with the pulse continuing for a long period. The fast rise time will guarantee that all thyristors start simultaneous conduction, and the long period will guarantee that each thyristor reaches its latching level, even with a reduced anode voltage.

A problem which can arise with parallel thyristors occurs if the load current falls to a low level; one or more of the thyristors may then turn off if their current drops below the holding level. Any subsequent rise in current could then overload the remaining thyristors. A continuous chain of gate firing pulses during the on-period would solve this problem.

6-7 SERIES CONNECTIONS

Where peak reverse (or forward) voltages of several thousand volts are required of the rectifying devices, it is necessary to connect the thyristors in series so as to share the voltage. Figure 6-15 illustrates the problems associated with a plain series connection. During the off-state, each thyristor will carry the same leakage current which, as shown in Fig. 6-15b, can lead to one thyristor taking most of the voltage and consequently breaking down. The other problem occurs during rapid turn-off, when a reverse current must flow to recover the storage change. As illustrated in Fig. 6-15c, differences between the thyristors means that, when one thyristor recovers, current ceases leaving the other thyristors unrecovered and hence unable to withstand the reapplication of forward voltage.

The problem of voltage sharing in the off-state is solved by connecting resistors, shown as R_1 in Fig. 6-16, across each thyristor. Provided the resistor current is somewhat greater than the thyristor leakage current, then the voltage sharing is independent of leakage values.

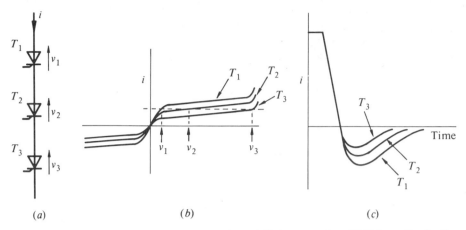

(a)   (b)   (c)

Figure 6-15 Voltage sharing in series connection. (*a*) Plain connection. (*b*) Voltage distribution in off-state with mismatched thyristors. (*c*) Illustrating differences in reverse recovery current requirements.

The connection of capacitors, shown as C in Fig. 6-16, across each thyristor will enable each thyristor to recover its off-state at turn-off. When the first thyristor recovers, its capacitor voltage builds up, demanding a charging current which can only flow through the other thyristors, thus enabling them in turn to regain their blocking state.

The capacitors across each thyristor create a problem at turn-on, as the capaci-

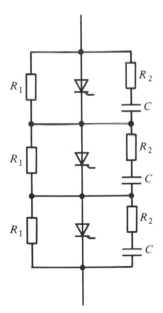

Figure 6-16 Equalization of voltages in series connection.

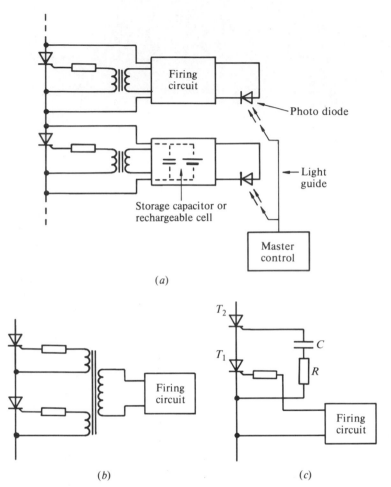

Figure 6-17 Alternative firing arrangements. (*a*) Optical link. (*b*) Transformer coupling. (*c*) Master-slave firing.

tor charge would be rapidly dumped into the thyristor with an excessive dI/dt, probably exceeding the thyristor rating. To control the dI/dt rate, a resistor R_2, as shown in Fig. 6-16, is placed in series with each capacitor C. The capacitors serve an additional function of equalizing the dv/dt rates across the series string during transient disturbances. When several thyristors are connected in series, a saturable inductor is added in series with the string to prevent excessive rates of current rise.

Simultaneous firing of the thyristors is required at turn-on when several thyristors are connected in series. As the gates of the individual thyristors can be at a potential of several thousand volts to the others, means other than transformer coupling have to be used to connect each gate to the master firing control. To overcome the insulation problem, light-guide optical systems are adopted for the link as shown in Fig. 6-17a. The supply to each thyristor firing module is obtained from a capacitor (or rechargeable cell) charged during the thyristor off-periods from across the thyristor itself.

Where only two or three thyristors in series are required, it may be more economic to use isolated secondaries of a transformer as shown in Fig. 6-17*b*.

Alternatively, a master-slave arrangement can be utilized as shown in Fig. 6-17*c*; the first thyristor is turned on, capacitor *C* then discharges into the gate of the next thyristor, turning it on. This can mean excessive voltage transients, so its use is not extended to series strings of many thyristors.

6-8 ELECTROCHEMICAL

Applications of electricity to electrochemical plant require direct current associated with a low voltage. The processes include electroplating, extraction, and refining of metals by electrolysis; production of elements such as chlorine and hydrogen; and electrochemical forming (machining) techniques.

Electroplating is concerned with the depositing of a metal on to a workpiece immersed in a suitable electrolyte by making it the cathode, the deposited metal being obtained from an anode made of that material. Figure 6-18 shows diagrammatically the process of electroplating. In electroplating, current densities are low, with large spacing of the electrodes to ensure even plating. The amount of metal deposited is proportional to the level of current and time of its flow. Most plating processes can tolerate ripple in the current, hence typically three-pulse supplies can be used. The voltage requirements are low, say 5 to 50 V, so normally the half-wave rectifier circuits are used. Load voltage and current control can be by a regulating (tap-changing) transformer to a diode rectifier or by the use of thyristors.

Extraction of metals such as aluminium and magnesium by electrolysis requires large quantities of electrical power. Aluminium is produced by the reduction of aluminium oxide (alumina) by passing direct current through an electrolytic bath (or pot) containing the alumina dissolved in cryolite. The aluminium is precipitated at the carbon cathode, sinking to the bottom of the pot. Normally each pot requires 5 V but, when the alumina concentration drops, the voltage requirement can rise

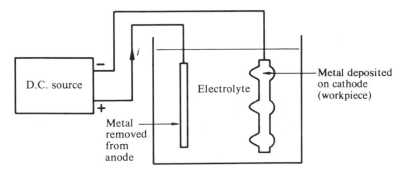

Figure 6-18 Electroplating.

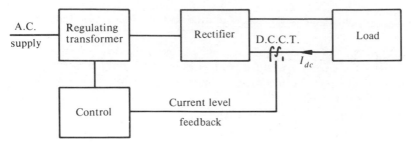

Figure 6-19 Feedback to maintain constant load current.

to 30 V to maintain the current level. Several pots are connected in series, so that typically the rectifier supplies 800 V, 70 kA to a potline.

The supply to the aluminium smelters will be from 12- or 24-pulse rectifiers, having several diodes in parallel for each arm. To maintain full-load current, it is necessary to supply the rectifier from an on-load tap-changing transformer, with fine control by use of saturable reactors or thyristors in the transformer input lines.

A closed-loop feedback system as shown in Fig. 6-19 would be used, so that voltage variations in the load (potline) can be compensated by change in the input voltage. A d.c.c.t. (d.c. current transformer) would sense the load-current level, any reduction in the feedback signal (i.e. the load current) automatically raising the input voltage, so bringing the load current back to its desired setting.

One method of measuring large direct currents is to use a d.c. current transformer as illustrated in Fig. 6-20. An alternating voltage is applied to the coils of two cores, which are connected in series opposition. During one half cycle, the load current will magnetize the core of one unit in the same direction as the a.c. source, hence taking the core into saturation with no induced voltage (zero $d\Phi/dt$) at its coil. During this half cycle, the coil current in the other unit will rise until an ampere-turn balance is reached, as in this unit the load-current m.m.f. is in oppo-

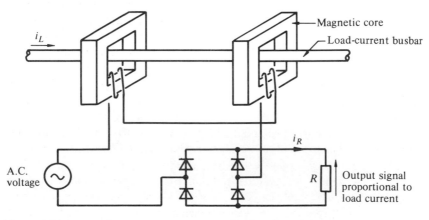

Figure 6-20 Simple direct-current current transformer (d.c.c.t.).

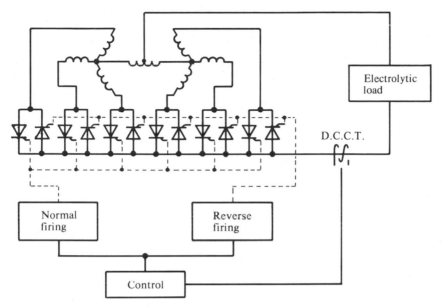

Figure 6-21 Typical circuit for periodic current reversal.

sition to the a.c. coil m.m.f. The alternating source current will remain fixed at the load-current level (times turns ratio) so that the core is out of saturation and able to change its flux so as to support the alternating source voltage. Therefore, the current i_R always corresponds to the load current i_L. In the other half cycle, the roles of the two cores reverse. The rectifier is necessary so that the current i_R is direct. The d.c.c.t. has the usual advantages of the current transformer of imposing negligible burden on the load circuit, together with electrical isolation.

The large-scale production of chlorine, hydrogen, oxygen, caustic soda requires typically 50 kA at 50 V, requiring half-wave connections for high efficiency, with 12- or 24-pulse outputs so as to reduce the harmonic loading on the supply system.

Many electrolytic processes (such as copper refining) produce a film of hydrogen at the cathode which increases the cell (pot) voltage. This hydrogen is most easily removed by reversing the current for a few seconds every few minutes. The use of inverse-connected thyristor converters (as shown in Fig. 6-21) enables easy periodic current reversal. The control logic must ensure that normal forward current has ceased, with all forward thyristors off, before the reverse thyristors are fired, and vice versa.

Electrochemical forming (machining) is in effect the reverse of electroplating. The principle is illustrated in Fig. 6-22, showing how material is removed from the work (anode) and is carried away in the high velocity stream of electrolyte thus avoiding plating on to the tool (cathode). Material is removed proportionally to the current density, which is of the order of 1 A/mm², with voltages of the order 5 to 20 V to maintain the current. The gap is typically 0.2 mm, with a metal removal rate of 2 mm per minute. The process is used for hard materials, difficult to machine

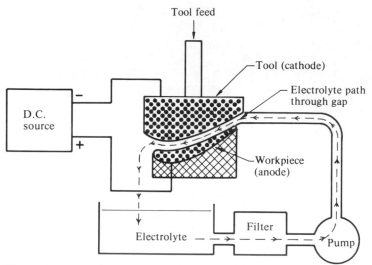

Figure 6-22 Electrochemical forming.

conventionally, for difficult three-dimensional shapes, or where the workpiece must not be distorted or stressed during machining.

The d.c. supply to electrochemical forming equipment is at a low voltage, say less than 20 V, at currents well into the kiloampere range. For this application, a diode rectifier fed from a regulated transformer would be used. Voltage regulation could be by inverse-parallel thyristors in the supply lines to the transformer.

The above examples of electrochemical applications show that the demand is for low-voltage, high-current supplies, demanding frequently the half-wave (single-way) rectifier connections, with parallel connection of diodes or thyristors of the highest current ratings associated with water or oil cooling.

6-9 H.V.D.C. TRANSMISSION

An application which requires strings of high-voltage thyristors in the converters is that of high-voltage, direct-current (H.V.D.C.) transmission.

Direct-voltage transmission lines are much more economical than alternating-voltage transmission lines, but the ease with which alternating voltages can be altered in level by transformers, coupled with generator and motor considerations, makes the three-phase a.c. system the best overall, both economically and technically. However, for long overland or underwater power transmission, it is economical to link two a.c. systems by an H.V.D.C. transmission line.

Referring to Fig. 6-23a, the principle is to rectify the a.c. power to a voltage level of say ± 200 kV, transmit power over a two-cable (pole) line to a converter operating in the inverting mode, hence feeding power into the other a.c. system. Typically, each converter would be 12-pulse as shown in Fig. 6-23b, the centre

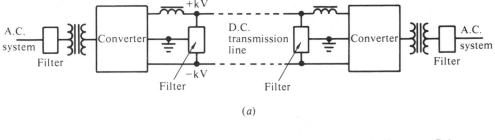

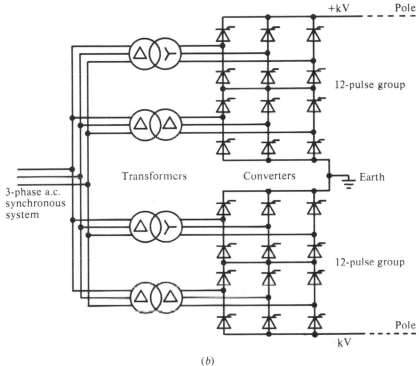

(a)

(b)

Figure 6-23 H.V.D.C. transmission. (*a*) Overall layout. (*b*) Converter terminal.

being earthed with one line at $+ 200$ kV and the other at $- 200$ kV to earth for a ± 200 kV system.

The two a.c. systems linked by the H.V.D.C. transmission line must each be synchronous, that is, contain generators so as to maintain the a.c. frequency constant. The two a.c. systems can be at different frequencies. Filters must be included, as shown in Fig. 6-23*a*, so as to attenuate the harmonics generated by the converters. Power flow can be in either direction, as each converter can operate in either the rectifying or the inverting mode.

The principle of the converter shown in Fig. 6-23*b* was described in Chapters 2 and 3, but each thyristor valve shown in the diagram is in fact composed of several thyristors connected in series, so as to support the very high voltages across the valves when non-conducting.

6-10 SWITCHED MODE POWER SUPPLIES

In the low to medium power range, say upwards from 50 W, a d.c. source of supply is often required which contains negligible a.c. ripple yet can be controlled in magnitude. For this application switched mode power supply circuits are used which have a layout as shown in Fig. 6-24.

The principle of the circuit shown in Fig. 6-24 is that a d.c. source of voltage is obtained from a rectifier, the voltage as shown in Fig. 6-24a being partially smoothed by the capacitor C_1. The switch is turned on and off rapidly to give, as shown in Fig. 6-24b, a voltage which is chopped between the source level and zero. This voltage is fed into an LC filter network which then smooths the chopped wave-form to give a level voltage to the load, as shown in Fig. 6-24c.

At first sight the circuit of Fig. 6-24 appears unnecessarily complex. Could not the rectifier output be smoothed to negligible ripple and so avoid the switching element? Filter theory is discussed in Sec. 7-4, where it is explained that the physical size of the capacitor C_1 would need to be very large and also that its d.c. level would be determined by the a.c. voltage magnitude. By switching the rectifier output on and off very rapidly control of the output voltage is obtained, the magnitude being a function of the on/off ratio. The a.c. ripple frequency fed into the $L_2 C_2$ filter is very high; hence the value of L_2 and C_2 can be correspondingly low but still give high a.c. ripple attenuation. Equation (7-27) shows that the attenuation is inversely proportional to the product of the frequency squared and $L_2 C_2$, assuming $\omega^2 LC \gg 1$; hence the higher the switching rate the smaller are the filter components. The principle of the filter is that the inductor presents zero impedance to the d.c. component and near infinite impedance to the a.c. components, whilst the capacitor is charged to the mean level of the chopped voltage. The diode is

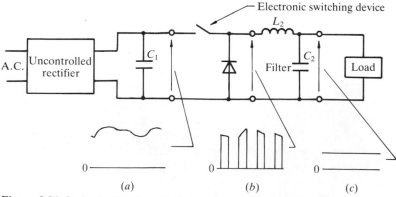

Figure 6-24 Basic circuit principle of a switched mode power supply.

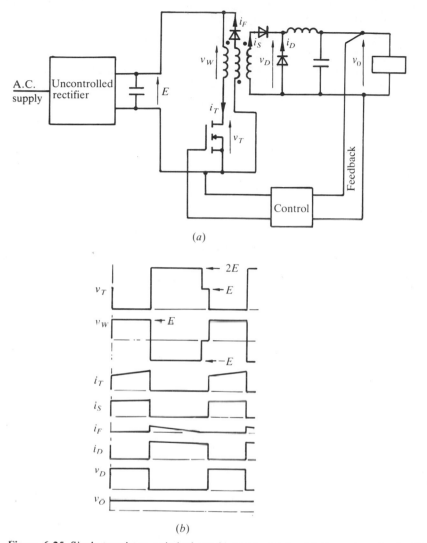

(a)

(b)

Figure 6-25 Single-transistor switched mode power supply. (a) Converter circuit. (b) Waveforms taking a turns ratio of 1/1/1.

essential to provide a path for the continuous current in the inductor when the switch is open. The magnitude of the load voltage can be controlled to any desired level.

The switched mode power supply circuit is essentially a d.c. to d.c. converter with control of the output voltage magnitude. Like all power electronic equipment the semiconductor devices are used in their switching mode in order to maximize efficiency. If power transistors are used as the switching devices the frequency of chopping would be limited to approximately 40 kHz, but by using power MOSFETs this frequency can be raised to 200 kHz or above, so giving considerable savings in component sizes. By introducing a transformer at the high-frequency interface between the two d.c. elements of source and load, it is possible to change voltage levels and to give the isolation which is often required in electronic equipment. Further, it is then possible to have two or more secondaries so that a range of d.c. output voltages are possible. Figure 6-25 shows such an arrangement and because the power MOSFET is switching in the higher voltage primary the switching losses are small. The ferrite core transformer will be small at the very high frequencies used.

Explanation of the circuit in Fig. 6-25 is best given by reference to the wave-forms. When the MOSFET is turned on it carries the rising magnetizing current in addition to the reflected secondary current. When the MOSFET is turned off the transformer secondary current commutates to the other diode, but the stored magnetic energy of the transformer core demands a current-carrying winding, being provided as shown by i_F. The core magnetic flux collapses at the same rate as it rose, so in order to reset the flux back to zero the off-time of the MOSFET must be equal to, or greater than, the on-time. The on/off ratio of the MOSFET will be set by the control circuit which has coupled to it a feedback loop sensing the output load voltage. The transformer turns ratio can, of course, be chosen to match the application. A disadvantage of this single-transistor circuit is that the flux in the transformer core is never reversed and there is a risk of the core drifting towards magnetic saturation.

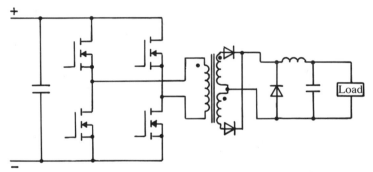

Figure 6-26 Basic bridge configuration for a switched mode power supply.

For the higher power levels it is necessary to use a full-wave version of the converter, as shown in Fig. 6-26. The transformer core flux alternates so that it is fully utilized magnetically. Pulse-width modulation of the MOSFETs gives control of the load voltage.

6-11 WORKED EXAMPLES

Example 6-1

A semiconductor contactor is used to connect a load of 5 Ω, 0.07 H to a 240 V, 50 Hz supply. Determine the initial firing angle to ensure no current transient. Plot the current waveform up to the second cycle if the initial firing is at the worst instant.

SOLUTION The analysis of the current is similar to that given in Examples 2-1(ii) and 2-2.

The steady-state a.c. component of current is $15.05 \sin (2\pi 50t - 1.347)$ A, with $t = 0$ at the voltage zero.

For no transient, the firing delay angle is at 1.347 rad $= 77.2°$ (or 257.2°).

The worst instant is at $77.2° + 90° = 167.2°$ (or 347.2°), when the d.c. component is $- 15.05 \, e^{-71t}$ A.

The total current $= 15.05 \cos 2\pi 50t - 15.05 \, e^{-71t}$ A, taking $t = 0$ at the instant of firing. The waveform is shown in Fig. 6-27.

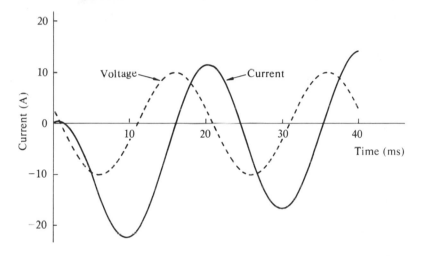

Figure 6-27

Example 6-2

A resistance heating load is controlled from a single-phase supply using a triac in the phase angle control mode. Determine the firing delay when the power is at (i) 80%, (ii) 50%, (iii) 30% of its maximum.

SOLUTION Using Eq. (6-1), 100% power will be when $\alpha = 0°$, (i) $(\pi - \alpha + \frac{1}{2} \sin 2\alpha)/\pi = 80/100$, giving $\alpha = 1.057$ rad $= 60.5°$, (ii) from inspection, at half power, $\alpha = 90°$, (iii) $(\pi - \alpha + \frac{1}{2} \sin 2\alpha)/\pi = 30/100$, giving $\alpha = 1.896$ rad $= 108.6°$.

Example 6-3

A three-phase resistance load is to be controlled by three triacs from a 415 V supply.

If the load is 15 kW, determine the required rating of the triacs. If thyristors were used instead of triacs, determine their rating.

SOLUTION The r.m.s. line current = $15\,000/(\sqrt{3} \times 415) = 20.9$ A. Hence the required triac r.m.s. current rating = 20.9 A. In the off-state, differences in leakage current could result in the line voltage appearing across the triac, hence the required peak off-state voltage = $415\sqrt{2} = 587$ V.

If thyristors were used, they would conduct for only one half cycle of the sinewave, hence their required r.m.s. current rating = $20.9/\sqrt{2} = 14.8$ A. The voltage requirement would be identical to that for the triac.

Example 6-4

Plot a curve showing the variation of power with firing angle delay for three-phase resistance loads with (i) fully-controlled, (ii) half-controlled circuits.

SOLUTION (i) Figure 6-4a applies with Eqs. (6-3)–(6-5). (ii) Figure 6-4b applies with Eqs. (6-6)–(6-8).

Taking the load power to be 100 at zero firing delay ($\alpha = 0°$), the required curves are shown plotted in Fig. 6-28.

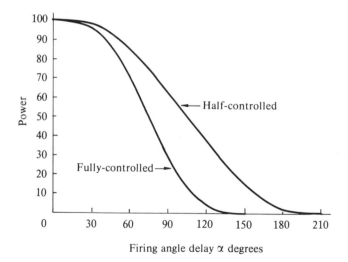

Figure 6-28

Example 6-5

A single-phase load of resistance 9 Ω, in series with an inductance of 0.03 H, is controlled by inverse-parallel-connected thyristors using phase angle delay. Taking the a.c. supply to be 240 V, 50 Hz, and neglecting thyristor voltage drops, determine the load-current waveform and load power at firing delay angles of (i) 0°, (ii) 60°, and (iii) 90°.

SOLUTION The circuit and waveforms shown in Fig. 6-9 apply to this problem.
(i) With zero firing delay, normal a.c. steady-state theory applies. Load impedance = $13.03/46.3°$ Ω. Load current = $26.04 \sin(2\pi50t - 0.808)$ A, taking the supply voltage as reference. Load power = $240 \times 29.04/\sqrt{2} \times \cos 46.3° = 3052$ W.
Note that a train of firing pulses will be required as the thyristor will not turn on until after 46.3°.
(ii) Using similar reasoning to that developed in solving Examples 2-1 and 2-2, the current can be considered as the above steady-state component, plus a decaying exponential component to satisfy the initial condition of zero current at the instant of firing.
Taking time $t = 0$ at the instant of firing ($\alpha = 60°$), then

$$i = 26.04 \sin(2\pi50t + \frac{\pi}{3} - 0.808) - 26.04 \sin(\frac{\pi}{3} - 0.808)e^{-(9/0.03)t}$$

$$= 26.04 \sin(2\pi50t + 0.239) - 6.16\,e^{-300t}\ \text{A}$$

The current will cease when $i = 0$ at $t = 9.19$ ms, that is, at an angle of 225.4° on the voltage wave. The negative half cycle will be the same shape.

$$\text{Mean load power} = \frac{1}{0.01} \int_0^{0.00919} 240\sqrt{2} \sin(2\pi50t + \frac{\pi}{3}) [26.04 \sin(2\pi50t$$

$$+ 0.239) - 6.16\,e^{-300t}]\,dt = 2494\ \text{W}.$$

(iii) Taking $t = 0$ at the instant of firing ($\alpha = 90°$), then by similar reasoning to above,

$$i = 26.04 \sin(2\pi50t + 0.762) - 17.99\,e^{-300t}\ \text{A}$$

The current will cease when $i = 0$ at $t = 7.33$ ms, that is, at an angle of 221.9° on the voltage wave.

$$\text{Mean load power} = \frac{1}{0.01} \int_0^{0.00733} 240\sqrt{2} \sin(2\pi50t + \frac{\pi}{2}) [26.04 \sin(2\pi50t$$

$$+ 0.762) - 17.99\,e^{-300t}]\,dt = 1190\ \text{W}.$$

Example 6-6

A three-phase star-connected load of resistance 9 Ω with an inductance of 0.03 H per phase is controlled by inverse-parallel-connected thyristors using phase angle

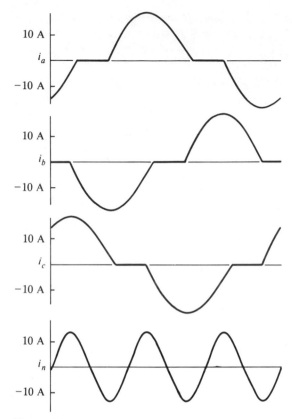

Figure 6-29

firing delay. Taking the supply at 415 V, 50 Hz, plot the neutral current waveform at a firing delay angle of 90°.

SOLUTION The data for this problem is identical per phase to that in Example 6-5 (iii). The neutral current is the addition of the three-phase currents which are mutually displaced 120°, as shown in Fig. 6-29. Note that the neutral current is at a frequency three times that of the supply.

Example 6-7

A three-phase delta-connected load of resistance 27 Ω, in series with an inductance of 0.09 H per phase, is controlled by inverse-parallel-connected thyristors using phase angle delay. Taking the supply as 415 V, 50 Hz, plot the line current for the first cycle after switch-on, at a firing delay angle of 90°.

SOLUTION This problem has a load equivalent to that of the previous example without the neutral connection.

Take the start as when thyristor T_1 in Fig. 6-4a is fired, which by reference to Fig. 6-5 shows thyristor T_6 is fired to take the return current. Line C is floating. By reference to Fig. 6-30, v_{ab} sets up i_{ab} and a current of half this value in the other two arms of the delta load.

Phase impedance = $39.1/0.808$ rad Ω. $T = L/R = 1/300$.

At $0°$ with $t = 0$, $v_{ab} = 415\sqrt{2} \sin (2\pi50t + \dfrac{2\pi}{3})$ V, giving $i_{ab} = 15.01 \sin (2\pi50t +$

$1.286) - 14.4\,e^{-300t}$ A, and $i_{ca} = -i_{ab}/2 = i_{bc}$. The line current $i_a = i_{ab} - i_{ca} = 22.52 \sin (2\pi50t + 1.286) - 21.6\,e^{-300t}$ A, this equation applying up to $t = 3.33$ ms (at $60°$) when thyristor T_2 is fired, connecting in line C. At this instant $i_{ab} = 5.58$ A, $i_{ca} = -2.78$ A and $i_{bc} = -2.78$ A. At $60°$ with $t = 0$, recalculating current equations and taking account of the initial current values in each arm, gives

$v_{ab} = 415\sqrt{2} \sin (2\pi50t + \pi)$ V
$i_{ab} = 15.01 \sin (2\pi50t + 2.333) - 5.28\,e^{-300t}$ A

$v_{bc} = 415\sqrt{2} \sin (2\pi50t + \dfrac{\pi}{3})$ V
$i_{bc} = 15.01 \sin (2\pi50t + 0.239) - 6.33\,e^{-300t}$ A

$v_{ca} = 415\sqrt{2} \sin (2\pi50t + \dfrac{5\pi}{3})$ V
$i_{ca} = 15.01 \sin (2\pi50t + 4.428) + 11.62\,e^{-300t}$ A
$i_a = i_{ab} - i_{ca} = 26 \sin (2\pi50t + 1.810) - 16.9\,e^{-300t}$ A

The current in line B $(= i_{bc} - i_{ab})$ is decaying until after $t = 1.00$ ms ($18°$) it is zero, when thyristor T_6 turns off, leaving line B floating. Hence at ($60° + 18°$) another set of conditions apply. Only v_{ca} is present, with initial currents $i_{ab} = 3.20$ A, $i_{bc} = 3.20$ A, $i_{ca} = -6.40$ A. Taking $t = 0$ at this new angle of $78°$, equations are:

$v_{ca} = 415\sqrt{2} \sin (2\pi50t + 5.550)$ V
$i_{ca} = 15.01 \sin (2\pi50t + 4.742) + 8.60\,e^{-300t}$ A
$i_{ab} = i_{bc} = -i_{ca}/2$
$i_a = 22.52 \sin (2\pi50t + 1.600) - 12.90\,e^{-300t}$ A

This will continue until $120°$ when thyristor T_3 is fired, connecting line B in, after which current in line A decays until thyristor T_1 turns off. The equations now to the line current i_a, taking $t = 0$ at the start of each period, are:

$120°$ to $140°$, $i_a = 26 \sin (2\pi50t + 2.857) + 2.57\,e^{-300t}$ A
$140°$ to $180°$, $i_a = 0$
$180°$ to $201°$, $i_a = 26 \sin (2\pi50t + 3.904) + 17.96\,e^{-300t}$ A
$201°$ to $240°$, $i_a = 22.52 \sin (2\pi50t + 4.790) + 11.7\,e^{-300t}$ A
$240°$ to $261°$, $i_a = 26 \sin (2\pi50t - 1.332) + 15.08\,e^{-300t}$ A
$261°$ to $300°$, $i_a = 22.52 \sin (2\pi50t + 4.791) + 11.68\,e^{-300t}$ A
$300°$ to $321°$, $i_a = 26 \sin (2\pi50t - 0.285) - 2.90\,e^{-300t}$ A
$321°$ to $360°$, $i_a = 0$.

By the second half cycle, the waveform has to all practical purposes reached its steady-state condition. The line-current waveform is shown in Fig. 6-30.

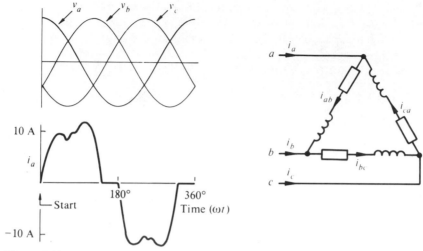

Figure 6-30

Example 6-8

The power in a resistor is to be controlled by a tapped transformer as shown in Fig. 6-11. Taking the load as $10\,\Omega$, the full transformer secondary as $100\,V$, with the tapping at $70.7\,V$, plot a curve of load power against firing delay angle. Determine the required thyristor ratings. Neglect losses.

SOLUTION Assume the thyristors T_3 and T_4 are fired at voltage zero, and the firing delay angle relates to thyristors T_1 and T_2, then the mean power is given by

$$\text{power} = \frac{V_{\text{rms}}^2}{R} = \frac{1}{10\pi}\left[\int_0^\alpha (70.7\sqrt{2}\sin\theta)^2 d\theta + \int_\alpha^\pi (100\sqrt{2}\sin\theta)^2 d\theta\right]$$

$$= 1000 - (500\alpha - 250\sin 2\alpha)/\pi\ \text{W}$$

The plot of mean load power against firing delay angle is shown in Fig. 6-31.

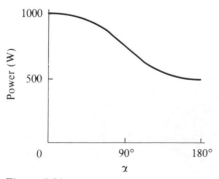

Figure 6-31

With the circuit connected, but no firing pulses to the thyristor gates, the peak voltages experienced by the thyristors are: thyristors T_1 and T_2, P.R.V. = P.F.V. = $100\sqrt{2} = 141$ V;

thyristors T_3 and T_4, P.R.V. = P.F.V. = $70.7\sqrt{2} = 100$ V.

At zero firing delay, thyristors T_1 and T_2 will carry a half cycle of sinusoidal current of maximum value $100\sqrt{2}/10$, that is, an r.m.s. value of 7.07 A. At $180°$ firing delay thyristors T_3 and T_4 will carry a half cycle of maximum value $70.7\sqrt{2}/10$, that is, an r.m.s. value of 5 A.

Example 6-9

If to the load of Example 6-8 is added a series inductance of 0.02 H, determine the load-voltage and current waveforms at a firing delay angle of $90°$. Take the supply frequency to be 50 Hz, and indicate precautions necessary regarding inhibition of the gate firing signals.

SOLUTION Using a.c. transient analysis as used in the previous examples, the following expressions are derived taking $t = 0$ at the start of the current flow for each thyristor.

Assume initially thyristor T_3 is fired at the supply voltage zero, then $i = 8.47 \sin (2\pi 50t - 0.561) + 4.51\, e^{-500t}$ A.

At $90°$ (after $t = 5$ ms), thyristor T_1 is fired, turning off thyristor T_3, now $i = 11.97 \cos (2\pi 50t - 0.561) - 2.6\, e^{-500t}$ A. Thyristor T_1 will not turn off until its current falls to zero at $t = 6.76$ ms, that is, at an angle $(90° + 121.7°) = 211.7°$.

Until thyristor T_1 ceases conduction, firing of thyristor T_4 must be inhibited, otherwise a short-circuit via thyristor T_1 and T_4 would occur across the tapped section of the transformer winding.

Thyristor T_4 can be fired at $211.7°$ when the load current becomes $i = 8.47 \sin (2\pi 50t + 3.134) - 0.06\, e^{-500t}$ A.

At $270°$, after $t = 3.24$ ms, thyristor T_2 is fired when the load current becomes $i = 11.97 \cos (2\pi 50t + 2.581) + 2.96\, e^{-500t}$ A, this current falling to zero at $t = 6.76$ ms, that is, at an angle of $(270° + 121.7°) = 391.7°$.

At this time thyristor T_3 can be fired, giving $i = 8.47 \sin (2\pi 50t - 0.007) + 0.06\, e^{-500t}$ A.

The circuit is now in the steady-state condition. The waveforms are shown in Fig. 6-32.

Example 6-10

An induction heater of nominal frequency 2500 Hz may be considered as being a resistance of 0.05 Ω in series with an inductance of 16 μH. Using the constant-voltage-fed inverter of Fig. 6-8a, estimate the required capacitor size and approximately sketch the load waveforms for an inverter voltage of 500 V. Estimate the heater power at 2500 and 2400 Hz.

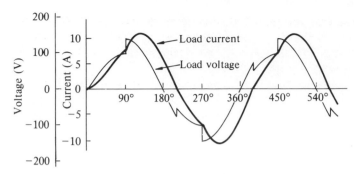

Figure 6-32

SOLUTION The minimum turn-off time for the thyristors will be the time of the diode conduction in the first cycle, which equals $\dfrac{1}{2}\left(\dfrac{1}{2500}-\dfrac{1}{f_L}\right)$, where

$$f_L \doteq \frac{1}{2\pi\sqrt{(LC)}},$$

the natural frequency of the load. Taking a turn-off time of $35\,\mu s$, $f_L = 3030\,\text{Hz}$, giving $C = 172\,\mu F$.

The damped load frequency from Eq. (4-7) is 3032 Hz but, as $\omega L \gg R$, the waveforms can be approximately derived by assuming an undamped load, that is, ignoring the resistance.

Taking the supply constant at 500 V, being switched at 2500 Hz, then the LC load sees a square wave. By ignoring the load resistance, the current must be symmetrical about the square-wave voltage to make the net power zero as shown in Fig. 6-33. Using Eqs. (4-8) and (4-9), noting that the initial value at the end of the half cycle is the negative of that at the start, we get an equation for the current of $i = 5004 \sin(2\pi3030t + 1.238)\,\text{A}$. The r.m.s. value of this current over the $200\,\mu s$ half cycle is 3815 A. Hence approximately the heater power $= 3815^2 \times 0.05 = 727\,\text{kW}$.

Note that the capacitor voltage $= 500 + 1528 \sin(2\pi3030t - 0.333)\,\text{V}$.

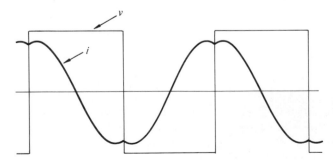

Figure 6-33

At an operating frequency of 2400 Hz, the half cycle square-wave time is extended in Fig. 6-33, whilst the current frequency is unchanged at 3030 Hz. The current equation is now $i = 4084 \sin (2\pi 3030 t + 1.158)$ A, having an r.m.s. value of 3144 A, with the heater power 494 kW.

It may be concluded that power control is possible by variation of the operating frequency.

The above calculations are approximate, but in any case the parameters of the heater will change over the melting cycle.

Example 6-11

If the induction heater of Example 6-10 is fed using the constant-current-fed inverter of Fig. 6-8b, estimate the required capacitor size, and approximately sketch the load waveforms for the inverter when the loading is 500 kW.

SOLUTION The current fed to the load by the inverter will be shaped as shown in Fig. 6-34, almost square-wave with an overlap period. The fundamental sinewave component of the current may be taken as having a maximum value of $4I/\pi$ (see Chapter 7). The thyristor firing must be advanced by (say) 35 μs to give adequate turn-off time. At 2500 Hz, 35 μs represents an angle of 31.5°. If overlap is (say) 5°, then the current is $(31.5° + 2.5°) = 34°$ in advance of the voltage, hence power

$$= 500 \times 10^3 = \frac{V \times 4I}{\sqrt{2}\pi} \cos 34°.$$

In order to develop 500 kW, the heater current must be $\left(\dfrac{500 \times 10^3}{0.05}\right)^{1/2} = 3162$ A, thus lagging the voltage by 78.75°, because it flows in a series combination of 0.05 Ω and 19 μH. At 2500 Hz, the impedance is 0.256 Ω, giving a load r.m.s. voltage of 810 V. For the inverter current to lead by 34°, the capacitor current must be 3517 A, giving an inverter output current of level value I (in Fig. 6-34) of 827 A.

With a current of 3517 A, voltage 810 V at 2500 Hz, the required capacitor is 267 μF.

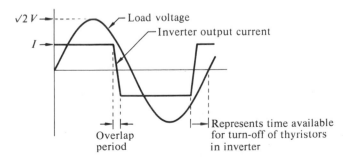

Figure **6-34**

In practice, to satisfy the conditions when the inverter switches, the voltage wave-form will be more like that of the current in Fig. 6-33, indicating a higher resonant load frequency, hence the above calculated values of capacitance, voltage, and current are only approximate, the practical value of capacitance probably being closer to 200 μF.

Example 6-12

A voltage doubler circuit is suppled by an a.c. voltage of maximum value 100 V at 50 Hz. If each capacitor is 100 μF and the load is 500 Ω, analyse the circuit from switch-on, allowing a constant diode volt-drop of 0.7 V, but otherwise assuming ideal components.

SOLUTION Referring to Fig. 6-12a, starting with the supply voltage zero and going negative, D_1 will conduct, connecting C_1 to the supply so that after one quarter cycle C_1 is charged as shown to $(100 - 0.7) = 99.3$ V. As the supply volt-age falls from its peak, D_2 will conduct so that the supply in series with C_1 sees a load of C_2 in parallel with the load resistance. In succeeding cycles, C_2 will have a charge, thereby delaying the time to when D_2 starts conduction.
In every cycle C_1 will be charged to 99.3 V.
Taking C_2 to be charged to E volts when D_2 starts conduction, the current as i_x in D_2, i_y in the load resistance, then with $t = 0$ at the start of D_2 conduction, the circuit equations are

$$\frac{1}{C_1} \int i_x dt + \frac{1}{C_2} \int (i_x - i_y) dt = -0.7 - 100 \cos(\omega t + \phi)$$

and
$$\frac{1}{C_2} \int (i_y - i_x) dt + 500\, i_y = 0$$

Putting in the values for C_1 and C_2, and the initial voltages on C_1 as 99.3 V, C_2 as E, with $\omega = 2\pi 50$, and $\cos \phi = \dfrac{99.3 - 0.7 - E}{100}$, the current equations are:

$$i_x = [0.0493 + 0.0005E + 0.000152 \cos \phi - 0.00159 \sin \phi]\, e^{-10t}$$
$$+ (-0.0499 \cos \phi + 1.5724 \sin \phi) \cos \omega t$$
$$+ (1.5724 \cos \phi + 0.0499 \sin \phi) \sin \omega t \text{ A}$$

Load voltage $= 500\, i_y = (49.3 + 0.5E - 0.05 \cos \phi - 1.59 \sin \phi)\, e^{-10t}$
$$+ (-49.95 \cos \phi + 1.59 \sin \phi) \cos \omega t$$
$$+ (1.59 \cos \phi + 49.95 \sin \phi) \sin \omega t \text{ V}$$

In the first cycle, E $= 0$, i_x falls to zero at $t = 9.66$ ms, leaving $E = 94.24$ V. The capacitor C_2 will now discharge into the load with an exponential discharge giving the load voltage as $94.24\, e^{-20t}$ V.

When this voltage has decayed to 71.33 V ($t = 13.93$ ms, $\cos \phi = 0.2727$), D_2 will again conduct due to the rising alternating voltage; i_x now flows for 6.14 ms leaving $E = 128.11$ V.

Next E decays to 95.59 V, then recharges to 140.11 V; in the next cycle, the values are 104.03 V, 144.32 V; and in the next cycle, the values are 106.97 V, 145.82 V. In practical terms, the circuit has now reached a steady-state condition, with the load voltage varying between 107 and 146 V each cycle.

Example 6-13

Two diodes are put in parallel in plain connection. One diode has an approximate characteristic of $v = [0.9 + (2.4 \times 10^{-4}i)]$ V, and the other diode $v = [1.0 + (2.3 \times 10^{-4}i)]$ V. Determine the current taken by each diode if the total current is (i) 500 A, (ii) 1000 A, (iii) 1500 A, (iv) 2000 A.

Determine the value of equal resistors which when placed in series with the diodes will in (iv) bring the diode current to within 10% of equal current sharing.

SOLUTION The voltage across each diode will be identical, so letting I_1 and I_2 be the respective diode currents,

$$0.9 + (2.4 \times 10^{-4})I_1 = 1.0 + (2.3 \times 10^{-4})I_2$$

and $I_1 + I_2 = $ total current.

For (i) $I_1 = 457$ A, $I_2 = 43$ A
    (ii) $I_1 = 702$ A, $I_2 = 298$ A
    (iii) $I_1 = 947$ A, $I_2 = 553$ A
    (iv) $I_1 = 1191$ A, $I_2 = 809$ A

In (iv) for 10% balance, the currents will be $I_1 = 1100$ A and $I_2 = 900$ A, then $0.9 + (2.4 \times 10^{-4} + R)\,1100 = 1.0 + (2.3 \times 10^{-4} + R)\,900$, giving $R = 0.215$ mΩ.

Example 6-14

A series string of three thyristors is designed to withstand an off-state voltage of 4 kV. Referring to Fig. 6-16, $R_1 = 20$ kΩ, $R_2 = 20$ Ω, and $C - 0.1$ μF.

Estimate the voltage unbalance during turn-off, the voltage distribution in the off-state, and the maximum discharge current from C when turn-on takes place.

Data for the three thyristors: T_1, leakage current 25 mA, hole storage charge 140 μC; T_2, 20 mA, 150 μC; T_3, 18 mA, 130 μC.

SOLUTION At turn-off, T_3 will turn off first with a stored charge of $140 - 130 = 10$ μC in T_1, and $150 - 130 = 20$ μC in T_2. The charge transferred to C (across T_3) is $10 + 10 = 20$ μC before T_1 turns off, and the remaining 10 μC on T_2 will be transferred equally to each capacitor across T_1 and T_3 when all the thyristors are then turned off.

Therefore charge on C across $T_3 = 20 + 10/2 - 25$ μC, charge on C across $T_1 = 10/2 = 5$ μC. Hence the unbalance is: the voltage across T_3 above that across $T_2 =$

$(25 \times 10^{-6})/(0.1 \times 10^{-6}) = 250$ V, and the voltage across T_1 above that across $T_2 = (5 \times 10^{-6})/(0.1 \times 10^{-6}) = 50$ V.

In the steady off-state at 4 kV, the leakage current, together with resistor R_1 current, will determine the voltage distribution. Taking the total string current as I mA at 4 kV, then the current in R_1 will equal I less the leakage current for that thyristor, hence $(I - 25)20 + (I - 20)20 + (I - 18)20 = 4000$, giving $I = 88$ mA, with T_1, T_2, and T_3 voltages as 1253 V, 1353 V, and 1393 V respectively.

At turn-on, the maximum capacitor voltage is 1393 V which, with the 20 Ω series resistor, will discharge initially at $1393/20 = 70$ A.

Example 6-15

An H.V.D.C. transmission system is rated at 500 MW, ± 250 kV, and the converters are arranged as shown in Fig. 6-23. Determine the r.m.s. current and peak reverse voltage of each thyristor valve.

SOLUTION The direct current = $(500 \times 10^6)/(250 \times 2 \times 10^3) = 1000$ A.

Each thyristor valve conducts 1000 A for one third cycle, therefore r.m.s. current from Eq. (2-10) is $1000/\sqrt{3} = 577$ A.

Each transformer feeds a 6-pulse group of mean voltage $250/2 = 125$ kV. Hence from Eq. (2-15) and Fig. 2-21b, the peak reverse voltage = $V_{\text{line(max)}} = 125 \times \pi/3 = 131$ kV.

SEVEN

HARMONICS

The earlier chapters have demonstrated that almost all the waveforms associated with power electronic equipment are non-sinusoidal, that is, contain harmonic components. The purpose of this chapter is to analyse the harmonic content of the various waveforms and discuss their effects as regards both supply and load.

In many applications it is desirable to reduce the harmonic content of the waveforms, and this may frequently be achieved either by filtering or by increasing the complexity of the converter circuitry.

7-1 HARMONIC ANALYSIS

Any periodic waveform may be shown to be composed of the superposition of a direct component with a fundamental pure sinewave component, together with pure sinewaves known as *harmonics* at frequencies which are integral multiples of the fundamental. A non-sinusoidal wave is often referred to as a *complex wave* and can be expressed mathematically as

$$v = V_0 + V_1 \sin(\omega t + \phi_1) + V_2 \sin(2\omega t + \phi_2)$$
$$+ V_3 \sin(3\omega t + \phi_3) + \ldots + V_n \sin(n\omega t + \phi_n) \qquad (7\text{-}1)$$

where v is the instantaneous value at any time t

V_0 is the direct (or mean) value

V_1 is the maximum value of the fundamental component

V_2 is the maximum value of the second harmonic component

V_3 is the maximum value of the third harmonic component

V_n is the maximum value of the nth harmonic component

ϕ defines the relative angular reference

$\omega = 2\pi f$, where f is the frequency of the fundamental component, $1/f$ defining the time over which the complex wave repeats itself.

Mathematically, it is more convenient to express the independent variable as x rather than ωt, and the dependent variable as y rather than volts or amperes. Then

$$y = f(x) = R_0 + R_1 \sin(x + \phi_1) + R_2 \sin(2x + \phi_2)$$
$$+ R_3 \sin(3x + \phi_3) + \ldots + R_n \sin(nx + \phi_n) \qquad (7\text{-}2)$$

or, alternatively, the series may be expressed as

$$y = f(x) = A + a_1 \sin x + a_2 \sin 2x + \ldots + a_n \sin nx$$
$$+ b_1 \cos x + b_2 \cos 2x + \ldots + b_n \cos nx \qquad (7\text{-}3)$$

Equation (7-3) is known as a Fourier series, and where $f(x)$ can be expressed mathematically, a Fourier analysis yields that the coefficients are:

$$A = \frac{1}{2\pi} \int_0^{2\pi} f(x)\, dx \qquad (7\text{-}4)$$

$$a_n = \frac{1}{\pi} \int_0^{2\pi} f(x) \sin nx \, dx \qquad (7\text{-}5)$$

$$b_n = \frac{1}{\pi} \int_0^{2\pi} f(x) \cos nx \, dx \qquad (7\text{-}6)$$

Equations (7-2) and (7-3) are equivalent, with

$$a_n \sin nx + b_n \cos nx = R_n \sin(nx + \phi_n) \qquad (7\text{-}7)$$

from which the resultant $\qquad R_n = (a_n^2 + b_n^2)^{1/2} \qquad (7\text{-}8)$

and the phase angle $\qquad \phi_n = \arctan \dfrac{b_n}{a_n} \qquad (7\text{-}9)$

The constant term of Eq. (7-4) is the mean value of the function, and is the value found in, for example, the calculation of the direct (mean) voltage output of a rectifier.

Certain statements and simplifications in the analysis of a complex wave are possible by inspection of any given waveform.

If the areas of the positive and negative half cycles are equal, then A is zero.

If $f(x + \pi) = -f(x)$, then there are no even harmonics, that is, no second, fourth, etc. In plain terms, this means the negative half cycle is a reflection of the positive half cycle.

If $f(-x) = -f(x)$, then $b_n = 0$, that is, there are no cosine terms.

If $f(-x) = f(x)$, then $a_n = 0$, that is, there are no sine terms.

Symmetry of the waveform can result in Eqs (7-5) and (7-6) being taken as twice the value of the integral from 0 to π, or four times the value of the integral from 0 to $\pi/2$, hence simplifying the analysis.

Where it is difficult to put a mathematical expression to $f(x)$, or where an analysis of an experimental or practical waveform obtained from a piece of equipment is required, graphical analysis can be performed.

Examination of Eq. (7-5) shows that a_n is twice the average value of $y \sin nx$ taken over one cycle. Setting up a number (m) of ordinates at equally spaced intervals, we get

$$a_n = \frac{2}{m} \sum_{r=1}^{r=m} y_r \sin nx_r \qquad (7\text{-}10)$$

Likewise,
$$b_n = \frac{2}{m} \sum_{r=1}^{r=m} y_r \cos nx_r \qquad (7\text{-}11)$$

The number of ordinates chosen must be above that of the harmonic order being derived — the higher the number of ordinates, the greater being the accuracy. (See Example 7-5 for the manner in which graphical analysis is used.)

The total r.m.s. value of the complex wave given in Eq. (7-1) is

$$V_{\text{rms}} = (V_{1(\text{rms})}^2 + V_{2(\text{rms})}^2 + V_{3(\text{rms})}^2 + \ldots + V_{n(\text{rms})}^2)^{1/2} \qquad (7\text{-}12)$$

Likewise for current,

$$I_{\text{rms}} = (I_{1(\text{rms})}^2 + I_{2(\text{rms})}^2 + I_{3(\text{rms})}^2 + \ldots + I_{n(\text{rms})}^2)^{1/2} \qquad (7\text{-}13)$$

The power flow in a section of circuit having a complex voltage with a complex current is given by

$$\text{total power} = \sum V_{n(\text{rms})} I_{n(\text{rms})} \cos \phi_n \qquad (7\text{-}14)$$

which says that the total power is the sum of the individual powers associated with each harmonic, including any direct (mean) term given by $V_0 I_0$. ϕ_n in Eq. (7-14) is the phase angle between the voltage and current components of the nth harmonic. The direction of the flow of the harmonic power can be determined by whether ϕ_n is less or greater than $90°$.

The distortion relating to a particular harmonic content in a waveform may be expressed as the relative magnitude of the r.m.s. harmonic voltage of order n to the r.m.s. amplitude of the fundamental.

The total harmonic distortion factor is the ratio of the r.m.s. value of all the harmonic components together to the r.m.s. amplitude of the fundamental.

Total harmonic distortion factor

$$= \frac{(V_{2(\text{rms})}^2 + V_{3(\text{rms})}^2 + V_{4(\text{rms})}^2 + \ldots + V_{n(\text{rms})}^2)^{1/2}}{V_{1(\text{rms})}} \qquad (7\text{-}15)$$

From Eq. (7-12) this can be written as

$$\frac{(V_{\text{rms}}^2 - V_{1(\text{rms})}^2)^{1/2}}{V_{1(\text{rms})}} \qquad (7\text{-}16)$$

7-2 LOAD ASPECTS

A d.c. load fed from a rectifier will in its voltage have a harmonic voltage content, the lowest order of which is the pulse-number of the rectifier. Harmonics at multiples of the pulse-number will also exist.

The controlled rectifier will (as seen in the waveforms developed in Chapter 2) have a higher harmonic content in the load voltage than when the rectifier is uncontrolled.

The harmonic voltages present in the voltage waveforms will inevitably give rise to harmonic current of the same frequency in the load. Although many of the waveforms illustrating rectifier performance in Chapter 2 were drawn with the assumption of level d.c. load current, in practice this assumption of infinite load inductance is not always justified, and harmonic currents do exist in the load waveform.

The effect these harmonics have on the load must be judged in respect of the individual application, but often they merely contribute to increased losses as, for example, if the load is a d.c. motor.

Loads requiring an alternating voltage fed from an inverter or cycloconverter as shown in Chapter 5 will, by virtue of their switching action, contain harmonic components.

Reduction of the harmonic content of the voltage to the load is either achieved by more complex circuitry, or with the use of filters such as those described in Sec. 7-4.

The fast switching action of the current can in certain circumstances result in the load transmitting air-borne radio interference signals which require suppression as described in Sec. 7-4-4.

The presence of harmonics can distort the supply waveforms and cause problems in respect to firing circuits which rely on the supply voltage zero as reference to the firing delay angle. Such distortion can cause slight differences in the firing delay from one device to the next, setting up a phenomenon known as *jitter*. The jitter frequency would be in the radio-frequency range, and hence worsen the radio interference problem.

The harmonic content in the load voltage of a p-pulse uncontrolled rectifier (neglecting overlap) can be determined by reference to Fig. 7-1.

The interval of each cycle in the waveform is 2π, the equation to the waveform over this period being $V_{\max} \cos \omega t$. Using Eq. (7-4),

$$A = \frac{1}{2\pi} \int_{-\pi}^{+\pi} V_{\max} \cos \omega t \, d(p\omega t) = \frac{p V_{\max}}{\pi} \sin \frac{\pi}{p} \qquad (7\text{-}17)$$

This is the equation for the mean voltage and is identical to Eq. (3-23) for $\alpha = 0$, $\gamma = 0$.
Using Eq. (7-5), $a_n = 0$, as $f(-p\omega t) = f(p\omega t)$.

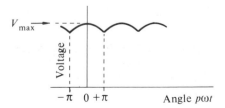

Figure 7-1 Waveform of a p-pulse uncontrolled rectifier without overlap.

Using Eq. (7-6),

$$b_n = \frac{1}{\pi} \int_{-\pi}^{+\pi} V_{max} \cos \omega t \cos mp\omega t \, d(p\omega t)$$

$$= \frac{p V_{max}}{\pi} \sin \frac{\pi}{p} \left[\frac{2}{(m^2 p^2 - 1)} \right] [- \cos m\pi] \qquad (7\text{-}18)$$

where $mp = n$, and m has values of 1, 2, 3, etc., and $-\cos m\pi$ only indicates a change of sign for each component.

Comparison of Eq. (7-18) to (7-17) indicates that the magnitude of the harmonic relative to the mean voltage is

$$\frac{2}{n^2 - 1} \qquad (7\text{-}19)$$

The lowest harmonic is when $m = 1$, that is, p times the input frequency. This statement is obvious from an inspection of the waveform.

7-3 SUPPLY ASPECTS

The switching action of the rectifying device inevitably results in non-sinusoidal current being drawn from the a.c. supply system. In essence, the a.c. supply delivers a sinusoidal voltage with power flow relating only to the fundamental (mains) frequency. The load then converts some of this power to higher frequencies, and transmits harmonic power back into the supply system. Hence a rectifying load acts in part as a harmonic generator.

With an uncontrolled rectifier, for every harmonic present in the direct voltage waveform, there are two harmonic components present in the three-phase alternating input current, one component at one higher-order, and the other at one lower-order frequency. Taking the case of an infinite load inductance, and consequently level load current, then the magnitudes of the harmonic current components are inversely proportional to their frequency.

These statements can be substantiated by assuming the following equations for the input quantities to a three-phase system, and deriving an expression for the total instantaneous power:

$$v_a = V \sin \omega t \qquad (7\text{-}20)$$

$$i_a = I \left\{ \sin \omega t + \frac{\sin [(n-1)\omega t]}{n-1} + \frac{\sin [(n+1)\omega t]}{n+1} \right\} \qquad (7\text{-}21)$$

Expressions for the b and c phases will be identical except for the substitution of $(\omega t - 2\pi/3)$ and $(\omega t + 2\pi/3)$ for ωt.

The total instantaneous power $= v_a i_a + v_b i_b + v_c i_c$

$$= VI(1.5 - \frac{3}{n^2 - 1} \cos n\omega t) \qquad (7\text{-}22)$$

This equation shows that a load voltage harmonic component of order n is related to two harmonic components in the supply current of orders $(n + 1)$ and $(n - 1)$, with respective magnitudes $1/(n + 1)$ and $1/(n - 1)$.

For the ideal p-pulse rectifier, taking level load current, no losses, and no overlap, then only harmonics exist of the order

$$r = mp \pm 1 \qquad (7\text{-}23)$$

(where $m = 1$, 2, 3, etc.) are present in the input current, and they are of magnitude $1/r$ times the fundamental. Equation (7-23) shows that the higher the pulse number p, the more harmonics will be eliminated, as $mp = n$, n being the order of the harmonics present in the voltage waveform and r the order of the input current harmonic.

The assumption of level load current is frequently not justified sufficiently for the above statements to be accurate, and in practice, particularly at large firing delays, the harmonic components differ from the simple relationship of $1/r$ to the fundamental.

Referring to Fig. 7-2, the supply network may be equated to its Thévenin equivalent. A load taking a current i will result in a voltage v different to the supply $V \sin \omega t$, by the voltage drop across the supply impedance $R + jX$. The value of the supply impedance will differ for each harmonic frequency; hence the voltage v will contain harmonic voltage components. If other loads or consumers are fed from this point of common coupling, then they will receive a non-sinusoidal voltage.

The penetration of the harmonics into the supply from the load will depend on how the components of the system react to each particular frequency. Capacitors connected across the lines for (say) power factor correction will absorb harmonic currents due to their reactance falling with frequency. Resonance within the system may occur at a harmonic frequency. It is possible for the harmonic voltages and currents to penetrate into all levels of the supply system, increasing losses, causing metering errors, and interfering with other consumers.

A problem of particular importance occurs in those circuits where a d.c. component is induced into the neutral of the supply system. This can lead to d.c. magnetization of the system transformers and overloading of the neutral conductor.

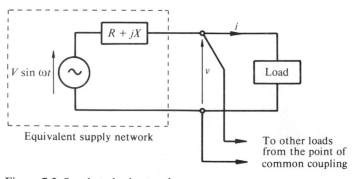

Figure 7-2 Supply to load network.

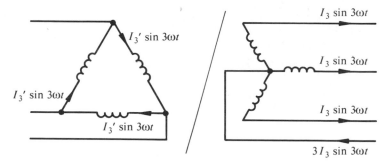

Figure 7-3 Flow of third-harmonic current in a three-phase system.

In respect to balanced three-phase systems, some general observations can be made. The triplen harmonics, that is, those of order $3m$, where $m = 1, 2, 3$, etc., are all in phase with each other at all instants as illustrated by

$$i_{a(3m)} = I \sin 3m\omega t$$

$$i_{b(3m)} = I \sin[3m(\omega t - 2\pi/3)] = I \sin 3m\omega t$$

$$i_{c(3m)} = I \sin[3m(\omega t + 2\pi/3)] = I \sin 3m\omega t$$

Hence $\qquad i_{a(3m)} = i_{b(3m)} = i_{c(3m)}$ \hfill (7-24)

As is obvious from Fig. 7-3, if $i_{a(3m)} = i_{b(3m)} = i_{c(3m)}$ then this triplen harmonic current can only flow in a star-connection with a neutral connection. In symmetrical component theory (Ref. 7) the flow of triple-frequency components is similar to a zero-sequence component with the neutral current being three times that in each line. The triple-frequency currents shown in the delta winding of the transformer of Fig. 7-3 can flow in the delta but do not appear in the input lines.

If the supply neutral is not connected to the load, then the absence of the neutral means that there can be no triplen harmonic components in the lines.

Harmonics of order $(3m - 1)$ can be expressed as

$$i_{a(3m-1)} = I \sin[(3m - 1)\omega t]$$

$$i_{b(3m-1)} = I \sin[(3m - 1)(\omega t - 2\pi/3)] = I \sin[(3m - 1)\omega t + 2\pi/3]$$

$$i_{c(3m-1)} = I \sin[(3m - 1)(\omega t + 2\pi/3)] = I \sin[(3m - 1)\omega t - 2\pi/3]$$

$$(7-25)$$

A harmonic of order $(3m - 1)$ is at a reversed (negative) sequence to that of the fundamental. For example, a fifth $(m = 2)$ harmonic will set up a reverse torque component in an induction motor if the input to the motor contains such a harmonic.

Harmonics of order $(3m + 1)$ can be expressed as

$$i_{a(3m+1)} = I \sin[(3m + 1)\omega t]$$

$$i_{b(3m+1)} = I \sin[(3m + 1)(\omega t - 2\pi/3)] = I \sin[(3m + 1)\omega t - 2\pi/3]$$

$$i_{c(3m+1)} = I \sin[(3m + 1)(\omega t + 2\pi/3)] = I \sin[(3m + 1)\omega t + 2\pi/3]$$

$$(7\text{-}26)$$

A harmonic of order $(3m + 1)$ such as the seventh $(m = 2)$ are of the same sequence as the fundamental.

The public electricity supply authorities impose limits to the level of harmonic current which can be taken by a consumer, at what is known as the *point of common coupling*. The point of common coupling is that point in the supply network at which the consumer is supplied from the authority, and to which other consumers are connected.

Limiting the harmonic current taken by a consumer is not the only criterion used by the supply authorities, and some individual loads may be judged by the harmonic distortion they cause in the voltage waveform at the point of common coupling.

Although only frequencies at a multiple of the fundamental have been considered, it is possible for sub-harmonic components to exist. These are variations in the waveforms at a frequency below the nominal mains frequency. Such frequency components can arise from the use of integral cycle control as shown in Fig. 6-2*e*, and may be a nuisance in causing voltage fluctuations at a low frequency. These cyclic voltage fluctuations cause severe problems to persons working under lamps connected to such a supply, by creating unacceptable flicker.

Two parallel loads, each drawing harmonic currents, are often out of phase in respect to the harmonics, and hence may tend to be self-cancelling.

A source of high-frequency interference can be caused by what is known as *jitter*. Jitter is said to occur when the firing delay angle from one cycle to the next is not precisely the same.

Radio interference can be quite severe from the harmonics in the supply system, as the whole system is acting as a high-frequency transmitting aerial in this respect. Coupled radio-frequency interference can occur where the power lines run parallel to communication lines, as in an electrified railway system. The signalling cables must be screened from such interference.

7-4 FILTERS

Filters serve the purpose of changing the harmonic content of a wave as it passes from the input to the output. The major areas of use are: in smoothing the voltage wave of a load fed from a rectifier, in reducing the harmonic content of an inverter output waveform, in preventing unwanted harmonic components being reflected into the a.c. system, and in eliminating radio-frequency interference.

7-4-1 Rectifier output smoothing

The basic filters for smoothing the load voltage of a rectifier load are shown in Fig. 7-4, illustrated with reference to a 2-pulse waveform. The inductance-only filter of Fig. 7-4b smoothes the current, hence resulting in continuous input current, and is similar to the inductive loads discussed in Chapter 2.

The capacitor-only filter of Fig. 7-4c acts by the capacitor being charged to the peak of the a.c. supply voltage, and then discharging exponentially into the load resistance at a rate dependent on the time constant RC. The supply current is of a pulse shape with a fast rise time due to the short charging period of the capacitor.

A combination of inductance and capacitance is shown in Fig. 7-4d, the capacitor acting to maintain the load voltage constant, while the inductor smooths the current, so resulting in an input current tending to be square-wave.

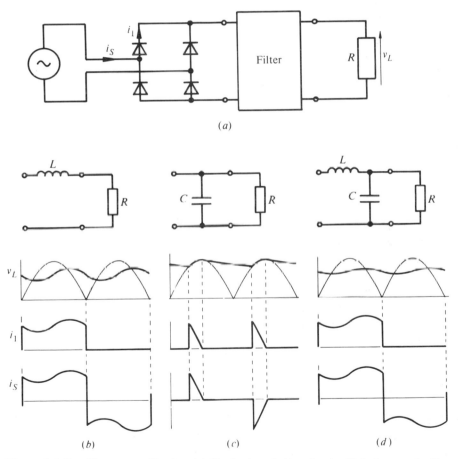

Figure 7-4 Rectifier output filtering. (a) Single-phase bridge circuit. (b) Inductor-only filter. (c) Capacitor-only filter. (d) Inductor-capacitor filter.

The performance of the LC filter can be obtained by considering the output of the rectifier to be its mean value plus the harmonic components, as given by Eq. (7-18), and then analysing the filter in respect of each harmonic frequency.

If the load resistance tends to infinity, then the ratio of the output voltage of the filter to its input voltage is

$$\frac{1/\omega C}{\omega L - 1/\omega C} = \frac{1}{\omega^2 LC - 1} \qquad (7\text{-}27)$$

Hence the higher-order harmonics are attenuated more than those of lower-order.

To obtain a given smoothing to a particular load, a cost balance must be made between the high filter cost of a low-pulse circuit, and the lower filter cost with a more expensive higher-pulse circuit.

A further reduction in the ripple voltage of the load can be obtained by a number of simple filters connected in cascade.

The ripple factor in the load voltage is defined as the ratio of the total r.m.s. value of all the alternating components to the mean value, giving

$$\text{ripple factor} = \frac{(V^2_{1(\text{rms})} + V^2_{2(\text{rms})} + \ldots + V^2_{n(\text{rms})})^{1/2}}{V_{\text{mean}}}$$

$$= \frac{(V^2_{\text{rms}} - V^2_{\text{mean}})^{1/2}}{V_{\text{mean}}} \qquad (7\text{-}28)$$

where V_{rms} is of the total wave.

7-4-2 Inverter output filtering

The inverters described in Sec. 5-4 have output waveforms of either the quasi-square waveform as in Fig. 5-21, or the pulse-width modulated waveform as in Fig. 5-23. In order to attenuate the harmonic content of these waveforms, it is necessary to pass them through a filter. Typically a reduction of the harmonic distortion factor (Eq. (7-15)) to 5% is adequate.

Various filter designs exist, but they are broadly variations of those shown in Fig. 7-5. The simple low-pass LC filter of Fig. 7-5a has a no-load output/input

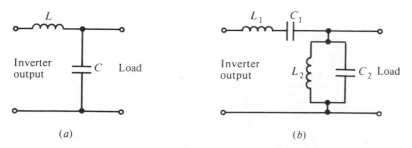

(a) $\qquad\qquad\qquad\qquad\qquad\qquad$ (b)

Figure 7-5 Two types of filter. (a) Low-pass filter. (b) Resonant-arm filter.

ratio as given by Eq. (7-27). If the inverter output is designed for elimination of the low-order harmonics such as those shown in Figs. 5-23 and 5-37, then this simple low-pass filter is adequate provided $\omega^2 LC$ is not excessively high at the fundamental frequency.

The resonant-arm filter (Fig. 7-5b) is more appropriate for situations where low-order harmonics are present. Both the series arm $L_1 C_1$ and the parallel arm $L_2 C_2$ are tuned to the inverter output frequency. The series arm presents zero impedance to the fundamental frequency, but finite increasing impedance to higher frequencies. The parallel arm presents infinite impedance at the fundamental frequency, but a reducing impedance to higher frequencies. Taking the fundamental frequency $\omega_0 = 1/\sqrt{(L_1 C_1)} = 1/\sqrt{(L_2 C_2)}$, making $C_1 = AC_2$ and $L_2 = AL_1$, and letting $\omega = n\omega_0$, where n is the order of the harmonic, then

$$\frac{V_{\text{out}}}{V_{\text{in}}} = \frac{1}{1 - \frac{1}{A}\left(n - \frac{1}{n}\right)^2} \tag{7-29}$$

for the nth harmonic component and open-circuit load.

The overall performance must be assessed with reference to the load when connected, and in relationship to transient responses, particularly where the load is suddenly changed or suffers a fault condition.

7-4-3 A.C. line filters

To attenuate the penetration of harmonics into the a.c. system from a rectifier load, harmonic filters can be connected to the neutral from each line.

The manner in which the harmonic currents are by-passed is to provide harmonic filters as shown in Fig. 7-6. For a 6-pulse system, tuned harmonic filters are provided for the 5th, 7th, 11th, and 13th harmonic components. For the higher-order harmonics, a high-pass filter is provided. Care must be taken to avoid excessive loss at the fundamental frequency. A practical problem is that of frequency drift, which may be as much as $\pm 2\%$ in a public supply system. Either the filters have to be automatically tuned or have a low Q-factor to be effective.

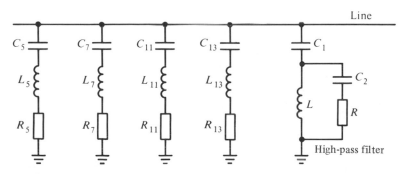

Figure 7-6 Harmonic line traps.

The use of filters as described above is normally limited to very large converters, such as those used in the H.V.D.C. transmission schemes described in Sec. 6-9.

7-4-4 Radio-interference suppression

The switching of thyristors, triacs, etc., gives rise to mains-borne harmonics in the radio-frequency range. The radio-frequency interference with communication systems can either originate from the devices themselves, or from the load equipment, or via radiation from the main supply lines. Screening of the equipment will eliminate most of the directly radiated interference, but not that from the mains, for which filters must be used.

National and international standards specify limits to the harmonics which equipment can induce in the lines, normally limits in the range of 0.15 to 30 MHz. At the higher frequencies, it is easily possible for standing-wave phenomena to arise (Ref. 9).

The harmonic voltage between lines is termed a *symmetrical voltage*, and those between either line and earth an *asymmetrical voltage*. Typically, a limit is specified of 2 mV at the supply terminals, and 10 mV at the load terminals for the range 5 to 30 MHz, the voltages being those measured across 150 Ω connected between lines or line to earth.

Many filtering arrangements exist, a few examples being shown in Fig. 7-7. In general it is easy to filter the symmetrical harmonic voltage as in Fig. 7-7a. With Fig. 7-7b any unbalance in the harmonic currents of the two lines will be attenuated by the coupled inductor, so reducing the asymmetrical voltages. Capacitors only, as shown in Fig. 7-7c, are sufficient for cases where harmonic distortion is less severe.

In general, the object of filtering is to provide a low-impedance path from the mains leads to the screen, and a high-impedance path in series with the mains leads.

7-5 WORKED EXAMPLES

Example 7-1

Determine the Fourier series for the load voltage of the following uncontrolled rectifiers taking the mean level to be 100 V: (i) 2-pulse, (ii) 3-pulse, (iii) 6-pulse, (iv) 12-pulse. Neglect overlap and losses.
Also determine the ripple factor for each configuration.

SOLUTION Using Eqs. (7-17) and (7-18), taking $V_{mean} = 100$, then

$$b_n = \frac{200}{(m^2 p^2 - 1)} (-\cos m\pi), \text{ and taking } n = mp \text{ with } m \text{ having values of } 1, 2, 3,$$

etc., we get the following series:
(i) $v = 100 + 66.7 \cos 2\omega t - 13.3 \cos 4\omega t + 5.7 \cos 6\omega t - 3.1 \cos 8\omega t$
     $+ 2.0 \cos 10\omega t - \ldots$

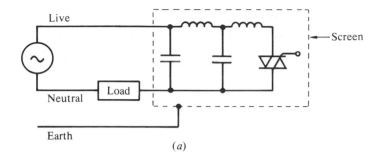

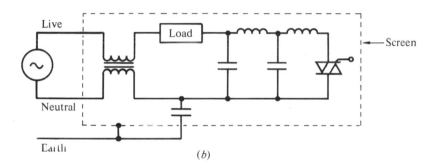

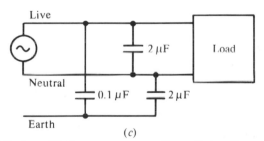

Figure 7-7 Various filtering arrangements for radio-interference suppression. (*a*) Low-pass filter. (*b*) Additional coupled inductor to suppress asymmetrical harmonics. (*c*) For apparatus with less severe harmonic generation.

(ii) $v = 100 + 25 \cos 3\omega t - 5.7 \cos 6\omega t + 2.5 \cos 9\omega t - 1.4 \cos 12\omega t + \ldots$
(iii) $v = 100 + 5.7 \cos 6\omega t - 1.4 \cos 12\omega t + 0.6 \cos 18\omega t - \ldots$
(iv) $v = 100 + 1.4 \cos 12\omega t - 0.35 \cos 24\omega t + \ldots$

Note that if a given order of harmonic is present, then its magnitude is the same in all the configurations.

The ripple factor is defined by Eq. (7-28) and, referring to Fig. 7-1 for a *p*-pulse output,

$$V_{rms} = \left[\frac{1}{\pi} \int_0^{\pi} (V_{max} \cos \omega t)^2 \, d(p\omega t) \right]^{1/2} = V_{max} \left[\frac{1}{2} + \frac{p}{4\pi} \sin \frac{2\pi}{p} \right]^{1/2}$$

Hence the ripple factors for the various configurations are
(i) 0.483, (ii) 0.183, (iii) 0.042, (iv) 0.01.

Example 7-2

Determine the relative low-order harmonic content compared to the mean level of the voltage output of a controlled 6-pulse rectifier when the firing delay angle is 30°, for the conditions of (i) neglecting overlap, (ii) assuming an overlap angle of 15°.

SOLUTION To determine a particular harmonic, the mathematics is made easier by treating the 300 Hz component as the fundamental, rather than adopt the approach of the previous example.
The waveform is similar to that shown in Fig. 2-22b and Fig. 3-10.
(i) Taking the period of 2π over one variation in the output, then the function is

$V_{max} \cos (x/6)$ between the limits $6\left(-\dfrac{\pi}{6} + \dfrac{\pi}{6}\right)$ to $6\left(\dfrac{\pi}{6} + \dfrac{\pi}{6}\right)$ for 30° ($\pi/6$ rad) delay.

Hence for the 300 Hz component,

$$a_1 = \frac{1}{\pi} \int_0^{2\pi} V_{max} \cos \frac{x}{6} \sin x \, dx = 0.1637 V_{max}$$

$$b_1 = \frac{1}{\pi} \int_0^{2\pi} V_{max} \cos \frac{x}{6} \cos x \, dx = 0.0473 V_{max}$$

giving $[a_1^2 + b_1^2]^{1/2} = 0.1704 V_{max}$ as the magnitude.
Using $\sin 2x$ and $\cos 2x$ for the 600 Hz component, its magnitude is calculated to be $0.081 V_{max}$.
Using Eq. (3-23), $V_{mean} = 0.827 V_{max}$; hence relatively the 300 Hz component is $(0.1704/0.827) = 0.21$, and the 600 Hz component is $(0.081/0.827) = 0.098$.

(ii) With an overlap of 15° ($\pi/12$ rad) the function will be

$\dfrac{\sqrt{3}}{2} V_{max} \cos [(x + \pi)/6]$, between $6\left(-\dfrac{\pi}{6} + \dfrac{\pi}{6}\right)$ to $6\left(-\dfrac{\pi}{6} + \dfrac{\pi}{6} + \dfrac{\pi}{12}\right)$,

and $V_{max} \cos \dfrac{x}{6}$, between $6\left(-\dfrac{\pi}{6} + \dfrac{\pi}{6} + \dfrac{\pi}{12}\right)$ to $6\left(\dfrac{\pi}{6} + \dfrac{\pi}{6}\right)$.

Hence for the 300 Hz component,

$$a_1 = \frac{1}{\pi} \left[\int_0^{\pi/2} \frac{\sqrt{3}}{2} V_{max} \cos [(x + \pi)/6] \sin x \, dx + \int_{\pi/2}^{2\pi} V_{max} \cos \frac{x}{6} \sin x \, dx \right]$$

$$= 0.06256 V_{max}, \quad \text{and further calculations yield}$$

$$b_1 = -0.1394 V_{max}$$

giving the magnitude as $0.1528 V_{max}$.

With $V_{mean} = 0.7511 V_{max}$, the relative 300 Hz component is 0.203.
Working in a similar manner, the relative 600 Hz component is 0.022.

Example 7-3

Taking a level load current and neglecting overlap and losses, determine the Fourier series for the input current to the bridge circuits of (i) 6-pulse, (ii) 12-pulse configurations.

SOLUTION The waveforms to be analysed for (i) are shown in Fig. 7-8a and b, being the input to the 3-phase bridge and the input to the delta primary transformer respectively, as shown in Fig. 2-21b. For (ii) the input current to the transformer (as shown in Fig. 2-25) is shown in Fig. 7-8c, which is the addition of the two waveforms in Fig. 7-8a and b.

(i) Inspection of the waveforms shows that there are no cosine terms, no even harmonics, and that there is quarter-wave symmetry.

For Fig. 7-8a,
$$a_n = \frac{4}{\pi} \int_{\pi/6}^{\pi/2} I \sin n\omega t \, d(\omega t) = \frac{4I}{n\pi} \cos \frac{n\pi}{6}$$

Hence the series is

$$i = 1.103 I [\sin \omega t - \frac{\sin 5\omega t}{5} - \frac{\sin 7\omega t}{7} + \frac{\sin 11\omega t}{11} + \frac{\sin 13\omega t}{13} - \frac{\sin 17\omega t}{17} - \ldots]$$

For Fig. 7-8b,

$$a_n = \frac{4}{\pi} \left[\int_0^{\pi/3} 0.577 I \sin n\omega t \, d(\omega t) + \int_{\pi/3}^{\pi/2} 1.155 I \sin n\omega t \, d(\omega t) \right]$$

$$= \frac{4I}{n\pi} \{0.577[1 + \cos(n\pi/3)]\}$$

Hence the series is

$$i = 1.103 I [\sin \omega t + \frac{\sin 5\omega t}{5} + \frac{\sin 7\omega t}{7} + \frac{\sin 11\omega t}{11} + \frac{\sin 13\omega t}{13} + \frac{\sin 17\omega t}{17} + \ldots]$$

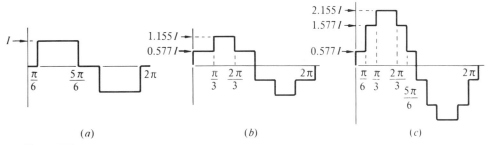

(a)　　　　　　(b)　　　　　　(c)

Figure 7-8

(ii) For Fig. 7-8c,

$$a_n = \frac{4}{\pi} \left[\int_0^{\pi/6} 0.577I \sin n\omega t \, d(\omega t) + \int_{\pi/6}^{\pi/3} 1.577I \sin n\omega t \, d(\omega t) \right.$$

$$\left. + \int_{\pi/3}^{\pi/2} 2.155I \sin n\omega t \, d(\omega t) \right]$$

$$= \frac{4I}{n\pi} \{0.577[1 + \cos(n\pi/3)] + \cos(n\pi/6)\}$$

Hence the series is

$$i = 2.205I[\sin \omega t + \frac{\sin 11\omega t}{11} + \frac{\sin 13\omega t}{13} + \frac{\sin 23\omega t}{23} + \ldots]$$

It is to be noted that the effect of the delta/star transformer is to reverse the sign of the fifth and seventh harmonics, so that the addition of waveforms (a) and (b) eliminates these harmonics from the 12-pulse supply current of (c).

The above derivations confirm the theory developed in relation to Eq. (7-23).

Example 7-4

Determine the value of the harmonic components, up to the seventh order, in the input current to a three-phase bridge rectifier, if the overlap angle is 15° and the load current can be considered level.

SOLUTION The general shape of the current being analysed is similar to that shown in Fig. 3-18. The positive half cycle starts at 30°, overlap is complete by 45°, the current is level until 150°, then there is a further period of 15° before the current falls to zero.

Using Eqs. (3-2) and (3-3) and taking $\omega t = x$, then during the growth period, $i = k(1 - \cos x)$, where the load current $I = k(1 - \cos 15°)$ hence $i = I(1 - \cos x)/(1 - \cos 15°)$.

During the decay period, $i = I - k(1 - \cos x)$

$$= I - [I(1 - \cos x)/(1 - \cos 15°)].$$

As the current is symmetrical and there is no neutral current, there are no even or triple-order harmonics.

$$a_n = \frac{2}{\pi} \left[\int_{\pi/6}^{\pi/4} \frac{I\{1 - \cos[x - (\pi/6)]\}}{0.03408} \sin nx \, dx + \int_{\pi/4}^{5\pi/6} I \sin nx \, dx \right.$$

$$\left. + \int_{5\pi/6}^{11\pi/12} \left\{ I - \frac{I[1 - \cos(x - 5\pi/6)]}{0.03408} \right\} \sin nx \, dx \right]$$

Similarly find b_n as above but with $\cos nx$ in place of $\sin nx$.
The magnitude of each harmonic is $\sqrt{(a_n^2 + b_n^2)}$.

Calculations give the fundamental as $1.729I$; the fifth harmonic as $0.33I$, that is, 19.1%; the seventh harmonic as $0.225I$, that is, 13.0%.

Compared to the analysis in Example 7-3 which ignores overlap, the harmonic values are slightly reduced.

Example 7-5

A three-phase diode bridge rectifier supplies a light inductance load. Determine the percentage level of the fifth- and seventh-order harmonics in the input line current, given the following measured instantaneous values taken at 5° intervals over the positive half cycle. Values are: 0, 11.0, 41.0, 90.3, 93.1, 95.9, 97.6, 98.5, 98.5, 97.5, 96.0, 94.0, 91.8, 89.7, 87.5, 90.3, 93.1, 95.9, 97.6, 98.5, 98.5, 97.5, 96.0, 94.0, 91.8, 79.0, 48.0, 0 A, continuing at zero until 180°.

SOLUTION It is only necessary to analyse the positive half cycle as there are no even harmonics. Graphical analysis has to be used as it is not practical to fit a mathematical equation to the waveform.

Equations (7-10) and (7-11) are used to determine the coefficients of the Fourier series.

A table can be drawn up as shown in Fig. 7-9. At each 5° interval, the value of $\sin nx$ and $\cos nx$ can be tabulated for each harmonic order. The product of the current value at that angle with $\sin nx$ (or $\cos nx$) is tabulated. Having tabulated

x	0°	5°	10°	15°	etc.	130°	135°	Sum	
i	0	11.0	41.0	90.3		48.0	0		
$\sin x$	0	0.087	0.174	0.259		0.766	0.707		
$i \sin x$	0	0.96	7.12	23.37		36.77	0	1752	$a_1 - 97.37$
$\cos x$	1	0.996	0.985	0.966		−0.643	−0.707		
$i \cos x$	0	10.96	40.38	87.22		30.85	0	654	$b_1 - 36.35$
$\sin 5x$	0	0.423	0.766	0.966		−0.940	−0.707		
$i \sin 5x$	0	4.65	31.41	87.22		−45.11	0	90	$a_5 = 4.996$
$\cos 5x$	1	0.906	0.643	0.259		0.342	0.707		
$i \cos 5x$	0	9.97	26.35	23.37		16.42	0	−394	$b_5 = -21.891$
$\sin 7x$	0	0.574	0.940	0.966		−0.174	−0.707		
$i \sin 7x$	0	6.31	38.53	87.22		−8.33	0	153	$a_7 = 8.509$
$\cos 7x$	1	0.819	0.342	−0.259		−0.985	−0.707		
$i \cos 7x$	0	9.01	14.02	−23.37		−47.27	0	−127	$b_7 = -7.06$

Figure 7-9

$i \sin nx$, it can then be summated. Over the half cycle, the number of 5° intervals is 36. The summation is divided by 36, multiplied by 2 as in Eq. (7-10) to give a_n. It is left to the reader to complete the gaps in Fig. 7-9 to confirm that the summation value is correct.

From Fig. 7-9 the peak value of the fundamental is $[97.37^2 + 36.35^2]^{1/2} = 103.9$ A, the peak value of the fifth harmonic is $[4.996^2 + 21.891^2]^{1/2} = 22.45$ A, and the peak value of the seventh harmonic is $[8.509^2 + 7.06^2]^{1/2} = 11.06$ A.

Relative to the fundamental, the percentage values are 21.6% and 10.64% for the fifth and seventh harmonics respectively.

It is worth noting that the fifth-harmonic value is greater than the 20% value found when level load current is assumed. The assumption of infinite load inductance predicts fifth-harmonic levels lower than those encountered in practice.

Example 7-6

Determine the harmonic content in the output voltage and input current of a single-phase half-controlled bridge rectifier, neglecting overlap and losses, and assuming the load current is level. Plot curves representing the harmonic content, taking the mean load voltage at zero firing delay angle as 100 V and the load current as 100 A.

SOLUTION The circuit and waveforms are shown in Fig. 2-10. From Eq. (2-8) for a mean voltage of 100 V at $\alpha = 0$, $V_{max} = 157$ V.

The easiest mathematical approach to analysing the voltage waveform is with the reference as shown in Fig. 7-10a, and then later relating ϕ to the firing delay angle α, and using $n = mp$ as defined for Eq. (7-18).

$$a_n = \frac{p}{\pi} \int_{-\phi}^{0} - V_{max} \sin \omega t \sin mp\omega t \, d(\omega t)$$

$$= \frac{pV_{max}}{2\pi} \left\{ \frac{\sin[(mp+1)\phi]}{mp+1} - \frac{\sin[(mp-1)\phi]}{mp-1} \right\}$$

$$b_n = \frac{p}{\pi} \int_{-\phi}^{0} - V_{max} \sin \omega t \cos mp\omega t \, d(\omega t)$$

$$= \frac{pV_{max}}{2\pi} \left\{ \frac{-2}{m^2 p^2 - 1} - \frac{\cos[(mp+1)\phi]}{mp+1} + \frac{\cos[(mp-1)\phi]}{mp-1} \right\}$$

The resultant r.m.s. nth harmonic $= [(a_n^2 + b_n^2)/2]^{1/2}$

For the single-phase bridge $p = 2$, hence n has values of 2, 4, 6, 8, etc., and $\phi = \pi - \alpha$, where α is the firing delay angle.

The harmonic values are plotted in Fig. 7-10b for the 2nd, 4th, and 6th orders, that is, 100 Hz, 200 Hz, and 300 Hz for a 50 Hz supply.

The easiest analysis of the current waveform is to take its reference as mid-way in the zero period as shown in Fig. 7-10c. The waveform then contains only sine terms. There are no even harmonics in this waveform.

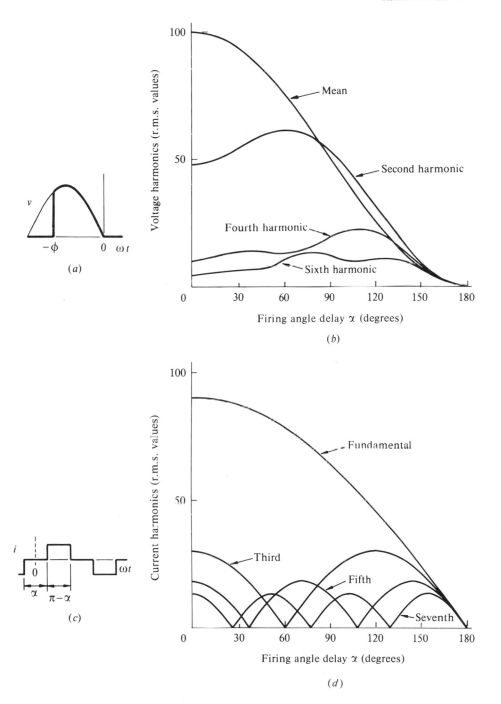

Figure 7-10

$$a_n = \frac{2}{\pi} \int_{\frac{\alpha}{2}}^{\pi - \frac{\alpha}{2}} 100 \sin n\omega t \, d(\omega t) = \frac{400}{n\pi} \cos \frac{n\alpha}{2}$$

$b_n = 0$, hence the r.m.s. value of the nth harmonic is $a_n/\sqrt{2}$.
For the fundamental $I_1 = 90 \cos (\alpha/2)$
for the third harmonic $I_3 = 30 \cos (3\alpha/2)$
for the fifth harmonic $I_5 = 18 \cos (5\alpha/2)$
for the seventh harmonic $I_7 = 13 \cos (7\alpha/2)$
These harmonic variations with firing delay are shown plotted in Fig. 7-10d.

Example 7-7

Determine the harmonic content in the output voltage of a three-phase half-controlled rectifier bridge up to the twelfth order, at firing delay angles of $60°$, $90°$, and $120°$. Take the peak value of the input voltage to be $100\,\text{V}$, neglect losses and overlap.
Assuming a level load current of $100\,\text{A}$, determine at the above conditions the harmonic content of the input current.

SOLUTION The waveforms are shown in Fig. 2-23 and Fig. 2-41, where it can be seen that at firing delays over $60°$ the load voltage has zero periods, hence the general formula developed in Example 7-6 relating to Fig. 7-10a can be used: $p = 3$, hence n has values of $3, 6, 9, 12$, etc.
At a firing delay angle of $60°$, $\phi = \pi - (\pi/3) = 2\pi/3$ rad, giving for $n = 3$

$$a_3 = \frac{3 \times 100}{2\pi} \left[\frac{\sin 4\phi}{4} - \frac{\sin 2\phi}{2} \right] = 31.0$$

$$b_3 = \frac{3 \times 100}{2\pi} \left[\frac{-2}{3^2 - 1} - \frac{\cos 4\phi}{4} + \frac{\cos 2\phi}{2} \right] = -17.9$$

The resultant r.m.s. third-order harmonic $= \left(\dfrac{31.0^2 + 17.9^2}{2} \right)^{1/2} = 25.32\,\text{V}$.

For $n = 6$, the resultant r.m.s. sixth-order harmonic $= 10.43\,\text{V}$.
For $n = 9$, the resultant r.m.s. ninth-order harmonic $= 6.70\,\text{V}$.
For $n = 12$, the resultant r.m.s. twelfth-order harmonic $= 4.96\,\text{V}$.
If the current waveform is analysed by taking a reference point midway between where I_1 ceases conduction and I_4 starts (Fig. 2-23c), then only sine components exist. At $\alpha = 60°$,

$$a_n = \frac{2}{\pi} \int_0^{2\pi/3} -100 \sin n\omega t \, d(\omega t) = \frac{200}{n\pi} [\cos \frac{n2\pi}{3} - 1] = \frac{-300}{n\pi}$$

giving r.m.s. harmonic components of: fundamental ($n = 1$), 67.5 A; 2nd, 33.8 A; 4th, 16.9 A; 5th, 13.5 A; 7th, 9.6 A; 8th, 8.4 A; 10th, 6.8 A; 11th, 6.1 A; 13th, 5.2 A. As there is no neutral connection, all triple-order harmonics are zero.

At a firing delay angle of 90°, $\phi = \pi - (\pi/2) = \pi/2$, giving r.m.s. harmonic values in the load-voltage waveform of: third, 33.76 V; sixth, 11.74 V; ninth, 6.75 V; twelfth, 5.69 V.

Adopting the same technique as above for determining the current harmonics at $\alpha = 90°$,

$$a_n = \frac{2}{\pi} \int_{\pi/12}^{7\pi/12} -100 \sin n\omega t \, d(\omega t) = \frac{200}{n\pi} \left[\cos \frac{n7\pi}{12} - \cos \frac{n\pi}{12} \right]$$

giving r.m.s. harmonic components of: fundamental, 55.13 A; 2nd, 38.99 A; 4th, 0 A; 5th, 11.03 A; 7th, 7.88 A; 8th, 0 A; 10th, 7.80 A; 11th, 5.01 A; 13th, 4.24 A. At a firing delay of 120°, $\phi = \pi - (2\pi/3) = \pi/3$, giving r.m.s. harmonic values in the load voltage waveform of: 3rd, 25.32 V; 6th, 10.07 V; 9th, 6.70 V; 12th, 4.91 V. The r.m.s. harmonic component values of the current are: fundamental, 38.99 A; 2nd, 33.76 A; 4th, 16.88 A; 5th, 7.80 A; 7th, 5.57 A; 8th, 8.44 A; 10th, 6.75 A; 11th, 3.54 A; 13th, 3.00 A.

Example 7-8

Derive a general expression for the harmonic content of the current taken by a single-phase pure resistance load supplied via a triac operating in phase angle control mode. Take the r.m.s. current at zero firing delay to be 100%, and neglect all losses and the supply impedence.

Plot curves showing the low-order harmonic content, power, and power factor against variation of firing delay angle.

SOLUTION The waveform being analysed is that shown in Fig. 6-2d, only odd harmonics being present. At any firing delay angle α, the nth harmonic component is

$$a_n = \frac{2}{\pi} \int_{\alpha}^{\pi} I_{max} \sin \omega t \sin n\omega t \, d(\omega t)$$

$$= \frac{I_{max}}{\pi} \left\{ \frac{\sin [(n+1)\alpha]}{n+1} - \frac{\sin [(n-1)\alpha]}{n-1} \right\}$$

$$b_n = \frac{2}{\pi} \int_{\alpha}^{\pi} I_{max} \sin \omega t \cos n\omega t \, d(\omega t)$$

$$= \frac{I_{max}}{\pi} \left\{ \frac{1 - \cos [(n-1)\alpha]}{n-1} - \frac{1 - \cos [(n+1)\alpha]}{n+1} \right\}$$

with the resultant harmonic component $= (a_n^2 + b_n^2)^{1/2}$.

For the fundamental when $n = 1$, the above expressions cannot be used, the components being analysed on their own, giving

$$a_1 = \frac{I_{max}}{\pi} \left[\pi - \alpha + \frac{\sin 2\alpha}{2} \right] \quad \text{and} \quad b_1 = \frac{I_{max}}{2\pi} [\cos 2\alpha - 1]$$

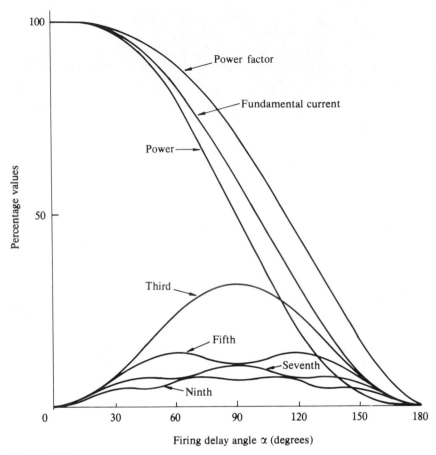

Figure 7-11

At zero firing delay, only a_1 has a value, it being I_{max}, hence in the calculations let $I_{max} = 100$ to find percentage values. The values up to the ninth harmonic are shown plotted in Fig. 7-11.

Equation (6-1) gives the power, taking 100% as the value at $\alpha = 0°$.

From Eq. (3-13), power factor = (power)/$(V_{rms}I_{rms}) = (I_{rms}^2 R)/(V_{rms}I_{rms})$, for a resistance of R. As R and V_{rms} are constant, then the power factor is proportional to I_{rms} which is in turn proportional to $\sqrt{(power)}$. Hence the power factor can be calculated as the square root of the per unit power. Values of power and power factor are shown plotted in Fig. 7-11.

Example 7-9

A pure resistance load of $2.4\ \Omega$ fed from a 240 V single-phase supply via a triac (or inverse-parallel thyristors) is operated in the integral cycle mode of control. Deter-

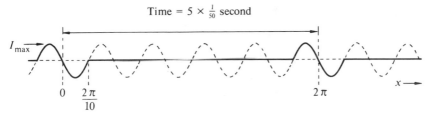

Time = $5 \times \frac{1}{50}$ second

Figure 7-12

mine the harmonic components in the supply current for (i) 1 cycle on, 4 cycles off; (ii) 5 cycles on, 1 cycle off.

SOLUTION The waveform to be analysed is similar to that shown in Fig. 6-2e. The waveform for (i) is shown in Fig. 7-12 with one cycle on, four cycles off, that is, the waveform repeats itself every five cycles, a rate of 10 Hz. The wave can be analysed by considering the current to have a fundamental frequency of 10 Hz, hence taking $n = 1$ for the 10 Hz component, $n = 2$ for the 20 Hz component, etc., with $n = 5$ being the 50 Hz component, the current being analysed is $i = I_{max} \sin 5x$. I_{max} for this problem = $(240\sqrt{2})/2.4 = 100\sqrt{2}$ A. Taking the reference as shown in Fig. 7-12, there are only sine terms ($b_n = 0$), hence

$$a_n = \frac{2}{\pi} \int_0^{2\pi/10} -100\sqrt{2} \sin 5x \sin nx \, dx = \frac{-100\sqrt{2}}{\pi} \times \frac{10 \sin 0.2n\pi}{5^2 - n^2}$$

A general formula for

$$a_n = -I_{max} \times \frac{2M}{\pi} \times \frac{\sin\left(\frac{N}{M}n\pi\right)}{M^2 - n^2}$$

can be derived where M is the number of cycles over which the pattern is repeated ($M = 5$ in above) and $N(= 1$ in above) the number of on cycles.
The above equation is indeterminate for $n = 5$ but putting $n = 5$ in the original equation yields $a_5 = -20\sqrt{2}$, that is, an r.m.s. value of 20 A. This is to be expected as the total power is one fifth of the maximum. Hence the fundamental current is one fifth of its continuous on-value of 100 A.
The supply frequency component expressed in general terms is $(N/M)I_{max}$.
Using r.m.s. values, being $1/\sqrt{2}$ of the maximum, then components for (i) are: 10 Hz, 7.8 A; 20 Hz, 14.4 A; 30 Hz, 18.9 A; 40 Hz, 20.8 A; 50 Hz, 20 A; 60 Hz, 17.0 A; 70 Hz, 12.6 A; 80 Hz, 7.8 A; 90 Hz, 3.3 A; 100 Hz, 0 A; 110 Hz, 1.9 A; 120 Hz, 2.5 A; 130 Hz, 2.1 A; 140 Hz, 1.1 A.
The waveform for (ii) repeats itself every 6 cycles, so its fundamental frequency = 50/6 = 8.33 Hz. By similar reasoning,

$$a_n = \frac{2}{\pi} \int_0^{\frac{2\pi}{12} \times 5} -100\sqrt{2} \sin 6x \sin nx \, dx = \frac{-100\sqrt{2}}{\pi} \times \frac{12 \sin 0.833n\pi}{6^2 - n^2}$$

where $n = 1$ is 8.33 Hz, $n = 2$ is 16.66 Hz, etc. At $n = 6$, 50 Hz, the r.m.s. current value is $100(5/6) = 83.3$ A.

Using r.m.s. values, the harmonic components for (ii) are: 8.33 Hz, 5.5 A; 16.66 Hz, 10.3 A; 25 Hz, 14.1 A; 33.33 Hz, 16.5 A; 41.66 Hz, 17.4 A; 50 Hz, 83.3 A; 58.33 Hz, 14.7 A; 66.66 Hz, 11.8 A; 75 Hz, 8.5 A; 83.33 Hz, 5.2 A; 91.66 Hz, 2.2 A; 100 Hz, 0 A; 108.33 Hz, 1.4 A; 116.66 Hz, 2.1 A; 125 Hz, 2.0 A.

Example 7-10

A resistance load of 19.6 Ω is fed from a 240 V, 50 Hz, single-phase supply which has an impedance of $(0.4 + j0.25)\,\Omega$. Determine the 250 Hz component of voltage appearing in the supply voltage waveform when the load is controlled by a triac operating in phase control mode, set to a firing delay angle of $90°$.

SOLUTION The r.m.s. current level when the control is fully on is $240/(19.6 + 0.4) = 12$ A, ignoring the effect of the supply reactance.

From Fig. 7-11 associated with Example 7-8, at a firing delay of $90°$, the fifth harmonic (250 Hz) component is 10.6%, that is, $12 \times 0.106 = 1.27$ A.

At 250 Hz, the supply impedance is $[0.4 + j(0.25 \times 5)]\,\Omega$, hence the r.m.s. 250 Hz voltage component at the supply terminals is $1.27 \times \sqrt{(0.4^2 + 1.25^2)} = 1.67$ V.

Example 7-11

For the circuit and conditions given in Example 7-10, determine the probable current and supply voltage harmonic components in the region of 5 kHz and 150 kHz.

SOLUTION To find the higher-order harmonics, the effect of the supply reactance cannot be ignored. Supply inductance $= 0.25/2\pi50$ H.

Using similar analysis to that given in Examples 2-1 (ii) and 2-2, the equation to the current is $i = 20\sqrt{2} \cos 2\pi50t - 20\sqrt{2}\, e^{-25\,133t}$ A, ignoring the small a.c. lag and having $t = 0$ at the firing delay angle of $90°$ on the voltage sinewave.

Letting $t = x/2\pi50$, the nth harmonic components are:

$$a_n = \frac{2}{\pi} \int_0^{\pi/2} 20\sqrt{2}(\cos x - e^{-80x}) \sin nx \; dx$$

$$b_n = \frac{2}{\pi} \int_0^{\pi/2} 20\sqrt{2}(\cos x - e^{-80x}) \cos nx \; dx$$

As only odd harmonics are present, we can find the 5.05 kHz component ($n = 101$). In the above equations, when $n = 101$, $a_n = 0.0474\sqrt{2}$, $b_n = 0.0614\sqrt{2}$, giving an r.m.s. value of 0.0775 A. The 5.05 kHz voltage component in the supply voltage waveform will be $0.0775 \times 0.25 \times 101 = 1.96$ V.

For the 150.05 kHz component, $n = 3001$, having an r.m.s. value of 0.113 mA, giving a supply voltage component of 0.0848 V. In practice, one could not ignore the effects of stray capacitance in the radio-frequency range, so any calculation such as this can be very inaccurate.

Example 7-12

A diode rectifier load takes a quasi-square-wave current of peak value 10 A from the 3-phase, 11-kV, 50-Hz, busbar feeder to a factory. In addition to the rectifier load, the busbars supply a star-connected load equivalent to 180 Ω in series with 0.3 H/ph. A star-connected capacitance of 1.75 μF/ph is connected to the busbars for power factor correction.

The 11 kV busbars are fed from an 800 kVA, 132 kV/11 kV transformer of 0.06 p.u. reactance and 0.01 p.u. resistance. The short-circuit impedance of the 132 kV system is 0.02 p.u. reactance, 0.005 p.u. resistance based on 800 kVA.

Determine the harmonic current and voltage levels in the various sections of the system up to the 23rd-order harmonic.

SOLUTION The system layout for this problem is shown in Fig. 7-13a. The rectifier may be considered as a harmonic generator, feeding current into the rest of the network, which reduces to the equivalent circuit shown in Fig. 7-13b.

Using a base of 800 kVA, and relating the per unit values to 11 kV, the base impedance $= 11^2/0.8 = 151.25\,\Omega$, giving the transformer impedance as $151.25(0.01 + j0.06) = (1.5125 + j9.075)\,\Omega = R_T + jX_T$ at 50 Hz. Likewise the system impedance $= 151.25(0.005 + j0.02) = (0.75625 + j3.025)\,\Omega = R_S + jX_S$ at 50 Hz. At 50 Hz, the load reactance $X_L = 2\pi50 \times 0.3 = j94.25\,\Omega$, and the capacitive reactance $X_C = 1/(2\pi50 \times 1.75 \times 10^{-6}) = -j1819\,\Omega$.

At the nth harmonic, the inductive reactance increases and the capacitive reactance decreases by the factor n.

The rectifier r.m.s. harmonic-current level can be obtained by using the expression developed in Example 7-3. The value for each harmonic is $I_n = \dfrac{4 \times 10}{n\pi\sqrt{2}} \times \dfrac{\sqrt{3}}{2} = \dfrac{7.8}{n}$ A, giving: 5th, 1.56 A; 7th, 1.11 A; 11th, 0.709 A; 13th, 0.600 A; 17th, 0.459 A; 19th, 0.410 A.

Referring to Fig. 7-13b, the current I_R divides into the three parallel arms of the circuit. Given three parallel impedances, Z_1, Z_2, and Z_3, the current I_1 flowing in Z_1 is given by

$$\frac{Z_2 Z_3}{Z_2 Z_3 + Z_1 Z_3 + Z_1 Z_2} \times \text{total current.}$$

To calculate the 5th harmonic values, frequency $= 250$ Hz, $n = 5$, $I_R = 1.56$ A. Impedance of the capacitance arm $= -j1819/5\,\Omega = Z_1$. Impedance of the load arm $= 180 + j5(94.25)\,\Omega = Z_2$.

Impedance of the supply arm $= (1.5125 + 0.75625) + j5(9.075 + 3.025)\,\Omega = Z_3$.

Current in the capacitor $= \dfrac{Z_2 Z_3}{Z_2 Z_3 + Z_1 Z_3 + Z_1 Z_2} \times 1.56 = 0.2735\underline{/175.2°}$ A.

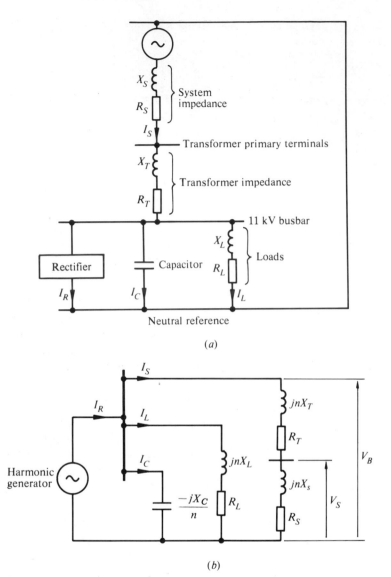

(a)

(b)

Figure 7-13 (a) System layout. (b) Equivalent circuit for each harmonic.

By like calculation, the load current $I_L = 0.1973\underline{/16.1°}$ A, and the supply current $I_S = 1.6443\underline{/-2.7°}$ A.

Harmonic phase voltage at the 11 kV busbars

$$= V_B = I_C \times X_C/5 = 0.2735 \times 363.8 = 99.5 \text{ V}.$$

Harmonic phase voltage at the 132 kV terminals

$$= V_S = 1.6443\,[0.756\,25^2 + (3.025 \times 5)^2]^{1/2} \times 132/11 = 298.8 \text{ V}.$$

n	1	5	7	11	13	17	19	23
frequency Hz	50	250	350	550	650	850	950	1150
I_R A	7.80	1.56	1.11	0.71	0.60	0.46	0.41	0.34
I_C A	3.49	0.27	0.46	1.77	22.3	1.11	0.77	0.50
I_L A	31.26	0.20	0.17	0.28	2.52	0.07	0.04	0.02
I_S A	37.16	1.64	1.40	2.20	19.83	0.58	0.32	0.14
V_B V	6351	99	119	293	3120	118	74	39
V_S V	78 139	299	356	880	9360	355	222	118

Figure 7-14

The components for the other harmonics are calculated in a similar manner. The values, together with those for the fundamental (50 Hz) frequency, are shown tabulated in Fig. 7-14.

At 650 Hz, the 13th harmonic frequency, the supply and load inductance resonate with the capacitor, hence the harmonic components are grossly amplified at this harmonic. In practice, the capacitor would be overstressed by the harmonic current, as its losses are frequency-dependent. At the higher frequencies, the capacitor presents such a low impedance that it absorbs most of the harmonic current.

Example 7-13

A 3-phase 50 Hz, bridge rectifier takes a quasi-square-wave current from the supply. Due to imperfections in the supply waveform and firing circuit, the positive half cycle is of length 120.1° and that of the negative half cycle 119.9°, with 60° zero periods. Taking the current to be level at a value I with instantaneous rise and fall, determine the harmonic content and compare to the case of a symmetrical wave.

SOLUTION Taking a reference about the centre of the positive half cycle, only cosine terms will be present, hence $a_n = 0$, and

$$b_n = \frac{2}{\pi} \left[\int_0^{60.05°} I \cos n\omega t \, d(\omega t) + \int_{120.05°}^{180°} -I \cos n\omega t \, d(\omega t) \right]$$

$$= \frac{4I}{n\pi} \sin (n90.05°) \cos (n30°).$$

Inspection of this equation shows that for all triple-order harmonics ($n = 3, 6, 9$, etc.), the value is zero.

For the second harmonic ($n = 2$) the value is $5.55 \times 10^{-4}I$, for the fourth harmonic ($n = 4$) the value is $5.55 \times 10^{-4}I$, for the fifth harmonic ($n = 5$) the value is $0.2205I$, for the seventh harmonic ($n = 7$) the value is $0.1575I$.

Compared to the perfect wave analysed in Example 7-3, we now have small even-order harmonics, but identical odd-order harmonics.

When $n = 1800$ (90 kHz), the value of $\sin(n90.05°) = 1$, giving significant levels of harmonics in the radio-frequency range.

Example 7-14

Analysis of the current into, and voltage at, the supply terminals to a controlled rectifier load gave the following phase values:

$i = 100 \sin(\omega t - 0.53) + 20 \sin(5\omega t + 0.49) + 14 \sin(7\omega t - 0.57)$
$\quad + 8 \sin(11\omega t + 0.45) + 6 \sin(13\omega t - 0.61) + 3 \sin(17\omega t + 0.41)$
$\quad + \sin(19\omega t - 0.65)\,\text{A};$

$v = 340 \sin \omega t + 10 \sin(5\omega t - 1.16) + 15 \sin(7\omega t - 2.32) + 30 \sin(11\omega t - 1.47)$
$\quad + 20 \sin(13\omega t + 1.31) + 8 \sin(17\omega t + 2.16) + 4 \sin(19\omega t + 1.01)\,\text{V}.$

Determine the power associated with each harmonic, its direction of flow, the net power delivered to the load, and the power factor of the current.

SOLUTION From Eq. (7-14) the power (P) associated with each harmonic is $V_{n(\text{rms})}I_{n(\text{rms})} \cos \phi_n$. If ϕ_n is less than $\pi/2$, power flow is into the load, if greater than $\pi/2$, the power flow is reversed.

For the fundamental component, $P_1 = \dfrac{100 \times 340}{2} \cos(0.53) = 14\,668\,\text{W}$; for the

fifth harmonic component, $P_5 = \dfrac{20 \times 10}{2} \cos(1.16 + 0.49) = -7.91\,\text{W}$; for the

seventh harmonic component, $P_7 = \dfrac{14 \times 15}{2} \cos(2.32 - 0.57) = -18.72\,\text{W}.$

Values for the other components are: 11th, $-41.06\,\text{W}$; 13th, $-20.53\,\text{W}$; 17th, $-2.14\,\text{W}$; 19th, $-0.18\,\text{W}$.

Note that only power at the fundamental frequency is being delivered to the load, acting in part as a harmonic generator feeding power back into the supply system at the harmonic frequencies.

The net power to the load is the sum of the above, giving

$$14\,668 - 7.91 - 18.72 - 41.06 - 20.53 - 2.14 - 0.18 = 14\,577\,\text{W}$$

Using Eq. (7-12),

$$V_{\text{rms}} = \left[\frac{340^2 + 10^2 + 15^2 + 30^2 + 20^2 + 8^2 + 4^2}{2}\right]^{1/2} = 242.2\,\text{V}$$

Using Eq. (7-13),

$$I_{\text{rms}} = \left[\frac{100^2 + 20^2 + 14^2 + 8^2 + 6^2 + 3^2 + 1^2}{2}\right]^{1/2} = 73.2\,\text{A}$$

$$\text{From Eq. (3-13), power factor} = \frac{14\,577}{242.2 \times 73.2} = 0.823$$

Example 7-15

A single-phase 50 Hz diode bridge rectifier is fed at 70.71 V. The load is 200 Ω, shunted by a capacitance of 400 μF. Neglecting all losses, determine the load voltage, its ripple factor, and the input current.

SOLUTION The waveform shapes are shown in Fig. 7-4c. While the diodes are conducting, the supply voltage is seen by the load, during which period the capacitor current $i_C = C\, dv/dt$. The load resistor current $i_R = v/R$.
When the diodes are non-conducting, the load current is given by the exponential equation $i_R = I\, e^{-t/RC}$.
Taking $t = 0$ at the supply voltage zero, when the diodes conduct,
$v = 70.71\sqrt{2}\ \sin 2\pi 50t$, therefore $i_C = 100 \times 2\pi 50 \times 400 \times 10^{-6}\cos 2\pi 50t = 12.57 \cos 2\pi 50t$, and $i_R = (100 \sin 2\pi 50t)/200 = 0.5 \sin 2\pi 50t$.
The diodes cease conduction when $i_R = -i_C$, that is, at $2\pi 50t = 1.611\ (= 92.28°)$, when $i_R = 0.499$ A. Taking $t = 0$ from this time $v = 200 \times 0.499\ e^{-t/0.08}$ V. The angle at which the diodes next conduct is where the decaying voltage meets the rising sinewave in the next half cycle. The angle can be calculated from $99.81\ e^{-12.5t} = 100 \sin(2\pi 50t - \pi + 1.611)$, from which $t = 0.008\,425$ s, an angle of 1.116 rad ($63,93°$) on the incoming sinewave. The load voltage varies between a maximum value of 100 V down to $100 \sin 63.93° = 89.83$ V.

$$\text{Mean load voltage} = \frac{1}{1/(50 \times 2)}\left[\int_{1.116/\omega}^{1.611/\omega} 100 \sin \omega t\ dt\right.$$

$$\left. + \int_{0}^{0.008\,425} 99.81\ e^{-12.5t}\ dt\right]$$

$$= 95.07\ \text{V}$$

$$V^2_{\text{rms}} = \frac{1}{1/(50 \times 2)}\left[\int_{1.116/\omega}^{1.611/\omega} (100 \sin \omega t)^2\ dt\right.$$

$$\left. + \int_{0}^{0.008\,425} (99.81\ e^{-12.5t})^2\ dt\right]$$

$$= 9047.4$$

Using Eq. (7-28), ripple factor $= \dfrac{(9047.4 - 95.07^2)^{1/2}}{95.07} = 0.0313$.

The supply current $= i_R + i_C$ during the diode on-period which is between $63.93°$ (1.116 rad) and $92.28°$ (1.611 rad) on the input voltage sinewave. Hence putting $t = 0$ at the start of the input current, its equation is $12.57 \cos(2\pi 50t + 1.116) + 0.5 \sin(2\pi 50t + 1.116)$ A, having a maximum value at $t = 0$ of 5.97 A, falling to zero at $t = 0.001\,575$ s, that is, after $28.35°$, and then remaining zero for $151.65°$.

Example 7-16

Given the same load values of mean voltage and ripple factor as in Example 7-15, design an LC filter for this purpose, and determine the a.c. supply requirements for the circuit. Neglect all losses.

SOLUTION The waveforms will be as shown in Fig. 7-4d. Taking continuous supply current, the mean voltage will be that given by a 2-pulse output, hence from Eq. (2-7) ($\alpha = 0$), $V_{mean} = 95.07 = 2V_{max}/\pi$, which gives $V_{max} = 149.3$ V, that is, a supply r.m.s. voltage of 105.6 V.

From Eq. (7-19), the amplitude of each harmonic component in the rectifier output can be determined as $2/(n^2 - 1)$ of the mean value, with $n = 2(100\,\text{Hz})$, $4(200\,\text{Hz})$, etc. The peak harmonic-voltage values are therefore: 100 Hz, 63.38 V; 200 Hz, 12.67 V; 300 Hz, 5.43 V; etc.

The major harmonic voltage appearing across the load will be the lowest, that is, 100 Hz, hence from Eq. (7-28) assume only the 100 Hz component exists. The ripple factor $= 0.0313 = V_{2(rms)}/V_{mean}$, giving $V_{2(rms)} = 2.98$ V at the load.

Referring to the circuit of Fig. 7-4d, we have a 100 Hz voltage of $63.38/\sqrt{2}$, into the LCR combination with 2.98 V across R. Using normal circuit analysis and taking $R = 200\,\Omega$, then $I_R = 2.98/200$, $I_C = 2.98/-jX_C$, $I_L = I_R + I_C$, $V_{in} = 2.98 + I_L(jX_L) = 2.98 + [0.0149 + j(2.98/X_C)]jX_L$.

To ensure continuous supply current, the ripple current must be less than the mean current of 95.07/200, giving an r.m.s. 100 Hz current of less than $95.07/(200\sqrt{2}) = 0.34$ A, (say) a value of 0.2 A. As most of the input voltage must appear across X_L, then $X_L = (63.38/\sqrt{2}) \times (1/0.2) = 224\,\Omega$, giving $L = 224/(2\pi \times 100) = 0.36$ H. From the above equations,

$$\left|\frac{63.38}{2}\right| = \left|2.98 - \frac{2.98X_L}{X_C} + j0.0149X_L\right|$$

using $L = 0.36$ H, then $X_C = 14.14\,\Omega$, hence $C = 10.17/2\pi100 = 112.55\,\mu$F.

The r.m.s. value of $I_L = (I_R^2 + I_C^2)^{1/2} = 0.211$ A, which is a peak value of 0.299 A. With a mean current of $95.07/200 = 0.475$ A, the input current level will vary between $(0.475 + 0.299) = 0.774$ A and $(0.475 - 0.299) = 0.176$ A.

With the above arrangement, the 200 Hz voltage at the load is given by $V_L = [Z/(Z + jX_L)]V_{in}$, where Z is the impedance of the parallel combination of R and C and V_{in} is the harmonic voltage component in the rectifier output. At 200 Hz, $V_{in} = 12.68/2V_L$, giving the load component as 0.142 V. At 300 Hz, $V_{in} = 5.43/2V_L$, giving the load component as 0.03 V.

An accurate calculation of the ripple factor from Eq. (7-28), using the 100, 200 and 300 Hz components, gives a value of 0.0314, hence it is accurate enough to consider only the lowest-order harmonic component.

To reduce the input current ripple, a higher value of L could be used. If (say) $L = 0.72$ H, then C would be $55.8\,\mu$F, with the maximum value of the current ripple 0.145 A.

Example 7-17

Derive a general expression for the fundamental and harmonic content of a quasi-square-wave output inverter. Plot curves showing the harmonic variation against width variation of the wave on-period.

SOLUTION The quasi-square-wave is that shown in Figs 5-20c and 5-21. Being symmetrical, there will be no even harmonics. If the wave is analysed by taking a reference at the centre of the on-period as shown in Fig. 7-15, with an on-period of θ, then only cosine terms are present, and

$$b_n = \frac{2}{\pi} \int_{-\theta/2}^{+\theta/2} E \cos nx \, dx = \frac{4E}{n\pi} \sin \frac{n\theta}{2}$$

giving the nth r.m.s. voltage as $\dfrac{2\sqrt{2}E}{n\pi} \sin \dfrac{n\theta}{2}$. The values including the fundamental $(n-1)$ are shown in Fig. 7-15.

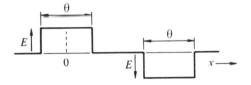

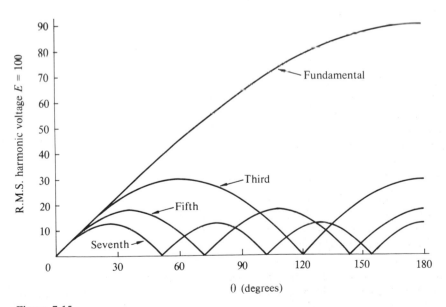

Figure 7-15

Example 7-18

Determine the low-order harmonic content of the selected harmonic reduction waveform shown in Fig. 5-27, if the angles at which the voltage reversal occurs are at 23.6° and 33.3°.

SOLUTION The waveform has quarter-wave symmetry and contains only sine terms, so taking the voltage level as E,

$$a_n = \frac{4}{\pi} \left[\int_0^{23.6°} E \sin nx \, dx + \int_{23.6°}^{33.3°} - E \sin nx \, dx + \int_{33.3°}^{90°} E \sin nx \, dx \right]$$

$$= \frac{4E}{n\pi} [1 - 2 \cos n23.6° + 2 \cos n33.3°]$$

giving a fundamental peak value of $1.068E$
   a third harmonic ($n = 3$) peak value of zero
   a fifth harmonic ($n = 5$) peak value of zero
   a seventh harmonic ($n = 7$) peak value of $0.317E$

Example 7-19

The output of a 50 Hz inverter is pulse-width-modulated by a triangular wave of frequency 500 Hz. Calculate the harmonic content of the wave when the reference wave is set to (i) its maximum value, and (ii) one half its maximum value.

SOLUTION The waveform for (ii) is identical to that shown in Fig. 5-45 with Example 5-14. For (i) the peak of the reference sinewave coincides with the peak of the triangular wave.
Plotting the waveforms to find the angles of switching, they are for (i) 27.6°, 49.7°, 56.9°, 90°; and for (ii) 31.3°, 42.0°, 63.9°, 80.9°.
Noting that the waveform contains only sine terms and odd harmonics, then taking the on-value to be E we get for (i),

$$a_n = \frac{4}{\pi} \left[\int_{27.6°}^{49.7°} E \sin nx \, dx + \int_{56.9°}^{90°} E \sin nx \, dx \right]$$

giving $a_1 = 1.0E$, $a_3 = 0.002E$, $a_5 = 0.033E$, $a_7 = 0.212E$, $a_9 = 0.182E$.
Similar calculations for (ii) give $a_1 = 0.5E$, $a_3 = 0.00E$, $a_5 = 0.001E$, $a_7 = 0.044E$, $a_9 = 0.361E$.
These calculations show that pulse-width modulation almost eliminates the low-order harmonics.

Example 7-20

The output of an inverter is pulse-width-modulated by a triangular wave of 24 times the inverter frequency. Calculate the harmonic content of the inverter output when the reference is set to one half maximum.

SOLUTION The timing waves are similar to those shown in Fig. 5-24c. The angles of firing can be found by drawing or by calculation. The first firing angle x is the solution to the equation

$$0.5 \sin x = 2 - (24x/\pi)$$

Calculations of this type yield the switching angles in the first quarter cycle to be 0.246, 0.280, 0.493, 0.558, 0.741, 0.834, 0.992, 1.106, 1.247, 1.373, 1.505 radians.

Using the same method to that used in Example 7-19, we get values of:
$a_1 = 0.500E$, $a_3 = 0E$, $a_5 = 0E$, $a_7 = 0E$, $a_9 = 0E$, $a_{11} = 0E$, $a_{13} = 0E$, $a_{15} = 0E$, $a_{17} = 0E$, $a_{19} = 0.001E$, $a_{21} = 0.0128E$, $a_{23} = 0.361E$.

These calculations show that the more segments into which the wave is divided, the greater the attenuation of the low-order harmonics.

Example 7-21

The output of an inverter is pulse-width-modulated by source alternation as shown in Fig. 5-26, by a triangular wave 11 times the reference sinewave, and set to give a maximum output. Determine the harmonic content of the inverter output voltage.

SOLUTION The output voltage is exactly as shown in Fig. 5-26, the output having quarter-wave symmetry, hence there are only odd-order sinewave components. The switching angles are at $0°$, $14.34°$, $37.73°$, $43.46°$, $73.29°$, $73.96°$.

Taking the source voltage to be E, then

$$a_n = \frac{4}{\pi} \left[\int_0^{14.34°} -E \sin nx \, dx + \int_{14.34°}^{37.73°} E \sin nx \, dx + \int_{37.73°}^{43.46°} -E \sin nx \, dx \right.$$

$$\left. + \int_{43.46°}^{73.29°} E \sin nx \, dx + \int_{73.29°}^{73.96°} -E \sin nx \, dx + \int_{73.96°}^{90°} E \sin nx \, dx \right]$$

giving values of $a_1 - 1.00E$, $a_3 = 0E$, $a_5 - 0E$, $a_7 = 0.018E$, $a_9 - 0.319E$, $a_{11} = 0.601E$, $a_{13} = 0.318E$, $a_{15} = 0.020E$, $a_{17} = 0.034E$, $a_{19} = 0.212E$.

Example 7-22

Design and analyse a simple LC low-pass filter for the inverter of Example 7-21, given that the output is 240 V, 50 Hz, the eleventh harmonic component is not to exceed 4%, and the load is resistive at 16 A.

SOLUTION The filter is that shown in Fig. 7-5a.

The ratio of the output to the input voltage of the filter with a load resistance R is

$$\frac{1}{(1 - \omega^2 CL) + j\omega L/R}.$$

The resonant frequency $1/2\pi\sqrt{(LC)}$ must be well above 50 Hz, but not a multiple of 50, say 140 Hz, giving $LC = 1.292 \times 10^{-6}$. The value of the load resistance is

$240/16 = 15\,\Omega$. The eleventh harmonic frequency is 550 Hz, with its value from the analysis of Example 7-21 as $240 \times 0.601 = 144.24\,\text{V}$ into the filter, to be attenuated to $240 \times (4/100) = 9.6\,\text{V}$ out of the inverter.
Therefore

$$\frac{9.6}{144.24} = \left| \frac{1}{1 - [(2\pi \times 550)^2 \times 1.29 \times 10^{-6}] + [j2\pi \times 550L/15]} \right|$$

giving L a value of 0.018 H, with $C = 1.292 \times 10^{-6}/0.018 = 72\,\mu\text{F}$.
The output/input ratio of the loaded filter at the other frequencies is: 50 Hz, 1.052; 150 Hz, 0.877; 250 Hz, 0.346; 350 Hz, 0.170; 450 Hz, 0.101; 550 Hz, 0.067; 650 Hz, 0.047.
On no-load, the filter acts as a simple LC circuit with the output across the capacitor. The output/input ratios of the unloaded ($R = \infty$) filter are: 50 Hz, 1.146; 150 Hz, 6.773; 250 Hz, 0.457; 350 Hz, 0.191; 450 Hz, 0.107; 550 Hz, 0.069; 650 Hz, 0.049.

Example 7-23

A 50 Hz square-wave output inverter is filtered by the resonant-arm filter shown in Fig. 7-5b. Determine suitable values for the filter if the harmonic distortion factor is to be less than 5% on open circuit. The load impedance is a pure resistance of $120\,\Omega$.

SOLUTION From Example 7-17, the harmonic content of a square-wave ($\theta = 180°$) inverter is proportional to $1/n$. Equation (7-29) relates to this circuit. The major harmonic is the third ($n = 3$). If (say) the third-harmonic component is 4%, then

$$\frac{4}{100/3} = \frac{1}{1 - \dfrac{1}{A}\left(3 - \dfrac{1}{3}\right)^2}, \quad \text{giving } A = 0.7619$$

$$2\pi 50 = \frac{1}{\sqrt{(L_1 C_1)}} = \frac{1}{\sqrt{(L_2 C_2)}}, \quad \text{and } A = \frac{L_2}{L_1} = \frac{C_1}{C_2}$$

The value of ωL_1 must be somewhat less than the load impedance so as to avoid excessive load voltage changes when the load varies, say ωL_1 is 30% of the load impedance.
Therefore $2\pi 50 L_1 = 0.3 \times 120$, giving $L_1 = 0.115\,\text{H}$, with from the above equations, $L_2 = 0.087\,\text{H}$, $C_1 = 88.1\,\mu\text{F}$, and $C_2 = 116\,\mu\text{F}$.
Using the above figures, the open-circuit fifth-harmonic component compared to the fundamental is 0.7%. For the seventh harmonic, it is 0.2%.

EIGHT

D.C. MACHINE CONTROL

A major use of the thyristor and other power semiconductor devices is in the control of electrical machines. The d.c. motor lends itself to speed adjustment much more easily than does the a.c. motor, so initial development of speed control systems using semiconductor devices concentrated on the d.c. machine.

Aspects of machine drives which are discussed in this chapter are: speed adjustment, starting, motoring and generating, braking, reversing torque levels, and general drive dynamics considerations.

8-1 BASIC MACHINE EQUATIONS

The d.c. machine (Ref. 10) consists of the field and an armature, the field being excited by d.c. windings, which set up a magnetic flux Φ linking the armature as shown in Fig. 8-1a. The function of the commutator and brushes is to ensure that the current in the armature conductors under any pole is in a given direction irrespective of the speed of the armature. The commutator acts as a mechanical frequency-changer to change the alternating current in a given armature conductor to the direct current in the brush lead.

The circuit diagram of a d.c. machine is shown in Fig. 8-1b. When rotating, an internal voltage (or back e.m.f.) E is generated in the armature by virtue of its rotation in a magnetic flux. The terminal armature voltage V will differ from E by the internal voltage drop. The load torque will give rise to a current in the armature, the current level being a function of the torque and magnetic flux values.

The basic machine equations are:

$$\text{terminal voltage } V = E + I_a R_a \tag{8-1}$$

$$\text{back e.m.f. } E = k_1 N \Phi \tag{8-2}$$

$$\text{torque } T = k_2 I_a \Phi \tag{8-3}$$

$$\text{flux } \Phi = k_3 I_f \tag{8-4}$$

$$\text{gross mechanical power } = TN = EI_a \tag{8-5}$$

where I_a and I_f are the armature and field currents respectively, R_a is the armature resistance, N is the speed(rad/s), and k_1, k_2, k_3 are constants of proportionality.

Equation (8-4) is only true when the magnetic circuit is below saturation.

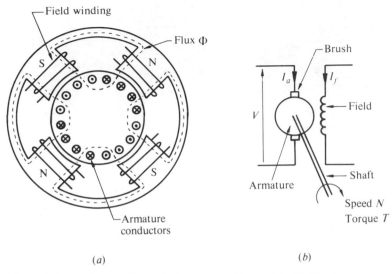

Figure 8-1 The d.c. machine. (*a*) Construction (four-pole). (*b*) Circuit representation.

Both the armature and field circuits have inductance which must be taken into account when the current is changing.

The current direction shown in Fig. 8-1*b* is for the motoring condition. As a generator, the torque will be reversed, current flow out of the positive terminal, Eq. (8-1) then being $V = E - I_a R_a$.

Figure 8-2 shows a separately excited d.c. motor in which the armature and field are supplied separately. As the torque is proportional to the product of armature current and flux, it is advisable to maintain the flux at its design level so as to minimize the armature current. For starting, R_2 is zero, and R_1 of such a value as to keep the armature current within safe limits. As the armature speeds up, the back e.m.f. rises from zero to a value proportional to the speed.

The resistors shown in Fig. 8-2 may be used to obtain speed adjustment. The

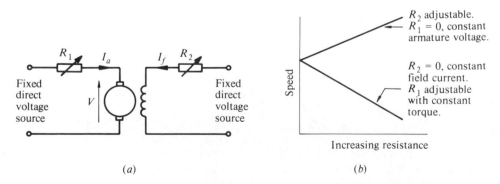

Figure 8-2 Speed adjustment by use of resistors. (*a*) Circuit. (*b*) Variation at fixed torque.

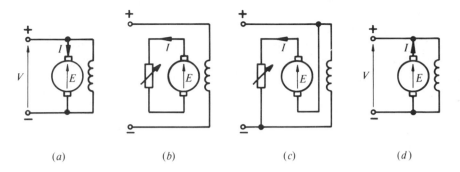

(a)   (b)   (c)   (d)

Figure 8-3 Braking modes for the d.c. machine. (*a*) Motoring $V > E$. (*b*) Resistive or dynamic. (*c*) Plugging. (*d*) Regenerative $V < E$.

presence of R_1 will cause a reduction in the armature voltage, and hence from Eq. (8-2) a reduction in speed. At fixed field current, the armature voltage is approximately proportional to the speed. A weakening of the field current by the inclusion of R_2 will reduce the flux, and hence from Eq. (8-2) increase the speed. Field weakening is limited in use as an increase in armature current occurs in order to maintain the load torque.

Electrical braking can be achieved in one of three ways illustrated in Fig. 8-3. Take the normal motoring condition in Fig. 8-3*a* as reference; remove the armature from the supply and place a resistor across the armature, so converting the machine to a generator as shown in Fig. 8-3*b*. *Plugging* is the term given to the technique of reversing the armature connections to the supply via a resistor as shown in Fig. 8-3*c*, the machine finally motoring in the reverse direction. If the terminal voltage is less than the internally generated voltage, then the current flow will reverse, and the machine generates as illustrated in Fig. 8-3*d*.

Instead of using the separate or shunt connection described earlier, it is possible to connect the armature and field in series as shown in Fig. 8-4. The characteristic of the series motor is one which is desirable for use in traction and high-speed drives.

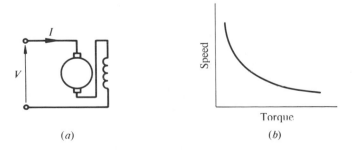

(*a*)   (*b*)

Figure 8-4 Series motor. (*a*) Connection. (*b*) Characteristic.

8-2 VARIABLE-SPEED DRIVES

In the previous section it was established that the speed of a d.c. motor could be adjusted by variation of the armature voltage. Figure 8-5*a* shows one layout where a diode rectifier can be used in conjunction with a voltage regulator. The voltage regulator may be a variable-ratio transformer or an induction regulator. Control of the voltage regulator controls the armature voltage and hence speed. Usually the field is supplied at its rated voltage via a diode rectifier, so that the flux is always at its optimum level.

The layout shown in Fig. 8-5*b* is the more usual, where a controlled rectifier is used to supply the motor armature. Any of the various rectifier configurations described in Chapter 2 can be used. If the armature current is continuous, then the voltage waveform at the motor will be identical to those developed in Chapter 2. Typically, a motor rated above 2 kW will have sufficient inductance to maintain continuous current. The speed of the motor is determined by its mean armature voltage, any oscillating torque produced by the harmonic voltage (current) components being heavily damped by the motor inertia. Hence, the motor speed is dependent on the firing delay angle of the rectifier. If required, the field can be supplied via a controlled rectifier.

Where the machine is required to generate, (say) if rapid braking is required, then the converter must be capable of operation in the inverting mode. Inversion was fully described in Sec. 3-3, where it was explained that the fully-controlled converter configurations are required.

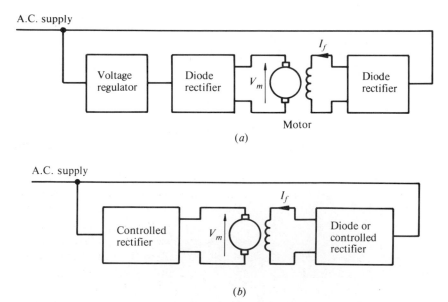

Figure 8-5 Basic variable-speed drives. (*a*) Voltage adjustment by regulator with diode rectifier. (*b*) Voltage adjustment by controlled rectifier.

Comparing the two systems shown in Fig. 8-5, that using diodes only is less flexible, has a longer response time to any call for speed change, but is advantageous regarding the power factor of the current drawn from the supply. The controlled rectifier is more flexible, has fast response time to any change, and can invert, but draws its supply current at a lagging power factor, and has a higher harmonic content in its armature voltage.

Traditionally, to start d.c. motors, resistors have been used in the armature circuit, which cut out as the motor gains speed. The object of the starting resistors is to keep the armature current within a safe level by matching the armature voltage to the internally generated voltage of the motor. With the controlled rectifier, the starting resistor is not required, as at starting the firing delay angle can be set to such a value that the armature voltage is only sufficient to overcome the armature resistance volt-drop. As the motor gains speed, the firing of the thyristors is advanced until the required speed is obtained.

Where the armature current is continuous, the waveforms are identical to those developed in the earlier chapters for inductive loads. However, with small motors supplied from single-phase sources, the armature current can be discontinuous, and here the armature voltage waveform differs from that described earlier.

Typically, a small motor will be supplied from a half-controlled single-phase bridge circuit as shown in Fig. 8-6a. It may be assumed that the inertia of the motor is large enough for the speed to remain constant over each cycle of the supply. With the field current at its rated value, the motor will have an internally generated voltage (back e.m.f.) E.

Referring to the waveforms in Fig. 8-6b and taking a firing angle as shown, once the thyristor is turned on, the motor voltage v_m is that of the supply, and the current rises at such a rate as to satisfy the parameters of the equivalent circuit shown in Fig. 8-6c.

Hence
$$v_m = L\frac{di_m}{dt} + Ri_m + E \qquad (8\text{-}6)$$

The instantaneous current will rise until the supply voltage falls to approximately the level of the back e.m.f. E, after which time it will decay. At the supply voltage zero, the commutating diode will take the armature current, until such time as the current has decayed to zero.

The motor armature voltage waveform has three sections: the supply voltage after a thyristor is fired, zero while the commutating diode conducts, and the motor internal voltage during the zero current period, as illustrated in Fig. 8-6b. In practice, the waveform is likely to contain oscillations at the time the diode turns off, due to protective capacitors and the tooth ripple voltage arising from the armature construction. In the absence of motor current, the firing of the thyristors can only take place after the time the instantaneous value of the supply voltage exceeds the motor internal voltage, as the thyristor will be reverse-biased up to this time.

The mean armature current is proportional to the mean motor torque, the harmonics only producing harmonic torques. The mean speed is proportional to the

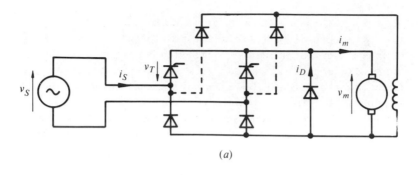

(a)

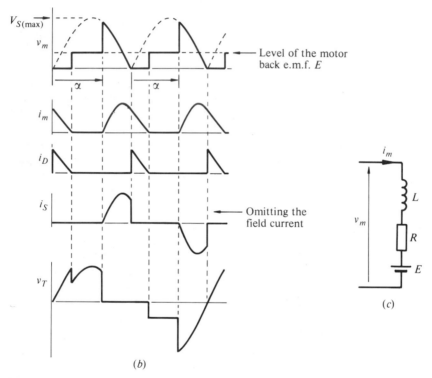

(b)

(c)

Figure 8-6 Small motor with discontinuous current. (*a*) Half-controlled single-phase circuit. (*b*) Waveforms. (*c*) Motor armature equivalent circuit.

mean armature voltage less the mean armature volt-drop. The mechanical power developed is EI_a, that is, the product of the mean current and internal voltage. Any harmonic power input to the motor does not emerge as mechanical power, but only contributes additional losses in the conductors and iron of the motor.

Because of the harmonic losses, a given motor rating will be lower if supplied from a thyristor drive than when compared to a supply from a level direct-voltage source. However, the current drawn by large motors is continuous and, with the

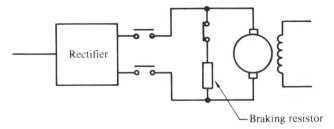

Figure 8-7 Dynamic braking.

higher inductance, the harmonic content of the current is small, hence resulting in much lower harmonic losses.

Where rapid slowdown is desired, a fully-controlled converter can be put in the inverting mode. Where regenerative braking is not possible, dynamic braking can be used. As illustrated in Fig. 8-7, when the stop button (not shown) is operated, a contactor places a resistor across the armature, while at the same time disconnecting the armature from the rectifier, either mechanically or electronically. Using a fixed resistor results in a linear drop in braking torque with speed, because the machine generates a voltage proportional to speed.

Various ways exist for reversing d.c. motor drives, the three major methods being shown in Fig. 8-8. If the field current is maintained in the same direction, then the motor can only be reversed by reversing the polarity of the voltage applied to the armature. Alternatively, the armature voltage can be maintained in polarity, and the field current reversed.

With a contactor as shown in Fig. 8-8*a*, a call to reverse the drive demands that the armature current is reduced to zero, and the contactor changed over; the current is then brought up, so reversing the torque and taking the motor into reverse rotation. The sequence of the events during reversal is as follows. Take the motor to be rotating in the forward direction, with the converter in the rectifying mode. The controls demand motor reversal. The firing angle is retarded, so dropping the current to zero, at which time the contactor is reversed. Referring to Fig. 8-9 we have passed from *a* to *b*, the drive is still rotating forwards, so the converter d.c. terminal voltage is reversed, hence the converter is in the inverting mode and regenerative braking is taking place. As the speed falls, the firing angle is advanced until at zero speed the converter goes into rectification, the drive then motoring in the reverse direction as shown in Fig. 8-9*c*.

During reversal, the firing angle is controlled so as to maintain the maximum rated current, thus ensuring reversal in the minimum time. The next section describes how this is arranged with the aid of feedback loops.

With contactor reversing, the drive may typically be reversed in 0.8 s, 0.2 s being spent in operating the contactor. This contactor time can be eliminated by using the double (anti-parallel) converter arrangement of Fig. 8-8*b*. Here immediately converter *A* has its current reduced to zero, converter *B* can commence conduction. The inductors shown will limit circulating current between groups if both

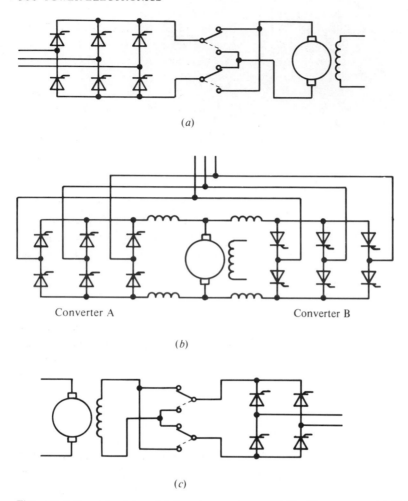

(a)

(b)

Converter A   Converter B

(c)

Figure 8-8 Reversing drives. (a) Armature contactor. (b) Double converter. (c) Field contactor.

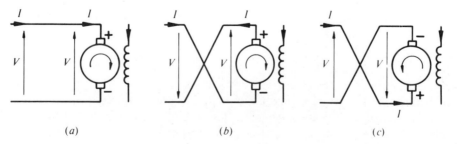

(a)   (b)   (c)

Figure 8-9 Modes during reversing. (a) Forward rotation, motoring, rectifying. (b) Forward rotation, generating, inverting. (c) Reverse rotation motoring, rectifying.

groups are fired simultaneously, but normally the firing of one group is inhibited while the other is conducting, so dispensing with the need for these inductors. As explained earlier, the incoming converter will invert during the period of regenerative braking, then rectify when the direction of rotation is reversed.

Reversal can be achieved by reversing the field current by contactor as shown in Fig. 8-8c, or alternatively with an additional converter in the anti-parallel connection. The high inductance of the field means that the converter must invert to remove the stored magnetic energy, thus reducing the current to zero; then the contactor is reversed, so that current can be built up in the opposite sense. This change is relatively slow and, until the field current is fully reversed, the torque is limited, due to the below rated flux conditions.

The power factor of the current drawn from the supply is directly related to speed in the case of the armature-fed d.c. motor drive. At (say) half full speed, the mean voltage will be halved, hence the firing delay angle will be 60°, the power factor of the fundamental component of the current being $\cos 60° = 0.5$.

8-3 CONTROL FEEDBACK LOOPS

It is clear from the earlier section that the speed of a motor can be controlled by the level of the mean output voltage of a rectifier. The rectifier can be considered as a power amplifier as shown in Fig. 8-10. The output voltage is determined by the angle of firing delay, this angle being set by (say) the level of input voltage to the firing circuit. The power associated with the input signal may only be milliwatts, whereas the rectifier output can be in the megawatt range.

The open-loop system of Fig. 8-10 can be converted to the closed-loop speed control system of Fig. 8-11 by the inclusion of a feedback signal and an amplifier (Ref. 11). The drive speed can be measured by a tacho-generator (permanent-magnet d.c. generator), whose voltage is proportional to speed. The difference between the input setting and the feedback level gives the error signal. Suppose the system is such that an error signal of 0.1 V gives a rectifier output of 200 V, the motor internal voltage is 200 V at 1000 rev/min, and the tacho-generator output is 10 V per 1000 rev/min. Using these figures, an input setting of 10.1 V gives a no-load speed of 1000 rev/min. If due to load torque the speed fell to 990 rev/min, then the error would be $(10.1 - 9.9) = 0.2$ V, the motor voltage rising to 400 V with very large motor current and acceleration of the drive back to near 1000 rev/min.

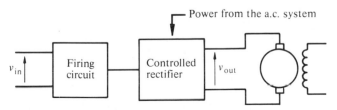

Figure 8-10 Rectifier as a power amplifier.

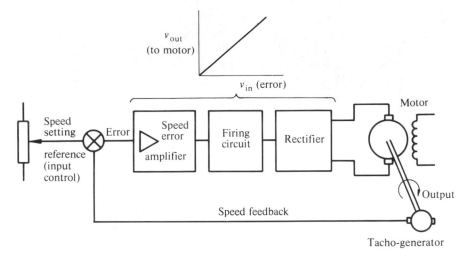

Figure 8-11 Closed-loop speed control.

Two observations on the above simple speed control system are that an error must alway exist, and (more importantly for thyristor drives) any change in motor speed will immediately give rise to excessive motor and thyristor currents which cannot be tolerated, hence means must be employed to limit the current level.

Automatic current limiting can be achieved by using a second feedback loop as shown in Fig. 8-12, The d.c. motor current can be measured by a d.c. current transformer (described in Sec. 6-8), or alternatively by current transformers in the a.c. input lines to the rectifier, connected as shown in Fig. 8-13.

The current limiting system of Fig. 8-12 includes two amplifiers, the first one

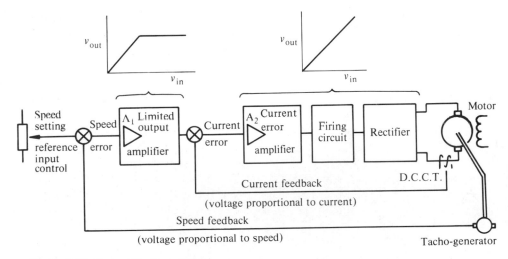

Figure 8-12 Current limiting system.

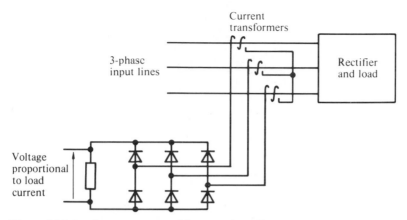

Figure 8-13 Load current measured from a.c. input lines.

A_1 having a limited output. To have an output from the rectifier there must be an input to the amplifier A_2, which is the difference between the output of A_1 and the current feedback signal. Hence the current feedback signal, that is, the load current, is limited to the maximum output of amplifier A_1, any tendency for the load current to exceed this value resulting in reduction of the current error signal, a reduction in the rectifier output, and consequently a reduction in the load current.

A clearer understanding of the manner in which the current limiting system operates can be gained by considering the starting situation. In Fig. 8-12 let the gain of A_1 be 100, limited to 10 V output, so that a current error signal of 0.1 V gives a rectifier output of 200 V; the current feedback signal is 10 V for 50 A load current, the speed feedback signal being 10 V per 1000 rev/min. If the input is set to 15 V, with initially zero speed feedback, the output of A_1 is limited to 10 V, and the load current will build up to near 50 A. Suppose that at a stationary condition and 50 A the motor armature voltage is 20 V (that is, an armature resistance of $0.4\,\Omega$), then the current error will be 0.01 V, with the current feedback signal equal to $(10 - 0.01) = 9.99$ V, that is, the load current is 49.95 A. For as long as the output of A_1 is limited, that is, the speed error is greater than $10/100 = 0.1$ V, the current will remain approximately 50 A throughout the acceleration period.

Taking the above data let us examine the conditions at a steady-load condition where the motor voltage is 300 V, and the motor current 20 A. For this condition, the current error signal is 0.15 V, the current feedback signal 4 V, and the output of A_1 4.15 V. The input to A_1 will be 0.0415 V, hence the tacho-generator feedback signal is $(15 - 0.0415) = 14.9585$ V, that is, a speed of 1495.85 rev/min. Ideally, the 15 V input setting should give an output of 1500 rev/min so there is a speed error of -0.277%. An increase in load torque would increase the speed error, probably bring the current limit into operation and increase the motor current to 50 A, so bringing the motor back near to its original speed.

In the current limit system, the current feedback senses a high current, and hence reduces the input voltage to the firing circuit, so retarding the firing angle in

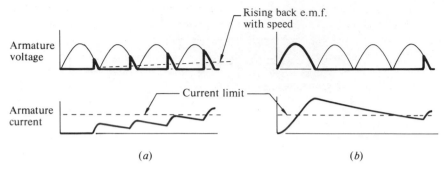

Figure 8-14 Initial armature current at start. (*a*) Controlled acceleration by a soft-start control. (*b*) Consequence of no soft-start control.

the next cycle. If, however, a large input change is made, and in the first period this change is permitted to fully advance the firing angle, a high current can result. Figure 8-14*a* illustrates the desirable starting conditions, with firing in the first period retarded so as to prevent excessive current build-up. Once the current reaches the set limit, the current feedback loop automatically ensures the correct firing angle is selected. Figure 8-14*b* shows the consequence of an initial full period, the excessive current giving a torque impulse to the motor shaft with possible mechanical damage.

To ensure that a sudden input change does not get transmitted directly to the error amplifiers, the circuit of Fig. 8-15 can be used, where a step change in the input is converted to a ramp, thus ensuring a gradual change in the firing angle.

An alternative to the tacho-generator feedback signal for speed control is:

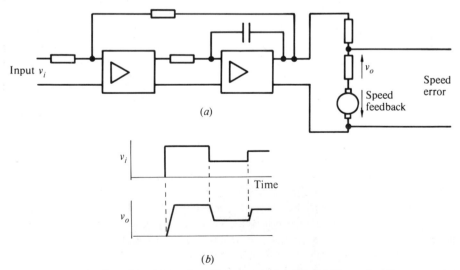

Figure 8-15 Softening the effect of sudden input changes. (*a*) Ramp function generator. (*b*) Relation of output to input voltage.

measure the armature voltage, measure the *IR* drop across a small resistor *R* in the armature line, subtract *kIR* from the armature voltage, so giving the internal generated back e.m.f. *E*. The value of *k* would be such as to make *kR* equal to the armature resistance. At fixed field current, *E* is directly proportional to speed. This is a cheaper method of closed-loop speed control and, although not so accurate, avoids having to couple a tacho-generator to the shaft.

Although speed has been given as the controlling feature in the application of closed-loop control, it is equally possible to convert the drive to one of constant torque, or constant power, by feeding back signals related to these quantities.

As the systems become more complex, the control features multiply. An example is the anti-parallel connection of two converters for reversing drives shown in Fig. 8-16, where, in addition to current and speed feedback, each firing circuit must be inhibited when the other converter is conducting.

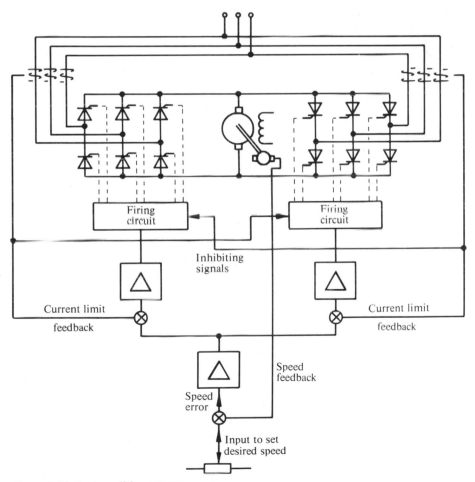

Figure 8-16 Anti-parallel converters.

It is evident from all the foregoing that the requirements of the control and firing circuits can be many and complex. Mention was made in Sec. 1-5-3, illustrated by Fig. 1-20, of the likely signals fed into the control circuits. It is not within the scope of this text to detail the electronic control systems, particularly as the requirements can be met in different ways. The introduction of digital electronics makes many of the earlier designs obsolete, and will enable superior solutions to be found to problems such as the effect of distortion in the supply voltage, or the accuracy of the firing pulse angles from one cycle to the next.

8-4 TRACTION DRIVES

Traditionally, the d.c. series motor has been used for traction drives; it has an ideal torque-speed characteristic of high torque giving fast acceleration at low speeds, coupled with high-speed motoring for cruising.

Except for a few low-frequency a.c. systems, most of the earlier electrified railway systems were fed by direct voltage, either via an overhead catenary, or by a third rail. For the high-density short-distance urban railway systems, d.c. has been retained. However, some long-distance inter-city railway electrification has in recent years been with single-phase alternating voltage fed from the public supply system at 50 Hz or 60 Hz, with voltage levels at the catenary of 25 kV.

In earlier and some present locomotives of the high-voltage a.c. railway electrification, the d.c. series traction motors were fed via a diode circuit, supplied from a step-down transformer, the primary of which is connected to the catenary pick-up. Low-speed control is by tappings on the transformer, so giving voltage control to the motor. At high speed, with full voltage applied, control is by field weakening on the motors.

An important consideration with the railway system is the harmonic current levels in the overhead catenary. The close and parallel proximity of the signalling cables to the railway lines does mean that careful screening of these cables has to be arranged so as to avoid interference. The single-phase nature of the supply results in the full range of harmonics being present, including all the triple-frequency components.

By the use of thyristors to control the traction motors in a railway locomotive, tappings on the transformer can be avoided, and a more accurate control of the motors is possible to utilize maximum wheel-to-rail adhesion without slipping. However, two prime considerations with thyristors are to avoid low power factor, that is, keep the catenary current to a minimum, and also to avoid excessive harmonics.

The accurate control and fast response times of the thyristor systems enable use to be made of separately excited motors rather than the series motor of the diode system. Figure 8-17 illustrates the basic layout of a typical locomotive arrangement. Four drive motors are shown, each being on a different axle. The

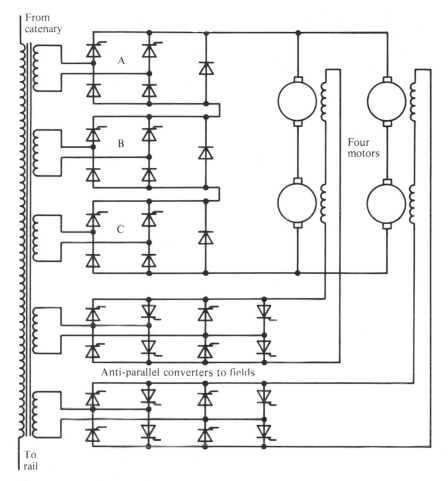

Figure 8-17 Basic layout of a thyristor locomotive arrangement.

armatures are connected in a series-parallel arrangement fed from a unidirectional rectifier, such that control is possible from zero to full voltage. The fields are supplied via fully-controlled circuits, so that field weakening can be used to obtain high speeds, and also to permit equal load sharing between the parallel groups. The anti-parallel field converters allow for easy drive reversal.

The armatures of the motors in Fig. 8-17 are fed from three series-connected half-controlled single-phase bridge rectifiers. For starting, firstly only bridge A is fired, firing being advanced as the speed and hence armature voltage builds up. When bridge A is fully conducting ($\alpha = 0°$), then bridge B is brought into conduction, until finally bridge C is fired, and its firing advanced until maximum armature voltage is reached. During the starting period, the field currents are set to a maximum. The object of the three bridges is to give a better power factor than would be possible with a single bridge; Example 8-14 illustrates this comparison.

Alternatively to the three half-controlled bridges of Fig. 8-17, two could be uncontrolled diode bridges brought into circuit via contactors in the secondary lines.

If regeneration were required, then the bridges would have to be fully-controlled, but normally dynamic braking is used for train retardation.

Diesel-electric locomotives use a system of the prime-mover being a diesel engine coupled to an electric generator, which in turn is coupled to the d.c. traction motors. If a d.c. generator is used, then clearly rectifying devices are not required, but in those locomotives where an alternator is used, the output has to be rectified. Voltage control to the motors is by field control on the generator, and the basic circuit of alternator to diode rectifier to the d.c. motors is a straightforward arrangement.

In both the a.c. and d.c. electrified railway systems, it is possible to generate on the locomotives a variable-frequency polyphase alternating voltage by the use of inverters or cycloconverters as described in Chapter 5. The traction motors would now be cage induction motors, the principle here being that which is described in the next chapter. Systems employing linear induction motors which develop the tractive thrust directly on to a fixed rail would be fed via inverters.

D.C. railway traction systems are usually fed at 600 to 700 V for third-rail pick-up, or 1500 V for overhead catenary. In both cases, the direct current is obtained by rectification of the public a.c. supply, usually by 12-pulse uncontrolled rectifiers. Any regeneration taking place within a given locomotive is taken up by any other locomotives which are motoring on the system.

The electric road vehicle is a particular case of a d.c. motor fed via a thyristor chopper circuit from a d.c. source (battery). The basic chopper circuit was described in Sec. 4-1. The underlying principle of chopper control is similar for both railway and road systems, and two typical circuits are shown in Fig. 8-18.

The manner in which the chopper functions was fully described in relation to Fig. 4-3, but it is appropriate to review the circuits shown in Fig. 8-18. In both circuits, thyristor T_2 is fired, so charging the chopper capacitor C_1 to the input voltage. Firing thyristor T_1 connects the motor to the input, and simultaneously causes the charge on the capacitor to reverse via the inductor. In Fig. 8-18b, thyristor T_3 has to be fired during the conducting period of T_1, a thyristor being preferred to a diode to prevent unwanted discharge of the capacitor via the battery as a result of load transients. In Fig. 8-18a, alternate firing of thyristors T_1 and T_2 controls the mean motor voltage, the resistance R being of a very high value to shunt any leakage current through T_2 and D_1, hence preventing discharge of the capacitor. A filter must be included in the railway systems to prevent high currents in the conductor rails.

The frequency at which the chopper operates is limited at the higher end by the limitations and losses in the chopper itself. To avoid undue ripple in the motor current, and hence pulsating torques, the frequency must be above a certain level. For railway use, the chopper frequency must be such as not to interfere with communication and signalling channels. Typically, frequencies around 1 kHz would be used.

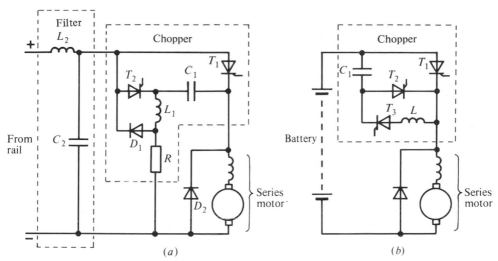

Figure 8-18 Chopper systems. (*a*) Typical for railway application. (*b*) Typical for road vehicle application.

Taking an ideal chopper as shown in Fig. 8-19, the motor voltage will be a series of on/off periods. Taking the on-period as t_1 and the off-period as t_2, the mean load voltage is

$$V_{\text{mean}} = V_S \frac{t_1}{t_1 + t_2} \qquad (8\text{-}7)$$

The motor current will rise and fall exponentially, pulses of current being taken from the supply. During the chopper off-periods, the inductive motor current will be diverted into the commutating diode. The equivalent circuit of the motor can be taken as in Fig. 8-19c, having inductance L, resistance R, and an internal back e.m.f. of E. In the series motor, E is dependent on the motor current, but may be considered approximately constant for small current changes.

The equation to the current during the on-period can be written from knowledge of transient response. The current starts at I_{min} (at $t - 0$), would if allowed to continue indefinitely reach $(V_S - E)/R$, and have an exponentially changing component starting at $\{[(V_S - E)/R] - I_{\text{min}}\}$. Neglecting all chopper losses, the equation to the motor current i_m in the on-period is

$$i_m = \frac{V_S - E}{R} - \left(\frac{V_S - E}{R} - I_{\text{min}}\right) e^{-(R/L)t} \qquad (8\text{-}8)$$

with at $t = t_1$, $i_m = I_{\text{max}}$.

By similar reasoning during the off-period, neglecting the diode loss,

$$i_m = -\frac{E}{R} + \left(\frac{E}{R} + I_{\text{max}}\right) e^{-(R/L)t} \qquad (8\text{-}9)$$

with at $t = t_2$, $i_m = I_{\text{min}}$.

IK — INN

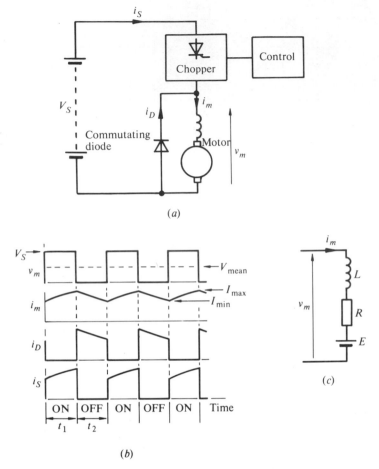

Figure 8-19 Chopper-fed motor. (*a*) Circuit. (*b*) Waveforms. (*c*) Approximate equivalent circuit.

The mean motor current will be given by

$$I_{\text{mean}} = \frac{V_{\text{mean}} - E}{R} \tag{8-10}$$

The chopper control would include current feedback to turn off the chopper in the event of overload. In normal steady-state conditions, the on-period would be determined by the speed demand. Starting would commence by the on-period being determined by the current limit, hence the on/off ratio would be low at the start of the on-period, automatically extending as the motor speed and its internal voltage build up.

In the railway system with (say) four traction motors, advantage can be taken to run two choppers as a two-phase system. The system is shown in Fig. 8-20*a*,

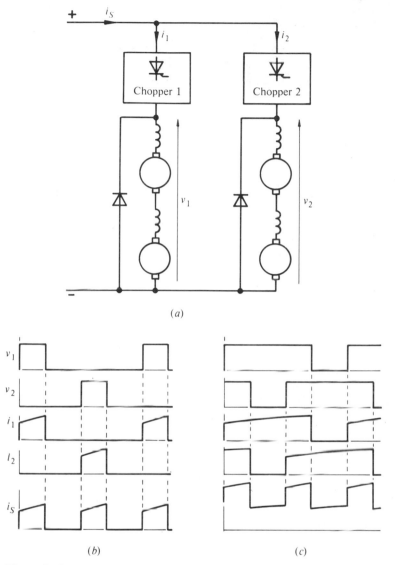

Figure 8-20 Two-phase chopper. (*a*) Basic circuit. (*b*) Low speed. (*c*) High speed.

where the timing of the start of the on-period of the two choppers is spaced as shown in the waveforms of *b* and *c*. It can be seen that the supply current's fundamental harmonic frequency is double that of either chopper, and at the higher output voltage (speed) the supply current is continuous.

Dynamic braking is possible with the chopper systems by changing the armature leads to a resistor and maintaining the field current. However, the need to preserve energy, particularly with the battery road vehicle, suggests regenerative braking with energy being returned to the d.c. source. Each of the chopper systems shown in Fig. 8-18 can be rearranged as shown in Fig. 8-21. Due to remanent flux, the direction of the current in the field must remain the same, whether motoring or generating, so as to maintain the machine's ability to self-excite. Hence, for generation, either the armature or the field connection must be reversed.

Referring to Fig. 8-21a, when thyristor T_1 is fired, current builds up in the field and armature via the short-circuit, with shaft power being converted to stored magnetic energy. Current feedback in the armature loop dictates the current level at which thyristor T_2 is fired to turn off thyristor T_1. Provided E is less than V_S, the only path for the current is now via diode D_2 and the battery, the collapsing current transferring magnetic energy from the motor inductance back into the supply. When the current has fallen to a low level, thyristor T_1 is again fired and the cycle repeated. At high speed, E may exceed V_S, in which case diode D_3 will conduct; however, the short-circuit now across the field will diminish i_f, and hence E will fall. Current feedback of both field and armature current will ensure optimum firing of thyristors T_1 and T_2.

The explanation of regeneration in the circuit of Fig. 8-21b is similar to that above; thyristor T_1 is fired, current builds up, transferring energy from the shaft into the magnetic field. When thyristor T_1 is turned off, the stored magnetic energy is transferred to the battery.

For the lower voltages associated with battery vehicles, such as fork-lift trucks, it may be more economical to use power transistors as the switching device rather than thyristors, avoiding the need for large and costly capacitors. The power transistor would be used in the switching mode, so the load waveforms would be identical to those described for the chopper.

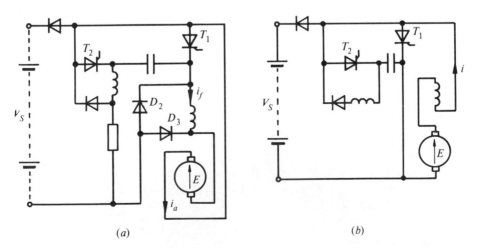

(a) (b)

Figure 8-21 Regenerative braking. (a) The circuit of Fig. 8-18a. (b) The circuit of Fig. 8-18b.

8-5 INDUSTRIAL APPLICATION CONSIDERATIONS

The role of the d.c. motor is to be found mostly in those applications where adjustable variable-speed drives are required, or where the particular characteristic of the series motor is required, such as in traction. Constant one-speed drives are best provided by a.c. motors, and it would be exceptional for a d.c. motor to be used in this application.

Although the material in this book clearly concentrates on the application of power semiconductor devices, it must be stressed that the older methods of control still command favourable consideration in many circumstances. If (say) speed control is required for very short periods only, the cost of the additional power loss of simple added armature circuit resistance will be less than the capitalized cost of the alternative power semiconductor system.

In some circumstances, a heavy load (such as a pump at a remote reservoir) will have to be driven by a motor control system which does not impose harmonics on the a.c. supply system. Excess voltage distortion would be induced in the inherently weak (high short-circuit impedance) supply to a remote area by thyristor drives, so precluding their use in such circumstances.

Traditionally, the four-quadrant drive, with speed and torque control in either direction, has been provided by Ward-Leonard systems. The absence of harmonics, and the ability of the Ward-Leonard system to buffer from the supply a sudden load transient, may make this the preferred drive to the four-quadrant drive illustrated in Fig. 8-8.

Drives requiring a set speed and/or a set torque have been discussed, but there is one class of drive which requires a constant power output from the motor. A reeling drive, that is, one in which a thread or wire is wound on to a drum, requires a constant force and linear speed condition for the wire, irrespective of the particular drum radius. The constant force and speed makes the requirement constant power. For the higher drum speeds when the winding radius is low, field weakening can be used on the motor, as the torque demand falls with increasing speed.

8-6 WORKED EXAMPLES

Example 8-1

A small separately excited d.c. motor is supplied via a half-controlled single-phase bridge rectifier. The supply is 240 V, the thyristors are fired at $110°$, and the armature current continues for $50°$ beyond the voltage zero. Determine the motor speed at a torque of 1.8 N m, given the motor torque characteristic is 1.0 N m/A and its armature resistance is 6 Ω. Neglect all rectifier losses.

SOLUTION The motor voltage waveform is as shown in Fig. 8-6b, with the supply voltage at the motor between $110°$ and $180°$, zero voltage for $50°$, and the motor back e.m.f. E from $50°$ to $110°$ (that is, a period of $60°$).

$$V_{mean} = \frac{1}{\pi} \left[\int_{110°}^{180°} 240\sqrt{2} \sin \theta \, d\theta + E\pi \frac{60}{180} \right] = 71.1 + 0.333E$$

Mean current $= 1.8/1.0 = 1.8$ A.

$E = V_{mean} - I_{mean}R = 71.1 + 0.333E - (1.8 \times 6)$; hence $E = 90.43$ V. From Eq. (8-5), $T/I = E/N$; therefore the voltage characteristic is 1.0 V/rad/s. Speed is $90.43/1.0 = 90.43$ rad/s $= 864$ rev/min.

Example 8-2

A small separately excited d.c. motor is fed from a 240 V, 50 Hz supply via a single-phase half-controlled bridge rectifier. Armature parameters are: inductance 0.06 H, resistance 6 Ω, torque (voltage) constant 0.9 N m/A (V/rad/s). If at a particular setting the firing delay angle is 80° and the back e.m.f. of the motor 150 V, determine an expression for the instantaneous armature current. Calculate the mean output torque and speed. Neglect thyristor and diode losses, and the supply impedance.

Recalculate the above quantities for back e.m.f.'s of 250 V and 90 V.

SOLUTION The current flow in the armature is in three sections. Referring to Fig. 8-22a, when the thyristor is fired, the supply is switched to the armature, giving a current i_1 starting from zero. At the voltage zero, the commutating diode conducts, that is, S_2 closes and S_1 opens, the current decaying as i_2 goes to zero. The third period is zero, assuming discontinuous current.

To find i_1, the total current is the sum of the steady-state a.c. component, the steady-state d.c. component, and an exponentially decaying component of such a value as to satisfy the initial condition of $i_1 = 0$ at $t = 0$.

A.C. impedance $= 6 + j2\pi 50 \times 0.06 = 19.78\underline{/1.263}\ \Omega$, d.c. impedance $= 6\ \Omega$; steady-state a.c. current $= (240\sqrt{2}/19.78) \sin (2\pi 50t + 1.396 - 1.263)$ A, steady-state d.c. current $= -150/6 = -25$ A. Therefore $i_1 = 17.16 \sin (2\pi 50t + 0.133) - 25 + Ae^{-t/T}$, where t is zero at the start of i_1 (i.e., 80° on the supply wave), the time constant $T = 0.06/6 = 0.01$ s. At $t = 0$, $i_1 = 0$, $\therefore A = 22.7$. Hence $i_1 = 17.16 \sin (2\pi 50t + 0.133) - 25 + 22.7e^{-100t}$ A. At voltage zero

$$t = \frac{180 - 80}{360} \times \frac{1}{50} = 0.005\ 55\ \text{s when } i_1 = 4.39\text{A}.$$

The current i_2 circulates in a closed circuit formed by R, L and E, starting at $t = 0$ with $i_2 = 4.39$ A. i_2 is a steady-state d.c. value of $-150/6$ plus a decaying exponential to satisfy the condition $i_2 = 4.39$ A at $t = 0$. Hence $i_2 = -25 + 29.39 e^{-100t}$ A, this current decaying to zero at $t = 0.001\ 618$ s, that is, at 29° after the voltage zero. The waveforms are shown in Fig. 8-22b.

The mean current can be found from the current expressions, or more readily by finding V_{mean} as in Example 8-1 and knowing $E = 150$ V, then $I_{mean}R = V_{mean} - E$, giving $I_{mean} = 3.22$ A. The mean torque $= 3.22 \times 0.9 = 2.89$ N m. The speed $= 150/0.9 = 166.7$ rad/s $= 1592$ rev/min.

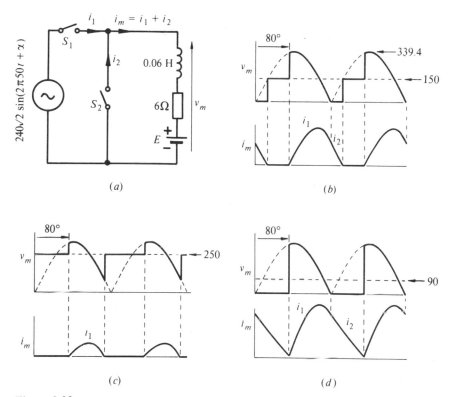

Figure 8-22

When the back e.m.f. $E = 250$ V, $i_1 = 17.16 \sin (2\pi 50t + 0.133) - 41.67 + 39.39$
e^{-100t} A. This expression will be zero when $t = 0.004\,74$ s, that is, $85.3°$ after the
thyristor is fired, so therefore the commutating diode does not conduct. The waveforms are shown in Fig. 8-22c. I_{mean} is 0.804 A, the torque 0.724 N m, and speed
2653 rev/min.

When the back e.m.f. $E = 90$ V, taking the current to be discontinuous and with
$t = 0$ at the start of each current component i_1 and i_2, we get $i_1 = 17.16 \sin$
$(2\pi 50t + 0.133) - 15 + 12.72 e^{-100t}$, and $i_2 = -15 + 23.65 e^{-100t}$ A. i_2 decays
to zero after 0.455 s, that is, $82°$ after the voltage zero, but the next thyristor is
fired at $80°$, so in fact the armature current is continuous with $i_1 = 0.263$ A at the
start of conduction from the supply. Hence the correct equations to the currents
are:

$$i_1 = 17.16 \sin (2\pi 50t + 0.133) - 15 + 12.99 e^{-100t} \text{ A}$$

and
$$i_2 = -15 + 23.80 e^{-100t} \text{ A}.$$

The waveforms are shown in Fig. 8-22d. V_{mean} is calculated without reference to
E, as the voltage waveform is the same as for passive loads, giving, by similar reasoning to above, I_{mean} as 6.13 A, the torque 5.52 N m, and speed 955 rev/min.

Example 8-3

For the motor and drive specified in Example 8-2, a closed-loop control is incorporated, and the control is set to maintain a constant speed of 1000 rev/min up to a
gross torque of 4 N m. Determine how the firing delay angle varies from no-load
to satisfy the constant-speed condition.

SOLUTION Back e.m.f. $E = 1000 \times (2\pi/60) \times 0.9 = 94.25$ V. At arcsin (94.25/
$240\sqrt{2}) = 163.9°$ the supply voltage equals the back e.m.f. At zero gross torque
this is the no-load firing angle but, in practice, current must flow to provide the no-
load loss torque.
Select a particular firing delay angle in advance of 163.9°, calculate the duration of
current flow in a similar manner to Example 8-2. Calculate V_{mean}, and from $I_{mean}R$
$= V_{mean} - E$, calculate I_{mean}, and hence the torque.
Calculations yield the following values: at a firing delay of 150°, the torque is 0.04
N m; 140°, 0.20 N m; 130°, 0.58 N m; 120°, 1.06 N m; 110°, 1.83 N m; 100°,
2.79 N m; 90°, 3.92 N m.
This and the previous example show how essential it is to have a closed-loop system
to maintain constant speed with varying load. With discontinuous current, the varying width of the zero-voltage period causes a large drop in speed at a fixed firing
angle.

Example 8-4

A separately excited d.c. motor is fed from a 240 V, 50 Hz supply via a single-phase
half-controlled bridge rectifier. Armature parameters are: inductance 0.05 H,
resistance 2 Ω, torque (voltage) constant 1.0 N m/A (V/rad/s). Plot the speed-gross
torque characteristic of the motor up to a torque of 12 N m at a fixed firing delay
angle of 90°.

SOLUTION An ideal motor would require zero torque at no load, hence zero input current, in which case the back e.m.f. would equal the peak input voltage,
which at 90° is $240\sqrt{2} = 339.4$ V, giving a speed of $339.4 \times 1.0 = 339.4$ rad/s. In
practice, there must be some frictional torque; hence the motor will not attain this
speed.
To determine points on the plot, select values of back e.m.f. E, then follow calculations as in Example 8-2 to determine the current expressions and hence the
torque. Calculations show that the current is discontinuous until E falls to 94.5 V,
when the current becomes continuous and the motor voltage waveform becomes
identical to that of a passive load.
With continuous current, $V_{mean} = 108.04$ V, hence I_{mean} can be determined from
$V_{mean} = E + I_{mean}R$; for example, when the torque is 10 N m, $I_{mean} = 10/1.0 =$
10 A, and $E = 88.04$ V. Speed $= E \times 1.0$ rad/s.
Values of speed against gross torque are shown plotted in Fig. 8-23.
Once the current is continuous, speed falls slightly with increasing torque, just

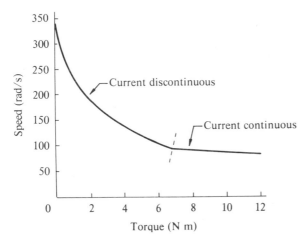

Figure 8-23

as if the motor were supplied from a battery source. In a closed-loop control system, the gain changes considerably between the two conditions of continuous and discontinuous current.

Example 8-5

For the conditions given in Example 8-2 with the back e.m.f. equal to 150 V, determine approximately the r.m.s. value of the armature current, and compare the armature resistance loss to that if the current were level.

SOLUTION The r.m.s. current can be calculated exactly by using calculus, or approximately by using the current values at $10°$ intervals, squaring these values, finding the mean, and then taking the square root. Using the current values at $85°$, $95°$, ... to $205°$, referring to Fig. 8-22b, we get

$$I_{rms} = [(0.83^2 + 2.48^2 + 4.00^2 + 5.31^2 + 6.31^2 + 6.96^2 + 7.20^2 + 6.98^2$$
$$+ 6.30^2 + 5.14^2 + 3.58^2 + 2.04^2 + 0.64^2)/18]^{1/2} = 4.24 \text{ A}$$

Armature loss $= 4.24^2 \times 6 = 108$ W.
With level current $I_{rms} = I_{mean} = 3.22$ A, giving an armature loss of $3.22^2 \times 6 = 62$ W. Additional loss $= 108 - 62 = 46$ W.

Example 8-6

A separately excited d.c. motor is fed from a 240 V, 50 Hz supply via a single-phase fully controlled rectifier. Armature parameters are: inductance 0.05 H, resistance $2\,\Omega$, torque (voltage) constant $1.0\,\text{N m/A}$ (V/rad/s). Neglecting rectifier losses, sketch the motor voltage and current waveforms when the firing delay angle is $60°$ and the motor is running at 150 rad/s. Determine the torque at this condition.

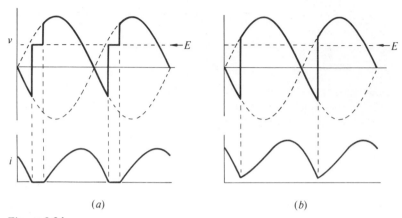

Figure 8-24

Determine the firing delay angle required if the load torque is increased by 50% but the speed is to remain unchanged. Plot the voltage and current waveforms.

SOLUTION Using the method developed in Example 8-2, the expression for the current is $i = 21.43 \sin (2\pi 50t - 0.397) - 75 + 83.29 \, e^{-40t}$ A, when the current is discontinuous with $t = 0$ at the instant of firing. In the fully-controlled circuit, the motor will see the supply voltage as shown in Fig. 8-24a, with i falling to zero at time 0.008 59 s, that is, 155° after the commencement of current flow. There is no voltage zero as in the half-controlled circuit.
The mean voltage is

$$V_{mean} = \frac{1}{\pi} \left[\int_{60°}^{(60°+155°)} 240\sqrt{2} \sin \theta \, d\theta + \frac{150\pi(180 - 155)}{180} \right] = 163.9 \text{ V}$$

$$I_{mean} = \frac{163.9 - 150}{2} = 7.0 \text{ A. Torque} = 7.0 \times 1.0 = 7.0 \text{ N m.}$$

A 50% increase in torque to $7 \times 1.5 = 1.5 = 10.5$ N m will it is found make the current continuous, so the waveforms are as shown in Fig. 8-24b, that is, the voltage is now as in a rectifier feeding an inductive load as described in Chapter 2. With torque $= 10.5$ N m, the mean current is $10.5/1.0 = 10.5$ A, and with unchanged speed $E = 150 \times 1.0 = 150$ V. $V_{mean} = E + IR = 150 + (10.5 \times 2) = 171$ V. From Eq. (2-7), $171 = (2/\pi)240\sqrt{2} \cos \alpha$, giving the firing delay angle $\alpha = 37.7°$.
Taking $t = 0$ at the instant of firing, the equation to the current is $i = 21.43 \sin (2\pi 50t - 0.786) - 75 + 92.04 \, e^{-40t}$ A, with $i = 1.87$ A at the instant of firing ($t = 0$ and $t = 0.01$ s).

Example 8-7

If the converter and machine of Example 8-6 are put to operate in the inverting mode for regeneration at a speed of 150 rad/s, determine the torque with a firing advance angle of $30°$. Sketch the voltage and current waveforms. Neglect converter losses and supply impedance.

Determine the firing advance if the torque is 11 N m, the speed remaining at 150 rad/s. Sketch the waveforms.

SOLUTION For continuous current flow at $\beta = 30°$, $V_{mean} = (2/\pi)240\sqrt{2} \cos 30° = 187$ V, but as this value is greater than E, which is 150 V, then the current must be discontinuous. The waveforms are as shown in Fig. 8-25a. The inverting condition may be drawn to the same reference as that for rectifying, in which case the converter voltage v_0 is mostly negative. However, considering the machine as a generator, it may be better to take the generator voltage v_G as the reference, in which case it is the reverse of v_0, as shown in Fig. 8-25c. Both waveforms are shown, one simply being the inverse of the other.

To calculate the current i taking $t = 0$ at the instant of firing, use the equivalent circuit of Fig. 8-25d. By similar reasoning to that used in Example 8-2, the equation to i is $i = 21.43 \sin (2\pi50t + \pi - \pi/6 - 1.444) + 150/2 + Ae^{-40t}$, where 1.444 rad is the current phase lag of L and R to 50 Hz, and A can be found from $i = 0$ at $t = 0$, hence $i = 21.43 \sin (2\pi50t + 1.174) + 75 - 94.77 e^{-40t}$ A, which falls to zero after a time of 0.006 59 s, that is, $118.6°$.

$$V_{mean} = \frac{1}{\pi} \left[\int_{-30°}^{88.6°} 240\sqrt{2} \sin \theta \, d\theta + \frac{\pi(180 - 118.6)150}{180} \right] = 142.1 \text{ V}$$

hence $I_{mean} - \dfrac{150 - 142.1}{2} = 3.95$ A, giving a torque of 3.95 N m.

Assuming that an increase in torque to 11 N m will make the current continuous, from Eq. (3-24) with $\gamma = 0$, $V_{mean} = (2 \times 240\sqrt{2} \cos \beta)/\pi = 150 - (11 \times 2)$, giving the firing advance angle $\beta = 53.7°$ with current $i = 21.43 \sin (2\pi50t + 0.761) + 75 - 89.65 e^{-40t}$ A, giving a current of 0.13 A at the instant of firing. The waveforms for this continuous-current condition are shown in Fig. 8-25b.

Example 8-8

A separately excited d.c. motor is supplied via a half-controlled three-phase bridge rectifier fed at a line voltage of 220 V, 50 Hz. The motor parameters are: inductance 0.012 H, resistance 0.72 Ω, armature constant 2 V/rad/s (N m/A). Neglecting rectifier losses and supply impedance, determine the speed-torque characteristic up to 60 N m at fixed firing delay angles of (i) $30°$, (ii) $90°$, (iii) $150°$. Sketch typical armature voltage and current waveforms.

SOLUTION At the lower torques, the current is discontinuous, and the voltage waveforms are as shown in Fig. 8-26. Calculations must be made like those in Example 8-2 to determine the duration of current flow:

a.c. impedance $= 0.72 + j2\pi50 \times 0.012 = 3.838\underline{/1.382}$ Ω

peak a.c. current component $- 220\sqrt{2}/3.838 = 81.06$ A

time constant $= L/R = 0.012/0.72 = 1/60$ s

(i) When the current is continuous, the armature voltage will be as shown in Fig. 2-41b and using Eq. (2-17), $V_{\text{mean}} = (3/2\pi) \times 220\sqrt{2} \times (1 + \cos \alpha) = 277.2$ V at 30° delay. Taking a back e.m.f. above this, the current must be discontinuous, say 280 V, that is, a speed of $280/2 = 140$ rad/s, the current equation after firing ($t =$

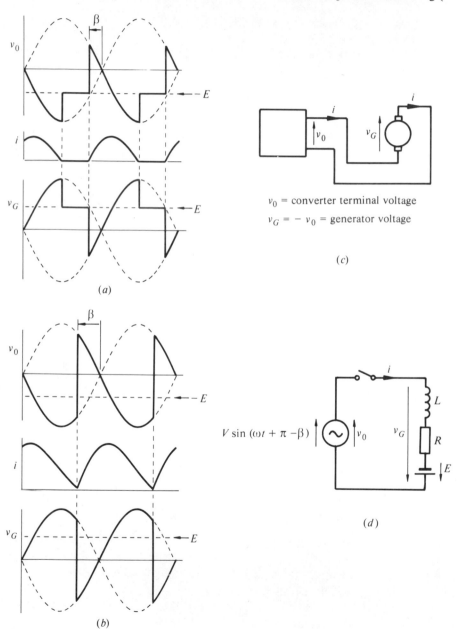

v_0 = converter terminal voltage

$v_G = - v_0$ = generator voltage

(c)

$V \sin (\omega t + \pi - \beta)$

(d)

(a)

(b)

Figure 8-25

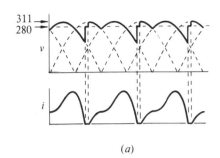

(a)

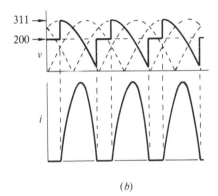

(b)

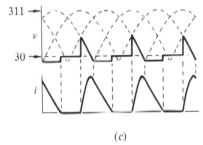

(c)

Figure 8-26

0) is $i = 81.06 \sin (2\pi 50t - 1.382 + \pi/2) - 280/0.72 + Ae^{-60t}$, where from $i = 0$ at $t = 0, A = 373.7$.

After $30°$ (time $= 1/600$ s), the voltage changes to another sinewave as shown in Fig. 2-41b and Fig. 8-26a, when the equation for the current is $i = 81.06 \sin (2\pi 50t - 1.382 + \pi/3) - 280/0.72 + Ae^{-60t}$, with $t = 0$ at the start of this period. From the earlier equation at $t = 1/600$, $i = 2.21$ A, which is the value of i at $t = 0$ in the second equation, giving $A = 414.7$.

This current will fall to zero at $t = 0.004\,69$ s, that is, after $84.4°$, giving a total period of current flow of $84.4 + 30 = 114.4°$.

The waveform for this condition is shown in Fig. 8-26a.

$$V_{mean} = \frac{1}{2\pi/3} \left[\int_{90°}^{120°} 220\sqrt{2} \sin\theta \, d\theta + \int_{60°}^{144.4°} 220\sqrt{2} \sin\theta \, d\theta \right.$$

$$\left. + \frac{280\pi(120 - 114.4)}{180} \right] = 282.4 \text{ V.}$$

From $V_{mean} = E + I_{mean}R$, $I_{mean} = (282.4 - 280)/0.72 = 3.34$ A, giving a torque of $3.34 \times 2 = 6.68$ N m.

Before the torque reaches 10 N m, the current is continuous and the voltage waveform is as shown in Fig. 2.41b. From $V_{mean} = E + I_{mean}R$, $I_{mean} = T/2$, $V_{mean} = 277.2$ V, and the speed is $E/2$, the speed is related to torque T as speed $= 138.6 - 0.18T$ rad/s. At $T = 10$ N m, speed $= 136.8$ rad/s; and when $T = 60$ N m, speed $= 127.8$ rad/s.

(ii) When the current is continuous, the armature voltage waveform will be as shown in Fig. 2-41d, with a mean value of 148.6 V. Following earlier reasoning, a back e.m.f. (speed) above this value must give discontinuous current. At (say) $E = 200$ V (100 rad/s), $i = 81.06 \sin (2\pi50t + 0.189) - 277.8 + 262.6 \, e^{-60t}$ A, this current falling to zero at $t = 0.004\,77$ s, that is, after 85.8°. This condition is shown in Fig. 8-26b where $V_{mean} = 205.1$ V, $I_{mean} = 7.14$ A, giving a torque of 14.29 N m.

(iii) When the current is continuous, the armature voltage waveform will be as shown in Fig. 2-41f, with a mean value of 19.9 V. At standstill (zero speed) when $E = 0$, $I_{mean} = 19.9/0.72 = 27.64$ A, giving a torque of 55.3 N m, that is, less than the stated limit of 60 N m.

Taking (say) a back e.m.f. of 30 V (15 rad/s), when the bridge conducts, $i = 81.06 \sin (2\pi50t + 1.236) - 41.66 - 34.89 \, e^{-60t}$ A. The voltage falls to zero after 30° ($t = 0.001\,66$ s), the current being 6.38 A, which is the value at the start of the zero-voltage period when the commutating diode conducts. The current equation is now $i = -41.66 + 48.05 \, e^{-60t}$ A, with $t = 0$ at the start of the zero-voltage period. The current decays to zero at $t = 0.002\,38$ s, that is, an angle of 42.8°. This waveform is shown in Fig. 8-26c. From the voltage waveform, the mean voltage is 31.7 V, giving $I_{mean} = 2.38$ A with the torque as 4.75 N m.

The speed-torque characteristics for the three conditions are shown in Fig. 8-27. The break-point in the characteristic is where the current changes from the discontinuous to the continuous condition.

Example 8-9

Determine the harmonic content up to the eighth order of the input current to the rectifier of Example 8-8 for the condition of 90° firing delay at a torque of (i) 30, (ii) 60 N m. Calculate also the power factor and distortion factor of the input current.

SOLUTION (i) Reference to Fig. 8-27 gives a speed of 75.84 rad/s, that is, a back e.m.f. E of 75.84×2 at a torque of 30 N m. During the period when the thyristor conducts, the equation to the motor current is $i = 81.06 \sin (2\pi50t + 0.189) -$

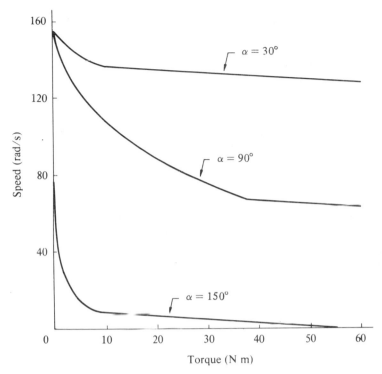

Figure 8-27

$210.7 + 195.9\ e^{-60t}$ A, taking $t - 0$ at the start of the thyristor conduction. From the reasoning in the development of Fig. 2-23c, the input waveform to be analysed is shown in Fig. 8-28a.

The waveform may be analysed for harmonic content mathematically, or by graphical analysis as in Example 7-5 by taking particular values of the current at different angles.

Analysis of the waveform gives r.m.s. values of: fundamental, 10.48; second, 8.44; third, zero; fourth 2.87; fifth, 0.84; sixth, zero; seventh, 1.2; eighth, 0.86 A.

The in-phase component of the fundamental current in r.m.s. terms is 6.47 A, hence the per phase input power is $(220/\sqrt{3}) \times 6.47 = 821.8$ W. The r.m.s. value of the input current is 13.89 A.

From Equation (3-13), power factor $= \dfrac{821.8}{(220/\sqrt{3}) \times 13.89} = 0.466.$

From Equation (3-16), distortion factor $= 10.48/13.89 = 0.755.$

(ii) The equation to the input current taking $t = 0$ at the start of thyristor current conduction is $i = 81.06 \sin (2\pi 50t + 0.189) - 176.3 + 172.4\,e^{-60t}$ A. This current waveform is shown in Fig. 8.28b.

Analysis of the waveform gives r.m.s. values of: fundamental, 18.48; second, 14.09; third, zero; fourth, 3.27; fifth, 1.69; sixth, zero; seventh, 2.23; eighth, 1.06 A.

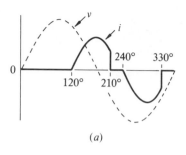

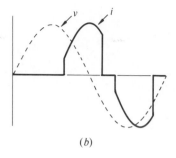

(a)   (b)

Figure 8-28

The in-phase component of the fundamental current is 11.79 A; hence the per phase power is $(220/\sqrt{3}) \times 11.79 = 1497.3$ W. The r.m.s. value of the input current = 23.72 A.

$$\text{Power factor} = \frac{1497.3}{(220/\sqrt{3}) \times 23.72} = 0.497$$

Distortion factor = $18.48/23.72 = 0.779$

Comparison of the above more practical waveforms with the idealized (infinite load inductance) waveforms analysed in Example 7-7 shows that, in practice, the low-order percentage harmonics are higher, but the higher-order harmonics are less, than in the idealized quasi-square waveforms.

Example 8-10

The motor detailed in Example 8-8 is supplied from the 220 V, 50 Hz supply via a fully controlled three-phase bridge converter. Neglecting all converter losses and supply impedance, determine and sketch the d.c. machine armature waveforms for the following conditions: (i) rectifying, 30° firing delay, 60 N m load torque; (ii) rectifying, 60° firing delay, speed of 75 rad/s; (iii) inverting, 30° firing advance, 60 N m torque; (iv) inverting, 60° firing advance, speed of 70 rad/s.

SOLUTION Following the reasoning developed in Example 8-8, the steady-state a.c. component of the armature current = $81.06 \sin(2\pi 50t - 1.382 + \phi)$, where $t = 0$ at the instant of firing and ϕ is the angle on the supply sinewave seen by the armature at $t = 0$.

The steady-state d.c. current component = $\pm E/0.72$, negative for motoring, positive for the generating condition.

The decaying transient component = Ae^{-60t}, the value of A being such as to satisfy initial conditions.

(i) Assuming the current to be continuous, from Eq. (2-16), $V_{\text{mean}} = (3/\pi)220\sqrt{2}$ cos 30° = 257.3 V. At 60 N m, $I_{\text{mean}} = 60/2 = 30$ A. Therefore $E = 257.3 - (30 \times 0.72) = 235.7$ V.

At the instant of firing, the supply sinewave is at 90°, hence $i = 81.06 \sin(2\pi 50t$

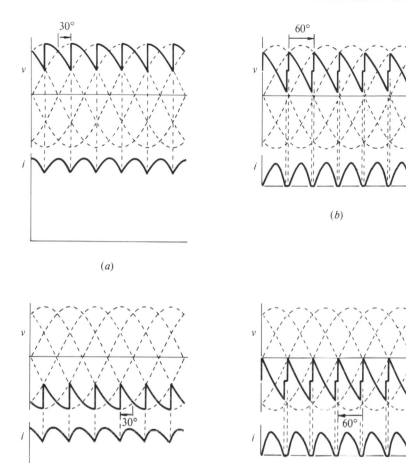

Figure 8-29

$+\ 0.189) - 327.4 + Ae^{-60t}$ A. The instantaneous value of the current will be the same at the start of the period as at the end (after $60°$, that is, $t = 1/300$), giving $A = 338.5$.

The waveforms are shown in Fig. 8-29a. For these conditions the speed is $235.7/2 = 117.85$ rad/s.

(ii) If the current were continuous, $V_{\text{mean}} = (3/\pi)220\sqrt{2} \cos 60° = 148.6$ V but, as $E = 75 \times 2 = 150$ V is greater, then the current must be discontinuous.

At the instant of firing $(t = 0)$, the supply sinewave is at $120°$ and $i = 0$, hence $i = 81.06 \sin (2\pi 50t + 0.712) - 208.3 + 155.4\ e^{-60t}$ A, this current falling back to zero at $t = 0.0032$ s $(57.6°)$.

The waveforms are shown in Fig. 8-29b. For these conditions $V_{mean} = 154.1$ V, $I_{mean} = (154.1 - 150)/0.72 = 5.64$ A, giving a torque of 11.28 N m.

(iii) Assuming the current to be continuous, then $V_{mean} = 257.3$ V, as in (i). At 60 N m, $I_{mean} = 60/2 = 30$ A, therefore $E = 257.3 + (30 \times 0.72) = 278.9$ V. At the instant of firing the supply sinewave is at $210°$, hence $i = 81.06 \sin(2\pi 50t + 2.283) + 387.4 + Ae^{-60t}$, giving $A = -422.2$, taking the current value to be the same at the start and finish of each period.

The waveforms are shown in Fig. 8-29c. For these conditions the speed is $278.9/2 = 139.45$ rad/s.

(iv) If the current were continuous, $V_{mean} = 148.6$ V, but, as $E = 70 \times 2 = 140$ V is less, then the current must be discontinuous.

At the instant of firing ($t = 0$) the supply sinewave is at $180°$ and $i = 0$, hence $i = 81.06 \sin(2\pi 50t + 1.759) + 194.4 - 274.1e^{-60t}$ A, falling back to zero at $t = 0.003$ s $(54°)$.

The waveforms are shown in Fig. 8-29d. For these conditions $V_{mean} = 136.5$ V, $I_{mean} = (140 - 136.5)/0.72 = 4.9$ A, giving a torque of 9.8 N m.

Example 8-11

The armature of a separately excited d.c. motor is fed from a 72 V battery via a chopper operating at 1 kHz. Determine the motor voltage and current waveforms at a speed of 90 rad/s, with a load torque of 5 N m, neglecting all chopper and motor losses. The motor armature has an inductance of 0.06 H and an armature constant of 0.5 V/rad/s (N m/A).

SOLUTION Figure 8-19 refers to this problem. At 92 rad/s the internal motor voltage E is $90 \times 0.5 = 45$ V, which is the mean output voltage of the chopper ignoring losses. The chopper on-time is a ratio of $45/72 = 0.625$. The chopper cycle time is $1/10^3 = 1$ ms. Hence the chopper is on for 0.625 ms and off for 0.375 ms.

At a torque of 5 N m, the mean motor current is $5/0.5 = 10$ A. When the chopper is on, the current will change at a rate such that $L \, di/dt = V_S - E$ (see Fig. 8-19), giving $di/dt = (72 - 45)/0.06 = 450$ A/s. When the chopper is off, the current will decay at the rate of $45/0.06 = 750$ A/s.

Ignoring the armature resistance, the current change is linear between the I_{min} and I_{max} values in Fig. 8-19b. Hence $I_{max} = I_{min} + (450 \times 0.625 \times 10^{-3})$ and also $I_{mean} = (I_{max} + I_{min})/2 = 10$. From these two equations, $I_{min} = 9.86$ A and $I_{max} = 10.14$ A. During the on-period, motor current $I = 9.86 + 450t$ A. During the off-period, the motor current $i = 10.14 - 750t$ A.

Example 8-12

If the motor of Example 8-11 has a resistance of 0.3 Ω, re-determine the solution to that example taking the resistance into account.

SOLUTION The mean value of the chopper output voltage will be higher by the

IR drop, therefore using Eq. (8-10), $V_{mean} = 45 + (10 \times 0.3) = 48$ V. Using Eq. (8-7), where $t_1 + t_2 = 1$ ms (see Fig. 8-19), then $t_1 = 48/12 = 0.667$ ms, and $t_2 = 0.333$ ms.

Using Eq. (8-8), during the on-period, $i = \dfrac{72 - 45}{0.3} - \left(\dfrac{72 - 45}{0.3} - I_{min} \right) e^{-(0.3/0.06)t}$,

which equals I_{max} at $t = 0.667 \times 10^{-3}$.

Using Eq. (8-9), during the off-period, $i = -\dfrac{45}{0.3} + \left(\dfrac{45}{0.3} + I_{max} \right) e^{-(0.3/0.06)t}$, which

equals I_{min} at $t = 0.333$ ms.

These two equations can be solved to give $I_{min} = 9.910$ A and $I_{max} = 10.177$ A.
During the on-period, $i = 90 - 80.09\, e^{-5t}$ A.
During the off-period, $i = -150 + 160.177\, e^{-5t}$ A.

Example 8-13

Describe conditions during run-up from the stationary condition if the system described in Example 8-11 is set to give maximum speed, the load torque being constant at 5 N m throughout the run-up period. Take the chopper to be set automatically to limit the motor current to 30 A, and take the armature and load inertia to be 0.2 kg m^2. Neglect chopper losses but include the armature resistance of 0.3 Ω.

SOLUTION Initially the motor is stationary, there is no internal voltage (back e.m.f.), hence the battery of 72 V is switched to a load of $R = 0.3$ Ω, $L = 0.06$ H.
The initial motor current is therefore $i = \dfrac{72}{0.3}(1 - e^{-(0.3/0.06)t})$ A. After one chopper cycle (0.001 s), $i = 1.2$ A, which is far below the set limit of 30 A; therefore the chopper remains fully on until 30 A is reached, given by a time t when $30 = 240$ $(1 - e^{-5t})$, that is, when $t = 0.0267$ s, that is, 27 chopper cycles.
The above assumes that the motor is still stationary, but once the motor current exceeds 10 A (5 N m) the load will receive an accelerating torque. Referring to the early part of the torque-time curve shown in Fig. 8-30, it can be assumed that the torque rises linearly with time.
From the equation, Torque = Inertia $\times\, d\omega/dt$, we can get the equation, $\int$ Torque $\times$ dt = Inertia $\times$ (change in speed). From the area ($\int T\, dt$) above the 10 N m line in Fig. 8.30, we get $[(15 - 5)/2] \times [(0.027 \times 2)/3] = 0.2 \times \delta\omega$, giving the change in

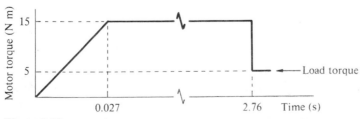

Figure 8-30

speed $\delta\omega = 0.445$ rad/s. At this speed, the back e.m.f. $= 0.445 \times 0.5 = 0.22$ V, which is negligible compared to the applied 72 V, hence justifying the earlier assumption.

The chopper is now automatically turned on and off, so as to maintain the 30 A limit. Therefore a constant accelerating torque of $(30 \times 0.5) - 5 = 10$ N m is maintained. The final maximum speed is $[72 - (10 \times 0.3)]/0.5 = 138$ rad/s. Hence the accelerating time is $(0.2 \times 138)/10 = 2.76$ s.

When up to full speed, acceleration ceases, and the motor current drops to the 10 A required by the load torque of 5 N m. Figure 8-30 shows the motor torque during the run-up period.

When the 30 A limit is initially reached (after 27 cycles) the chopper turns off. When off, the current will decay at a rate given by $30\,e^{-st}$, ignoring the small back e.m.f. At this time during the off-period $i = 30\,e^{-st}$ A, having fallen to I_{min} by the time the chopper is next turned on, the rising current now being $i = \dfrac{72}{0.3} - \left(\dfrac{72}{0.3} - I_{min}\right)e^{-st}$ A, turn-off again occurring at 30 A. With a chopper cycle time of 1 ms, this means the chopper is on for 0.13 ms, off for 0.87 ms, the value of I_{min} being 29.87 A.

As the motor accelerates, the motor back e.m.f. rises and the chopper is on for a longer period. At (say) half speed of $138/2 = 69$ rad/s, the back e.m.f. $E = 34.5$ V, the motor current when the chopper turns off is $i = -\dfrac{34.5}{0.3} + \left(\dfrac{34.5}{0.3} + 30\right)e^{-st}$ A, and when turned-on is $i = \dfrac{72 - 34.5}{0.3} - \left(\dfrac{72 - 34.5}{0.3} - I_{min}\right)e^{-st}$ A. From these two equations the value of I_{min} is 29.713 A and the chopper on/off times are 0.604/0.396 ms.

Example 8-14

In the locomotive layout shown in Fig. 8-17, take the transformer primary voltage to be 15 kV and each main secondary to be 300 V. If the total motor load is 1000 A, determine, at one half maximum output voltage, the r.m.s. current in the primary, and the harmonic content of this current. Neglect all losses, assume the motor current is level, and ignore the current taken by the motor fields.

Compare the above values with those obtained if only a single bridge (rather than three) were used.

SOLUTION At one half maximum output, bridge A will be fully conducting ($\alpha = 0°$), and bridge B at half maximum ($\alpha = 90°$), with bridge C not conducting. The load of 1000 A will be reflected into the primary as $1000 \times (300/15\,000) = 20$ A. The addition of the two bridge currents is shown in Fig. 8-31, the r.m.s value of the current being $\left(\dfrac{20^2 + 40^2}{2}\right)^{1/2} = 31.62$ A.

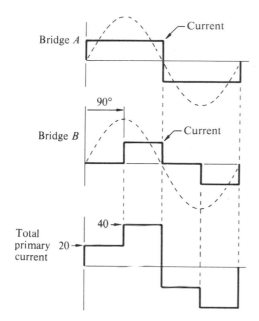

Figure 8-31

Using Eqs. (7-5) to (7-8) to find the harmonic content, there being no even-order harmonics,

$$a_n = \frac{2}{\pi} \left[\int_0^{\pi/2} 20 \sin nx \, dx + \int_{\pi/2}^{\pi} 40 \sin nx \, dx \right] = \frac{120}{n\pi}$$

$$b_n = \frac{2}{\pi} \left[\int_0^{\pi/2} 20 \cos nx \, dx + \int_{\pi/2}^{\pi} 40 \cos nx \, dx \right] = \pm\frac{40}{n\pi}$$

For the nth harmonic, the r.m.s. value is

$$\frac{1}{n\pi} \left[\frac{120^2 + 40^2}{2} \right]^{1/2} = \frac{28.47}{n} \text{ A}$$

If only a single bridge were used, at one half maximum output, the primary current would be $3 \times 20 = 60$ A for $90°$ to $180°$, and $270°$ to $360°$, during each cycle. Its r.m.s. value would be $\sqrt{(60^2/2)} = 42.43$ A and from Example 7-6 its r.m.s. harmonic component values at $\alpha = 90°$ would be $38.2/n$ A for the nth harmonic (no even components).

The above shows that using three bridges in series reduces the supply r.m.s. current, that is, improves the power factor.

NINE

A.C. MACHINE CONTROL

The construction of the cage induction motor is relatively simple compared to other machines, but it does not lend itself to speed adjustment so readily as does the d.c. motor. The major use of the power semiconductor device in a.c. motor control is centered on variable output frequency inverters. To obtain conquerable control characteristics comparable with those of the d.c. motor, a greater complexity of power and control equipment is required, although this additional cost has to be set against the lower cost of the cage induction motor.

This chapter explains the characteristics and manner in which power electronic equipment is employed in the control of cage induction motors and the other types of a.c. machines, such as the slip-ring induction and synchronous motors.

9-1 BASIC MACHINE EQUATIONS

In this section, basic characteristics (Ref. 10) are revised and developed with particular reference to how these characteristics are exploited in power electronic applications.

9-1-1 Synchronous Machine

The application of balanced three-phase currents to the three-phase winding of a machine stator sets up a magnetic flux in the air gap. This magnetic flux pattern remains constant in form throughout the alternating cycle, rotating one pole pair in one cycle. Figure 9-1a shows diagrammatically the flux pattern formed in a four-pole stator, drawn at an instant when the current in the A phase is at a maximum, the currents in B and C phases being reversed at half maximum. The current distribution at an instant $60°$ later in the electrical cycle, that is, when the current in C phase is reversed maximum and that in A and B phase one half positive maximum, results in the flux pattern shown in Fig. 9-1b. The pattern is clearly identical, but has moved on $30°$ mechanically. For the 4-pole (2 pairs of poles) machine, the rate of flux rotation is one revolution in 2 cycles.

The speed of rotation of the flux is known as the *synchronous speed*

$$N_{syn} = \frac{f}{p} \text{ rev/s} \tag{9-1}$$

where f is the frequency in Hz, and p the number of pairs of poles.

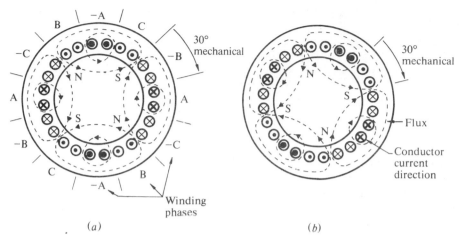

Figure 9-1 Flux set up by three-phase stator winding in a 4-pole machine. (*a*) At the instant when the current in phase A is a maximum. (*b*) At an instant 60° later in the electrical cycle.

Expressed in radian measure, the synchronous speed is

$$\omega_{\text{syn}} = \frac{2\pi f}{p} \text{ rad/s} \tag{9-2}$$

A simple motor which makes use of the rotating magnetic flux is the reluctance motor illustrated in Fig. 9-2. The flux set up by the stator follows a path via the asymmetrical rotor in such a manner as to minimize the length of the flux path in the air. Adopting the useful simile of considering the flux as elastic strings, it can be visualized that the rotor is dragged round with the flux. The higher the rotor torque, the greater is the displacement load angle, the rotor falling out of step with the rotating flux when the displacement load angle exceeds 90°. The speed of the rotor

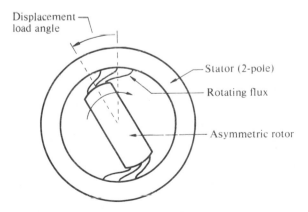

Figure 9-2 Illustrating the principle of the synchronous reluctance motor.

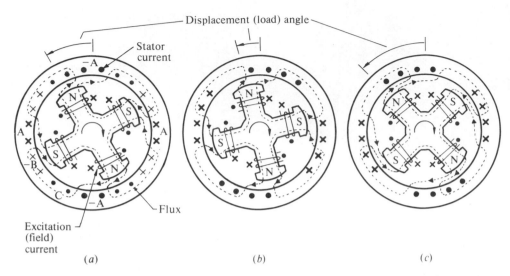

Figure 9-3 4-pole synchronous motor, shown at an instant when A phase voltage is a maximum. (*a*) Conditions when the stator current is at unity power factor. (*b*) Increased field excitation giving leading current in the stator. (*c*) Reduced field excitation giving lagging current in the stator.

is equal to the synchronous speed, and is hence directly proportional to the frequency of the stator supply.

The reluctance motor is electrically inferior to the normal excited synchronous motor regarding power factor and torque output for a given frame size. However, the simplicity of the reluctance motor does mean that it has some industrial applications.

The principle of the excited synchronous motor is that the magnetic flux set up by the d.c. excitation current locks in with the rotating magnetic flux set up by the stator, so that torque is produced, and the rotor runs at the synchronous speed given by Eq. (9-1). The basic layout of the synchronous motor is shown in Fig. 9-3, where it can be seen that the distorted air-gap flux pattern results in the rotor being dragged round at synchronous speed by the rotating flux.

A complete explanation of the synchronous machine can be found elsewhere (Ref. 10), but it is worth while examining in a simplified manner the salient features of the motor. Ignoring the stator winding resistance and leakage reactance, the actual strength of the magnetic flux in the air gap will always be related to the applied stator voltage, and hence this flux will be of constant magnitude. Figure 9-3 shows an instant where the A phase voltage is a maximum, so that the flux entering the stator is symmetrical about the A phase winding. Taking the condition where the stator current is at unity power factor as shown in Fig. 9-3*a*, the load torque will have caused the rotor to lag in space behind the stator flux. Following the closed path formed by a line of flux, no net m.m.f. contribution is made by the stator current.

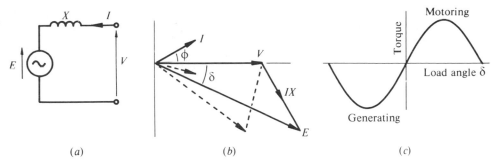

Figure 9-4 Synchronous motor characteristic. (*a*) Equivalent circuit neglecting resistance. (*b*) Phasor diagram at a leading power factor (lagging shown dotted). (*c*) Torque-load angle characteristic.

If a sudden increase is made in the excitation current, the flux magnitude will increase momentarily, the torque will increase, hence accelerating the rotor forward until balance is reached at a smaller displacement (load) angle. As the total flux cannot change, the increased field m.m.f. is balanced by the stator currents advancing in time as shown in Fig. 9-3*b*, so that the stator m.m.f. contribution within a closed flux line opposes the increased field excitation. The net result of increased excitation is to reduce the displacement angle and cause the stator current drawn from the supply to be at a leading power factor. The opposite effect results from a reduction of excitation, as illustrated in Fig. 9-3*c*, where the displacement angle increases and the stator currents are now drawn from the supply at a lagging power factor.

The simple equivalent circuit of the synchronous motor is shown in Fig. 9-4*a*, where E is the open-circuit voltage as a generator ignoring saturation effects, V is the terminal voltage, and X is known as the *synchronous reactance* (per phase). E is proportional to the field excitation current and, given values of E and V at a particular load condition, the stator current I will take up a magnitude and power factor such as are represented in the phasor diagram shown in Fig. 9-4*b*.

The electrical input power is $VI \cos \phi$ and, neglecting losses with suitable substitutions from the phasor diagram, this power can be derived as $\dfrac{VE}{X} \sin \delta$. Using line values for V and E,

$$\text{the total power} \ = \ \frac{VE}{X} \sin \delta \text{ watts} \tag{9-3}$$

$$\text{the mechanical power} \ = \ T \omega_{\text{syn}} \text{ watts} \tag{9-4}$$

$$\text{and the torque} = \frac{VE}{X \omega_{\text{syn}}} \sin \delta \text{ N m} \tag{9-5}$$

where δ is known as the *electrical load angle*, and this angle divided by the number of pairs of poles is the mechanical displacement angle shown in Fig. 9-3.

From Eq. (9-5), the torque is proportional to the sine of the load angle and, as can be seen from Fig. 9-4c, is a maximum at 90°. A reversal of load torque will automatically advance the rotor, change the sign of the displacement angle, and convert the machine to the generating mode. In practice, the reluctance torque produced by the saliency of the rotor structure and the effect of losses will modify the characteristic shown in Fig. 9-4c.

The synchronous motor can be braked by disconnecting the stator from the a.c. supply and connecting it to a resistor bank; the machine than acts as a generator. For starting from a fixed-frequency source, it is usual to fit a cage winding, or rely on eddy currents in the pole faces, so permitting the motor to start as if it were an induction motor. The excitation would be switched in at near synchronous speed, pulling the rotor into synchronism.

9-1-2 Cage Induction Motor

The principle of the cage induction motor is illustrated in Fig. 9-5. To set up the air-gap flux, magnetizing current must flow; this is shown in Fig. 9-5a as lagging the voltage by 90°. The movement of the rotating flux across the conductors induces a voltage in the short-circuited closed-cage rotor winding, hence causing current to flow. Because of the inductive nature of the winding, the induced currents will lag the voltage as shown in Fig. 9-5b. Also Fig. 9-5b shows the current flow in the stator which must flow by transformer action to balance the rotor current. The interaction of the rotor currents and flux is to produce a torque in the same direction as the rotating field. The total current is shown in Fig. 9-5c, being the addition of the currents shown in a and b.

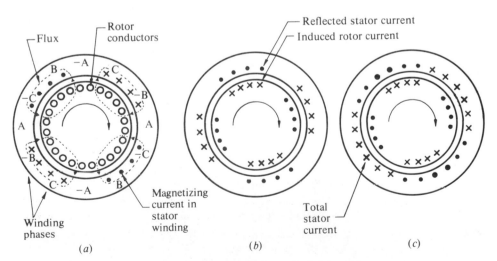

Figure 9-5 4-pole induction motor at instant when A phase voltage is a maximum. (a) Stator magnetizing current to set up the flux. (b) Showing how the induced rotor currents are reflected into the stator. (c) Showing the total instantaneous current distribution.

The rotor must always rotate at a different speed from the synchronous speed for a voltage, and hence current and torque to be induced in the rotor. The relative speed of the rotor to the synchronous speed of the stator flux is known as the slip s:

$$s = \frac{\text{synchronous speed} - \text{rotor speed}}{\text{synchronous speed}} = \frac{N_{syn} - N_{rotor}}{N_{syn}} \tag{9-6}$$

The frequency of the induced rotor current is equal to the slip times that of the stator frequency, giving

$$f_2 = sf_1 \tag{9-7}$$

where suffix 1 applies to the stator (primary), and suffix 2 applies to the rotor (secondary).

As the rotor speed approaches synchronous speed, the inductive reactance (proportional to slip frequency) is less, hence the current is closer in phase to the voltage, resulting in a better power factor in the stator. It is evident from Fig. 9-5 that the cage induction motor must always take a lagging power factor current.

The power transmitted across the air gap must be associated with the rotating flux. The power across the air gap into the rotor is the torque times synchronous speed, giving

$$\text{air-gap power transfer} = T\omega_{syn} \tag{9-8}$$

The output power (neglecting friction and iron loss) must be given by the same torque times the rotor speed, giving

$$\text{rotor output power} = T\omega_r = T(1-s)\omega_{syn} \tag{9-9}$$

The difference between these two powers is that lost in the conductors of the rotor circuit. It is worth noting from Eqs. (9-8) and (9-9) that

$$\text{efficiency} < (1 - \text{slip}) \tag{9-10}$$

If one ignores the stator winding resistance and leakage reactance, then the flux may be considered constant at all loads and proportional to the applied stator voltage.

Using values of rotor resistance R_2 and inductance L_2 per phase, referred (by the turns ratio squared) to the stator, then Fig. 9-6a represents the electrical conditions with both the reactance and the input voltage proportional to the slip. However, the same rotor current (referred value I_2) is given by the circuit of Fig. 9-6b, where the stator voltage V_1 is applied to a reactance independent of slip, but where the resistance is inversely proportional to slip. The power input to the circuit of Fig. 9-6b represents the total power transfer across the air gap, hence $I_2^2 R_2/s$ represents the addition of rotor power output and electrical $I_2^2 R_2$ loss.

Expressing the rotor speed ω_r in rad/s,

$$\omega_r = (1-s)\omega_{syn} \tag{9-11}$$

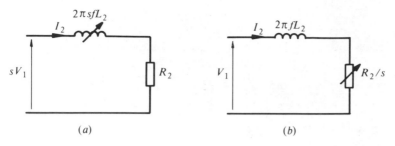

Figure 9-6 Equivalent rotor circuit of the cage induction motor. (*a*) True electrical conditions. (*b*) Representing the total power input to the rotor.

From the equivalent circuit of Fig. 9-6*b*, the power into the rotor is $I_2^2 R_2/s$, and from Eq. (9-8) this equals $T\omega_{\text{syn}}$, therefore

$$T\omega_{\text{syn}} = \frac{I_2^2 R_2}{s} \tag{9-12}$$

$$= \frac{V_1^2}{(2\pi f L_2)^2 + (R_2/s)^2} \times \frac{R_2}{s}$$

Hence   torque $T = \dfrac{sV_1^2 R_2}{\omega_{\text{syn}}[(2\pi s f L_2)^2 + R_2^2]}$ N m per phase $\qquad$ (9-13)

If we are to examine the torque-speed characteristic of a particular motor at different fixed supply frequencies, then the effect of the different frequencies must be considered.

In any magnetic circuit, the induced voltage is proportional to the flux level and the frequency (from $v = d\Phi/dt$), hence, to maintain optimum flux level in a machine, the ratio

$$\frac{\text{voltage}}{\text{frequency}} = \text{constant} \tag{9-14}$$

It is necessary to maintain the optimum flux at just below the saturation level, firstly, to make maximum use of the magnetic circuit, secondly, to minimize the current drawn from the supply to provide the torque (torque is proportional to current × flux).

For example, a motor rated at 400 V, 50 Hz needs to be supplied at 320 V for 40 Hz operation, or at 480 V for 60 Hz operation.

Take V_1 as the voltage at rated frequency f. At any other frequency kf, the rated voltage is kV_1, and the synchronous speed is $k\omega_{\text{syn}}$. Hence at any frequency, Eq. (9-13) becomes

$$\text{torque } T = \frac{skV_1^2 R_2}{\omega_{\text{syn}}[(2\pi s k f L_2)^2 + R_2^2]} \tag{9-15}$$

Taking a typical ratio of $2\pi f L_2/R_2$ as 5 at the rated frequency, typical torque-

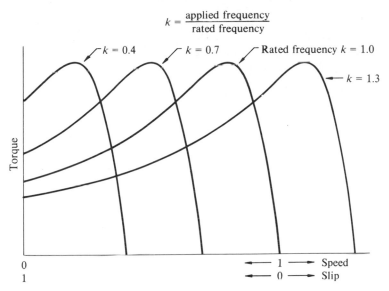

$$k = \frac{\text{applied frequency}}{\text{rated frequency}}$$

Figure 9-7 Induction motor torque-speed curves at different frequencies.

speed curves are shown in Fig. 9-7, where it can be observed that the shapes are similar, with the maximum torque value independent of frequency.

The running region of cage induction motors is normally with a small slip at just below its synchronous speed. An efficient means of speed adjustment is possible by variation of the supply frequency, provided the ratio of voltage to frequency is maintained resonably constant. At the very lowest speeds, the supply voltage has in practice to be increased slightly to compensate for the stator impedance voltage drop.

The current demanded by the induction motor for a direct-on-line start at fixed frequency has a magnitude of approximately six times the normal full-load current. With a fixed-frequency source, this starting current can only be reduced by voltage reduction. However, using the inverter drives described later, it is possible to start at low frequency, then raise the frequency to accelerate the motor. Reference to Fig. 9-7 shows the high starting torque possible with a low-frequency source. With a low-frequency start, the rotor inductive reactance is low, hence the induced rotor currents are much closer in phase to the voltage, so giving high torque with high power factor, and consequently minimum starting current magnitude.

If an induction motor is started at a frequency of k times its rated frequency, and the voltage is such as to maintain constant air-gap flux, then from Eq. (9-15) the starting torque T is

$$T = \frac{kV_1^2 R_2}{\omega_{\text{syn}}[(2\pi k f L_2)^2 + R_2^2]} \qquad (9\text{-}16)$$

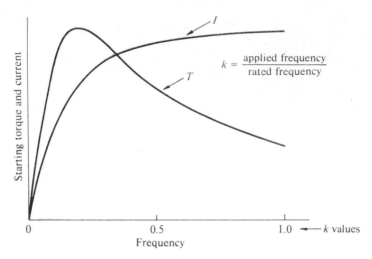

Figure 9-8 Starting values related to frequency for constant air-gap flux.

and the starting current I_2 from Fig. 9-6a is

$$I_2 = \frac{kV_1}{[(2\pi kfL_2)^2 + R_2^2]^{1/2}} \tag{9-17}$$

Using the same basis of $2\pi fL_2/R_2 = 5$, as used for Fig. 9-7, the starting torque and current values for different starting frequencies are shown in Fig. 9-8. It can be seen that a low-frequency start gives peak torque with low current.

If the induction machine is driven by external means above its synchronous speed, the polarity of the induced rotor voltage and current is reversed, the slip is negative, and the machine generates current at a leading power factor back to the a.c. system. The torque characteristic as a generator is a reflection of the motoring characteristic, as shown in Fig. 9-9; the same equations apply except for slip s being negative. The net result of a reduction in supply frequency to a revolving induction machine is to cause it to generate until the speed falls below the new synchronous condition.

A braking condition can be imposed on an induction motor by reversing the direction of the rotating magnetic field. As the field and rotor are then rotating in opposite directions, the slip is greater than unity, giving the characteristic shown in Fig. 9-9. The current during this form of braking exceeds in magnitude that at starting. The field direction is reversed by interchanging two of the three a.c. input conductors, or in an inverter-fed motor by changing the phase rotation sequence.

An effective form of induction motor braking is to disconnect the motor from the a.c. system and inject current from a d.c. source as illustrated in Fig. 9-10a. The current distribution is identical to an instant in the a.c. cycle when one phase is at its maximum. The magnetic field is now stationary, so the slip is directly proportional to the speed, giving the braking torque characteristic shown in Fig. 9-10b.

If the voltage, at fixed frequency, to an induction motor is reduced, the slip is

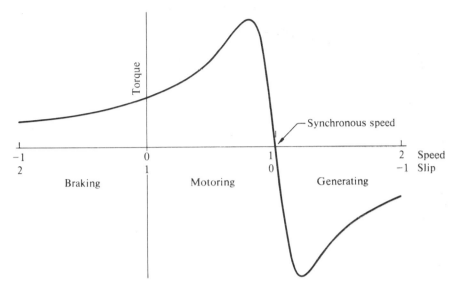

Figure 9-9 Complete torque-speed characteristic at fixed frequency.

increased in order to maintain the same torque. Examination of Eq. (9-13) shows the torque to be proportional to the square of the voltage. The effect is shown in Fig. 9-11, where it can be seen that a narrow range of speed adjustment is possible by voltage reduction. The disadvantages are those of loss of efficiency [see Eq. (9-10)], the increase in rotor losses leading to possible overheating, and the reduction in the maximum torque.

The complete equivalent circuit for an induction motor must include the stator winding resistance R_1, the leakage reactance $2\pi f L_1$, and the magnetizing components. Such an equivalent circuit is shown in Fig. 9-12a, with the approximate equiva-

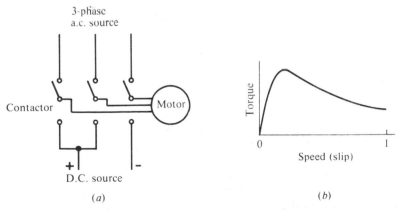

Figure 9-10 Braking by d.c. injection. (a) Circuit. (b) Characteristic.

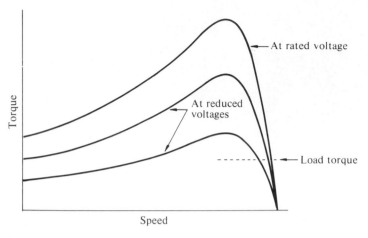

Figure 9-11 Torque-speed characteristic at reduced voltage, fixed frequency.

lent circuit shown in Fig. 9-12*b*, where the magnetizing arm has been moved to the input terminals.

The equivalent circuits of Fig. 9-12 assume linear constant values for the various parameters, but in practice this is not so. The resistance values are temperature-dependent and their a.c. values will be above the d.c. values due to skin effect, which in turn is dependent on the particular slip value. The values of the leakage inductances L_1 and L_2 depend on the flux conditions, with saturation of the teeth at heavy loads making these parameters variable and non-linear. Finally, the magnetizing arm includes friction and windage losses, which are themselves variables with speed, as are the magnetic losses, which are a function of slip and frequency. However, the equivalent circuits do give a good measure of understanding and prediction of performance over the machine's operating range.

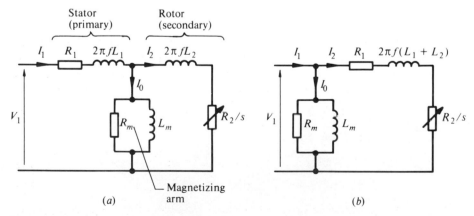

Figure 9-12 Equivalent circuits for induction motor. (*a*) With magnetizing arm correctly located. (*b*) Approximate equivalent circuit.

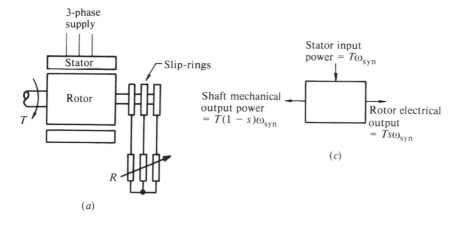

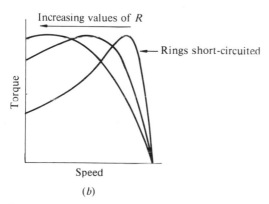

Figure 9-13 Slip-ring induction motor. (a) Arrangement. (b) Characteristic with added rotor resistance. (c) Power balance.

9-1-3 Slip-Ring Induction Motor

The slip-ring induction motor has a wound rotor carrying a three-phase winding similar to that of the stator. The ends of the winding are taken to slip-rings. If the slip-ring external connections are open-circuited, no current can flow, and the voltage at the slip-rings when stationary will be given by the stator voltage times the turns ratio. If the shaft is rotated, then the voltage at the slip-rings is reduced in proportion to the slip, as is its frequency.

The slip-ring circuit relationships are:

$$\text{open-circuit slip-ring voltage} = \text{open-circuit standstill voltage} \times \text{slip} \quad (9\text{-}18)$$

$$\text{slip-ring frequency} = \text{stator frequency} \times \text{slip} \quad (9\text{-}19)$$

If resistors are connected to the slip-rings as shown in Fig. 9-13a, current can flow, the level being dependent on the torque. Figure 9-13b shows the well-known characteristics given with added rotor resistance, showing how added resistance can improve starting conditions and give speed adjustment. The added resistance improves the starting torque by bringing the rotor current closer in phase to the induced voltage, while at the same time reducing current by increase in secondary impedance.

Neglecting all internal motor losses, the power balance of the motor is illustrated in Fig. 9-13c. As far as the stator is concerned, its power input is the power crossing the air gap, and hence is proportional to the torque. The rotor mechanical output is the same torque times the rotor speed, hence is less than the stator input by the slip. The difference between these two powers emerges as electrical power from the slip-rings. Control of the slip-ring voltage controls the slip, and thereby the speed of the rotor. Later sections of this chapter will discuss the manner in which power semiconductor circuits can be utilized in the slip-ring circuit to give efficient adjustment of speed.

If power is taken out of the slip-rings of the motor, then only speeds below synchronous speed are possible. However, if the rotor power flow is reversed, with electrical power flow into the rotor via the slip-rings, the slip will be negative, with speeds above synchronous speed.

9-2 MOTOR SPEED CONTROL BY VOLTAGE REGULATION

As explained in relation to Fig. 9-11, it is possible to get some measure of cage-induction-motor speed adjustment by reduction of the supply voltage, with consequential increase in slip. Thyristors may be used in the fully-controlled arrangement shown in Fig. 9-14, where firing delay of the thyristors removes sections of the supply voltage, so reducing the r.m.s. value of the voltage to the motor. Waveforms associated with such voltage regulation were shown in Figs. 6-5 and 6-6 for a resis-

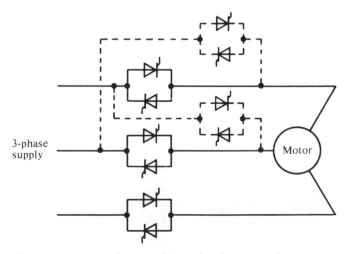

Figure 9-14 Speed adjustment by supply voltage control.

tance load; the effects of inductance (as must be the case with a motor load) were demonstrated in Examples 6-6 and 6-7. It is possible to use the half-controlled configuration shown in Fig. 6-4*b*, but the current waveforms are now asymmetrical as shown in Fig. 6-7, hence leading to even harmonics in the motor waveforms. Reversal is achieved by employing extra thyristors, as shown dotted in Fig. 9-14, to reverse the supply sequence.

Apart from the serious handicap of reduced efficiency with increased slip [see Eq. (9-10)], other disadvantages are associated with voltage regulation control. The increased rotor losses with falling speed means the torque has to be restricted so as to avoid excessive rotor heating. This form of speed adjustment is only practicable for loads such as (say) a fan, where the load torque falls considerably with decreasing speed. At particular firing angles, it is possible to have large harmonic components in the waveforms, so setting up reverse torques with harmonics of the order $(3m - 1)$ [see Eq. (7-25)]. The second harmonic associated with the half-controlled configuration is particularly onerous, therefore normally only the fully-controlled configuration is used, where the fifth order is the lowest harmonic component present.

Although the use of voltage reduction is limited as a means of speed control it is, however, used as a means of permitting soft-starting of the cage induction motor. Direct-on-line starting of the cage induction motor results in an initial current of typically six times the rated current of the motor. Voltage reduction in the form of star-start, delta-run, connection of the windings does result in the possibility of transient current at the instant of the change-over from star to delta, due to the rotating magnetic flux freezing in a stationary position at the instant of star disconnection and then almost certainly being in the wrong position at the instant of delta connection. Using the voltage regulation circuit shown in Fig. 9-14 it is possible to delay the firing angle at starting to restrict the current, and then advance the firing angle so as to accelerate the motor yet keep the current magnitude limited. The advantages of the gradual increase in voltage to give soft-starting is twofold; firstly it maintains the current within acceptable limits and secondly it permits control of the accelerating torque, avoiding any snatching of the shaft which may overstress drive components.

A further use of voltage regulation is to permit more efficient operation of the motor when it is underloaded, which is described later in Sec. 9-11.

9-3 CONSTANT-VOLTAGE INVERTER DRIVES

The major means of controlling the speed of a cage induction motor is to supply it from a variable-frequency inverter as illustrated by the basic circuit of Fig. 9-15*a*. The inverter operation was described in Chapter 5, Sec. 5-5, and the presence of the capacitor in Fig. 9-15*a* is to emphasize that the d.c. feed to the inverter is at constant voltage, that is, the capacitor holds the voltage sensibly constant over each cyclic change in the inverter. The battery source is shown variable to emphasize that over several cycles the voltage can change to match the inverter load requirements.

Each thyristor in Fig. 9-15a will receive firing pulses for 180°, so giving the voltage waveforms shown in Fig. 9-15b. The motor does not present a simple load to the inverter, as it reacts differently to each harmonic in the voltage waveform. Typically the on-load current waveform is as shown in Fig. 9-15b. These waveforms relate to a quasi-square-wave output inverter.

Equation (9-14) shows that to maintain optimum magnetic-flux conditions in the induction motor, the ratio of voltage to frequency must be kept constant. Hence, any variation in the inverter output frequency must be accompanied by a voltage change. With the quasi-square-wave inverter, the voltage adjustment can be achieved by varying the level of the d.c. source voltage to the inverter.

Two common arrangements for obtaining the necessary voltage adjustment are shown in Fig. 9-16. The section between the rectifier and the inverter is known as the *d.c. link*. A controlled rectifier can be used, as it gives a fast response to any

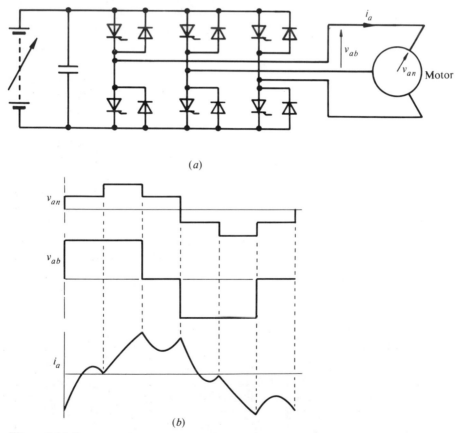

(a)

(b)

Figure 9-15 Constant-voltage quasi-square-wave inverter feeding an induction motor. (a) Basic inverter circuit. (b) Basic on-load waveforms.

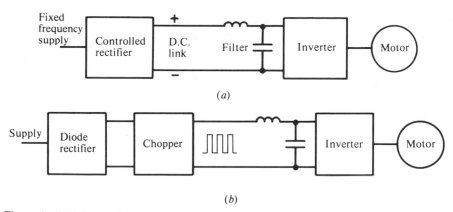

(a)

(b)

Figure 9-16 Voltage adjustment with quasi-square-wave inverter. (a) By controlled rectifier. (b) By chopper regulator.

control demand; but it suffers from the major disadvantage of all controlled rectifiers of lagging power factor on the a.c. supply. However, regeneration into the supply mains is possible with the fully-controlled rectifier, and the arrangement is relatively cheap. The other arrangement involves a diode rectifier giving a fixed d.c. busbar voltage, which is then fed to a chopper such as shown in Fig. 4-3. The chopper on/off action adjusts the mean level of the inverter input voltage, the chopper output being filtered before reaching the inverter. An advantage of the diode rectifier is that a battery can be placed across the fixed busbar to maintain the inverter supply in the event of a mains failure.

A low-voltage input to the inverter can present a problem if the commutation circuits of the inverter rely on this voltage. To overcome this problem the capacitors associated with commutation can be charged by a voltage derived from the fixed mains input. An example of a possible commutation circuit is shown in Fig. 9-17. The commutation capacitor is charged to the rectified mains voltage. When the commutating thyristor is fired, the inverter is short-circuited, the capacitor-inductor oscillates, and the thyristor turns off after a half cycle. The oscillation continues via the inverter thyristors and diodes, whereby the thyristors are turned off.

Many other commutating circuits are used, often being basically those shown

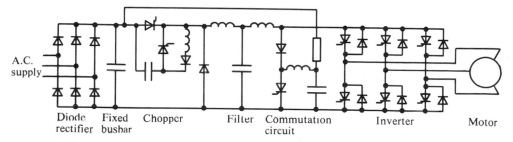

Figure 9-17 An example of a chopper-inverter showing the commutation circuit.

in Figs. 4-19 and 4-22. Chapter 4 detailed the many aspects of d.c. line commutation, the final choice of circuit often being a compromise between cost and commutation losses. With low-frequency quasi-square-wave output, the commutations per second will be low but, with high-frequency pulse-width modulation, the commutations per second can be very high, so that careful attention has to be given to the commutation losses.

The various forms of inverter control were described in Chapter 5, but it is worth reminding ourselves of the various systems and how the constant voltage/ frequency relationship is achieved. Figure 9-18 shows the major methods of control, illustrating how the fundamental voltage is halved at the half frequency. The quasi-square wave requires the inverter input voltage to be adjustable, but the harmonic content of the wave is invariable. The other three systems all have fixed input voltage level to the inverter, but the relative off-times are increased, with the relative harmonic levels in the waveforms changing as the fundamental frequency varies.

The quasi-square-wave inverter is unsuitable for powering an induction motor below around 5 Hz, due to harmonic torques making the drive unstable at low speed. For low-speed operation down to zero speed, the pulse-width modulation system is used. Above (say) 100 Hz, the commutation losses associated with the high rate of switching required for pulse-width modulation prohibits its use, so the quasi-square-wave inverter is used for high speeds. The power transistor inverters with their lower switching losses may be favoured for very high speed drives.

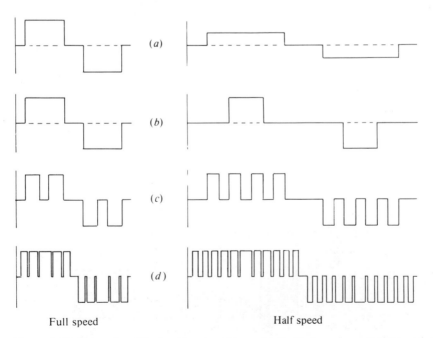

Full speed   Half speed

Figure 9-18 Illustrating different inverter systems. (*a*) Quasi-square wave. (*b*) Phased inverters. (*c*) Notched waveforms. (*d*) Pulse-width modulation.

A feature of the induction motor drive must be to keep the inverter voltage/ frequency relationship fixed over most of its operating range, so as to maintain constant air-gap flux. At the lowest frequencies (voltages), it is necessary to increase the voltage slightly to maintain the air-gap flux constant, as the stator impedance drop becomes a proportionally larger component of the total motor voltage.

The inverter designer can either choose to control the inverter frequency with voltage adjustment following, or control the output voltage with frequency following. A synchronous motor will demand frequency control, but with the induction motor it is usual to control the inverter output voltage.

The power entering the inverter is the product of the direct voltage and current. The shaft output power of the motor is the product of torque and speed. As the speed is related to the frequency (assuming a small slip) and hence to the inverter input voltage, the inverter input current is directly related to torque. Hence, control of the d.c. link current directly controls torque. The reactive components of the motor current circulate within the inverter, and so do not affect the mean d.c. link current.

Figure 9-19 shows a block diagram of the voltage-controlled induction motor. The input setting is a voltage (calibrated to speed) which is compared to the link voltage. The difference (error) is amplified so that the firing signals automatically raise the link voltage until it matches the input settings. However, in a like manner to that described for the d.c. motor (Fig. 8-12), feedback of the link current level limits the link voltage so that the current limit is not being exceeded.

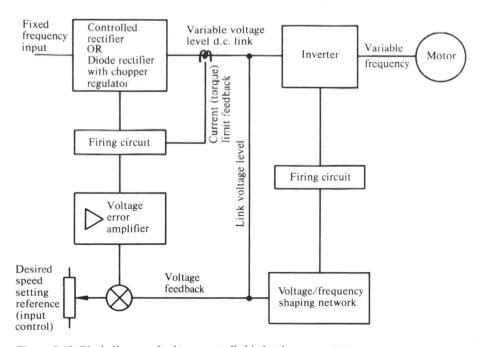

Figure 9-19 Block diagram of voltage-controlled induction motor drive.

The sequence of events at (say) starting, with the input in Fig. 9-19 set to the desired speed, is as follows: with zero link voltage feedback, the link voltage automatically rises to the level required to achieve the current limit; the link voltage is low so, via the voltage/frequency network, the inverter frequency is low. Reference to Fig. 9-7 shows that, at low frequency, the motor experiences almost maximum torque. As the motor accelerates, the link voltage must rise so as to maintain the limited link current, thus the motor frequency rises throughout its accelerating period, with the motor characteristic operating near to its maximum torque capability and favourable power factor. Eventually the link voltage reaches its set level, when the link current falls to a level to match the load torque.

In the steady state, at a fixed control setting, the inverter voltage is constant, so giving constant frequency, hence the torque-speed characteristic will be such that the slip will increase with load torque. A slip of 0.01 at light load may increase to 0.04 at full load, giving a speed regulation of 3%. It is possible to include slip compensation related to the current level, so that the link voltage rises with load, thus increasing the frequency, and so maintaining speed to an accuracy less than 1%. A further refinement is to measure the speed by a tachogenerator, compare this speed to the input setting, convert the difference to slip frequency, and increase the inverter frequency by this amount.

The induction motor can be easily reversed by changing the sequence of firing in the inverter, so reversing the sequence of the three-phase supply to the motor. To avoid sudden torque changes within the motor, a ramp network is included in the control circuitry, so that any sudden input control change is introduced slowly into the power circuit.

Regenerative braking is possible if the motor is overhauled by the load, the energy being returned to the d.c. link. This raises the d.c. link voltage by increasing the charge on the d.c. link capacitor. Unless this energy can be returned to the fixed-frequency source by using a converter in the inverting mode, regeneration is very limited. The energy can be dissipated into a resistor placed across the d.c. link,

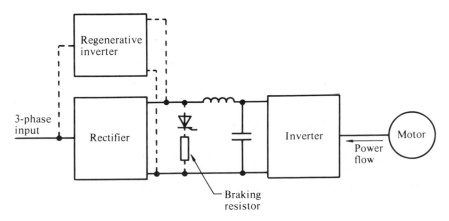

Figure 9-20 Two methods of regenerative braking.

which is a form of dynamic braking. These two methods of braking are illustrated in Fig. 9-20.

A very accurate open-loop control system can be obtained by the use of synchronous motors as the final drive, with the inverter input control setting the frequency. In block diagram form, the system is illustrated in Fig. 9-21 which in many ways is similar to Fig. 9-19. In the frequency-controlled arrangement, the input voltage directly sets the oscillator and hence the inverter frequency. The input setting also determines the link voltage, so maintaining the voltage/frequency relationship. As explained in Sec. 9-1-1, the synchronous motor can only produce a torque if its speed is matched to the supply frequency. Hence, in inverter systems, the frequency must start from zero and rise at a rate such that the torque can accelerate the rotor to this rate, so that the motor speed matches the increasing frequency. Provided the load angle of a synchronous motor stays within 90°, the system will remain stable during the run-up period. The ramp unit shown in Fig. 9-21 delays the rise of voltage resulting from any step input, and allows the inverter frequency to rise slowly, so maintaining stability during the accelerating period. Typically, the ramp unit can be set to give linear acceleration or deceleration rates between 3 and 90 seconds, standstill to full speed.

The reluctance and permanent-magnet synchronous motors run precisely at the inverter frequency, hence the accuracy of the drive depends on the oscillator frequency. This is an example of an accurate open-loop control system. A single inverter feeding several synchronous motors in parallel ensures each motor is running at exactly the same speed, an essential requirement on many production line processes. Transient load changes provide problems if they exceed the pull-out torque

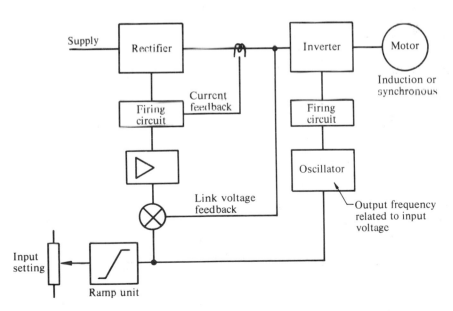

Figure 9-21 Frequency-controlled inverter drive.

of the synchronous motor, e.g., loads which the induction motor could handle by increased slip, the transient load energy being provided by loss of stored inertial energy.

The basic inverter circuits shown earlier in Fig. 9-16 control the motor voltage magnitude by adjustment of the d.c. link voltage. The adaptation of the pulse-width modulation strategy for the inverter, as explained in Sec. 5-4, Fig. 5-24, means that it is possible to control both voltage magnitude and frequency within the inverter, allowing the d.c. link voltage to be maintained constant. The emergence of the gate turn-off thyristor and MOSFET has enabled switching within the inverter to be both faster and more efficient, thereby eliminating the need for the commutating circuits required by the conventional thyristor. The variable d.c. link voltage system involved extra losses because thyristors were used in the rectifier rather than diodes, or where a chopper is used the losses within the chopper devices. The use of pulse-width modulation results in an overall reduction of the losses because only a diode rectifier is required; where gate turn-off thyristors are used there is an absence of commutation component losses.

The simplicity of the power component requirement is shown in Fig. 9-22, illustrated with gate turn-off thyristors in the inverter. The snubber circuit across each thyristor is for protection against voltage transients, its function being more fully explained in Chapter 10.

Typical waveforms associated with pulse-width-modulated waves are shown in Fig. 9-23. The frequency ratio is the ratio of the inverter switching frequency to the inverter output, that is, to the motor frequency. The current waveforms clearly show the reduction in harmonic current compared to the quasi-square-wave inverter waveform shown in Fig. 9-15, as well as the desirability to raise the frequency ratio.

A problem with inverter-fed cage induction motors is that instability may arise, particularly where sub-harmonics are present in the voltage waveforms. To avoid sub-harmonics the frequency ratio of the carrier wave to the modulating wave has to be in multiples of three. There are technical limits to the rate at which the inverter devices can be switched, so a technique known as *gear changing* may be used. By this technique the frequency ratio reduces at discrete intervals as the output frequency rises until possibly at the highest output frequency the waveform may be quasi-square wave. Gear changing does itself create problems of a transient

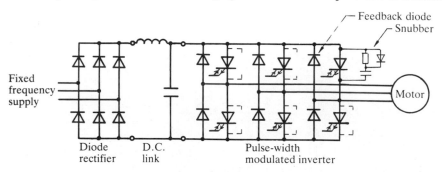

Figure 9-22 Pulse-width-modulated inverter system using gate turn-off thyristors.

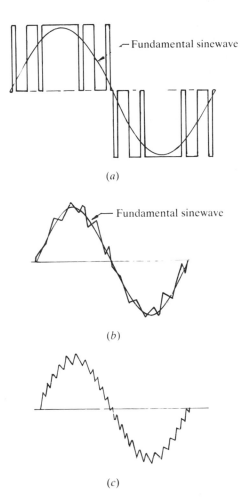

Figure 9-23 Waveforms of three-phase pulse-width-modulated inverter. (*a*) Ideal line voltage with low frequency ratio. (*b*) Typical motor line current with low frequency ratio. (*c*) Illustrating motor line current with higher frequency ratio.

nature at the change-over positions. It must be stated that satisfactory performance can be obtained using a fixed inverter switching frequency of (say) 1 kHz, allowing an asynchronous frequency ratio to occur as the inverter output frequency changes. Like many engineering problems the solution is a compromise between conflicting interests.

The choice of the switching frequency with pulse-width-modulated inverter-fed motors is also a compromise between conflicting considerations. High switching frequency within the inverter will increase the inverter switching losses, but reduce the harmonic content of the current waveform, and hence reduce the motor conductor losses giving a smoother torque. However, the magnetic circuit of the motor when having to respond to a high-frequency voltage component will show increased magnetic losses and become a source of acoustic noise. Likewise rapid switching of the inverter devices can and does generate high levels of acoustic noise which can be a nuisance in certain quiet locations.

It is clear that technically the variable-speed cage induction motor drive is more complex than the controlled rectifier-fed d.c. motor drive. However, the constant improvement in the switching devices and the advance of the microprocessor control systems has made the variable-speed pulse-width-modulated inverter-fed cage induction motor an increasingly attractive proposition in areas which were traditionally those of the d.c. motor drive.

The control strategy of determining the switching instants within each cycle was described in Sec. 5-4, Fig. 5-24, where the instants were determined by the intersection of a higher frequency triangular wave to a reference sinewave which corresponded to the desirable inverter output. This analogue form of control has the problem of accurately defining the two waves, balancing the three phases to an equal $120°$ displacement, and then ensuring the switching times are not so close as to not allow the minimum on or off times required by the devices. Microprocessors are now almost universally used to determine the switching instants. The switching instants can be determined in a like manner to the analogue method by generating a triangular wave from small up/down steps and then comparing this to a sampled and hold sinewave. Built into the microprocessor unit could be minimum on or off times, optimization of voltage magnitude to motor load, soft-starting, acceleration and braking times.

Microprocessor-based control systems can perform several functions:

pulse-width modulation control,
monitoring for fault conditions,
drive features such as acceleration or braking times,
torque-speed characteristics,
loss minimization,
closed-loop control.

These functions could be hardware or more likely a combination, allowing the user to program the inverter to suit his/her own application.

9-4 CONSTANT-CURRENT INVERTER DRIVES

The constant-current source inverter was fully described in Sec. 5-6, where it was explained that by constant current what was meant was that the current from the d.c. link changed negligibly over the period of each inverter cycle. The application of this inverter to the cage induction motor drive is illustrated in Fig. 9-24. The converter which is shown supplying the d.c. link could equally well be a chopper following a diode rectifier, if regeneration is not required. Compared to the constant-voltage inverter, the capacitor in the d.c. link is not present in the constant-current inverter, the inductor being the dominant feature.

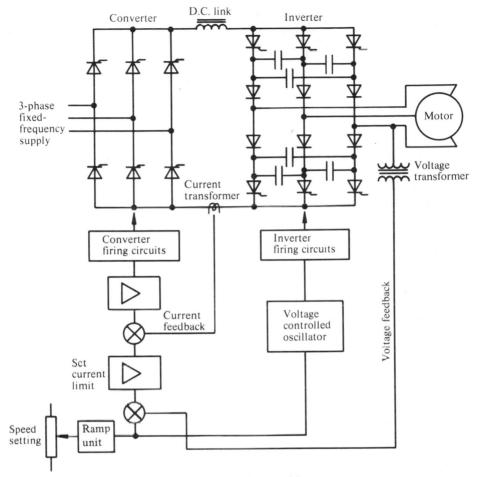

Figure 9-24 Control scheme for constant-current inverter drives.

The controls shown in Fig. 9-24 function in a similar manner to those described for the constant-voltage inverters. The input sets the inverter frequency and hence the speed, with the voltage adjusted to match the frequency. The current (and hence torque) is controlled during the acceleration periods.

In the constant-voltage inverter, the reverse-connected diodes prevent d.c. link voltage reversal, so any regeneration has to involve current reversal. In the constant-current inverter, such problems do not exist, as an overhauling load can cause the motor to generate, reversing the d.c. link voltage, the current direction remaining unchanged. With a reversed voltage, the converter can be placed in the inverting mode by firing angle delay, thereby returning energy to the supply. The system can, without any additional power circuitry, operate in any of the four quadrants, that is, operation is possible in either direction of rotation by the inverter sequence, with either motoring or generating torque.

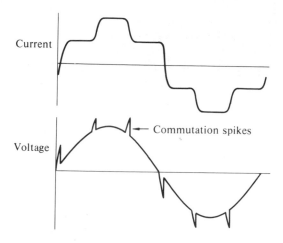

Figure 9-25 Motor waveforms when fed from constant-current inverter.

The stepped nature of the motor-current waveforms does preclude operation at very low speeds. The drive is very suitable for applications where violent changes in shaft torque must be avoided, because the very nature of the d.c. link inductor prevents sudden current changes. Another advantage of the constant-current inverter is that a fault such as a short-circuit at the motor terminals does not damage the inverter, as the current remains constant. The constant-current inverter is more complex in its control circuitry, and careful attention has to be given to match the inverter inductor and capacitors to the motor. Instability can occur in the constant-current inverter drives. Typical motor waveforms are shown in Fig. 9-25.

To help stabilize the drive, a closed-loop system can be used which controls the slip. A block diagram of such a system is shown in Fig. 9-26. Here any rise in link

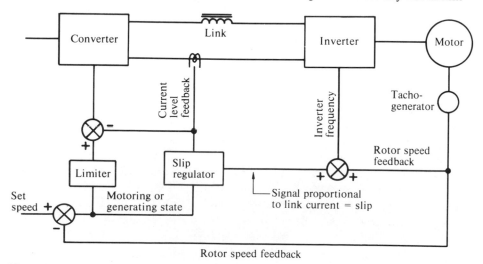

Figure 9-26 Controlled-slip constant-current inverter drive.

current forces a certain slip on the system, so dictating the difference between the rotor speed and related inverter frequency. The system regulates the current rather than the voltage.

The principle of regeneration is to delay the firing of the inverter thyristors, hence dropping the inverter frequency such that the motor is running at a super-synchronous speed and generating as illustrated by the torque-speed characteristic shown in Fig. 9-27.

When regenerating the link voltage is reversed and the energy can be passed back into the main fixed-frequency a.c. supply via the mains converter operating in the inverting mode. The commutating capacitors in the constant-current inverter are appropriately charged for commutation because when (say) thyristor T_3 is fired (see Fig. 5-35), thyristor T_1 is turned off, but because the load is inductive the capacitor between T_3 and T_1 is discharged via diode D_1 and then appropriately recharged when thyristor T_5 is fired. Care has to be taken to match the inverter capacitors to the link inductor and to the motor parameters.

Compared to the constant-voltage inverter the constant-current inverter has the advantages of: using conventional thyristors with simple commutation by using capacitors only, the output can be short-circuited momentarily without fusing occurring because the link inductor prevents sudden current change, regeneration is relatively easy, sudden dips in the mains fixed-frequency input are not directly

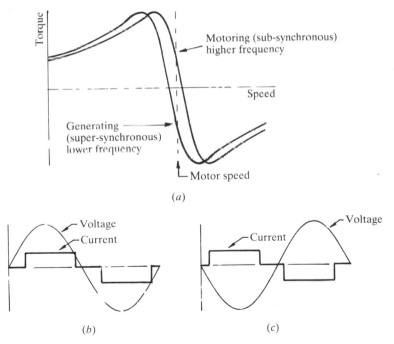

(a)

(b)   (c)

Figure 9-27 Illustrating regeneration. (a) Inverter frequency reduced. (b) Basic motoring line current with its phase voltage. (c) Generating current and voltage relationship.

reflected to the motor, it can supply a number of motors in parallel, and has high efficiency.

The increasing use of the gate turn-off thyristors, power transistors, and MOSFETs in the constant-voltage inverters does mean that the constant-current inverter with its inherent disadvantages of torque pulsations at low speed, bulky capacitors and inductor, and poorer speed regulation is not widely used for the lower power drives.

The use of pulse-width modulation techniques with the constant-current inverters has not been adopted commercially because of the long commutation times within the inverter. Its use would be restricted to low output frequency in order to eliminate the low-order harmonics in the current waveform.

9-5 MOTOR SPEED CONTROL VIA THE CYCLOCONVERTER

The principle and operation of the cycloconverter was explained in Sec. 5-1, where it was shown that a low-frequency supply can be directly synthesized from a higher-frequency source by suitable switching of the cycloconverter elements. A major limitation of the cycloconverter is that its output frequency is limited to (say) one third of the input frequency, possibly slightly better for a higher-pulse configuration. If the input is 50 or 60 Hz, the maximum output frequency is around 20 Hz, the net result being that the cycloconverter application is limited to low-speed drives. If the power source is (say) 400 Hz, then clearly higher speeds are possible.

Figure 9-28 shows that both voltage and frequency levels are directly controllable in the cycloconverter. However, although the cycloconverter is technically attractive for some applications, its use is severely limited on economic grounds compared to the inverter schemes. The cycloconverter is expensive in the number of thyristors and the extensive electronic control circuitry which is required. The system is inherently capable of braking by regeneration back into the a.c. source, thus four-quadrant operation is possible.

Technically, cycloconverter use is limited by the low-output-frequency range and by the quite severe harmonic and power factor demands made on the a.c. supply system. The waveforms associated with the cycloconverter were shown in Figs. 5-4 to 5-14.

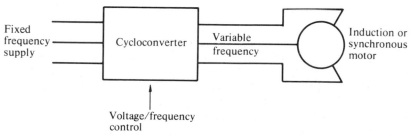

Figure 9-28 Cycloconverter control.

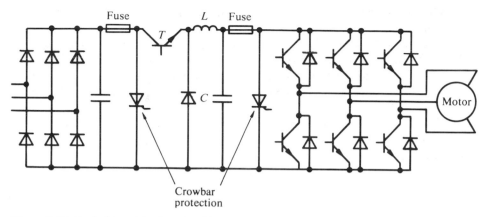

Figure **9-29** Typical transistor inverter drive.

9-6 TRANSISTORIZED INVERTER DRIVES

The power aspects of the transistorized inverter as shown in Fig. 9-29 are the same as for the thyristor inverter in regard to the manner in which the motor is controlled. The transistors are used in the switching mode, so the output can be either quasi-square wave or pulse-width modulated. The circuit shown in Fig. 9-29 is quasi-square wave with the transistor T acting as a chopper to control the d.c. link voltage. Compared to thyristors, the chopping rate can be much higher, resulting in smaller components being required for the LC filter. Alternatively, the transistor T can be dispensed with if the pulse-width modulation form of control is used.

Protection is covered in detail in Chapter 10, but it is worth mentioning here that the overload capability of the transistor is somewhat less than that of the thyristor; one way to protect them is by the use of a crowbar protection thyristor, as shown in Fig. 9-29. Once excessive currents are detected in the inverter, the thyristor is fired, short-circuiting the inverter and operating the protection associated with the diode rectifier. Alternatively, or more likely concurrently, all base drive to the transistors is removed, so permitting transistor turn-off.

Compared to the thyristor inverter, the obvious advantage in using transistors is that the commutation circuits are dispensed with. Also, the switching losses are less, so it is possible to operate the inverter at much higher frequencies. The transistor inverter is smaller and lighter in weight than its thyristor counterpart. A disadvantage is the requirement for continuous base drive during the transistor on-period but, by using Darlington pairs, current gains of 400 are possible. Another disadvantage is in the voltage rating, which is somewhat less than the thyristor. It is fairly safe to predict that the use of power transistors will expand into fields formerly dominated by the thyristor, as ratings and cost improve.

A further development in inverters which dispense with the commutating elements required for the conventional thyristor is to use the gate turn-off thyristor, an illustration of which is shown in Fig. 9-22.

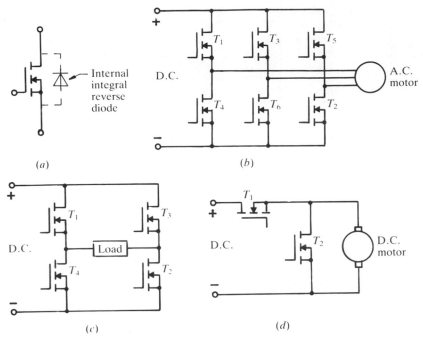

Figure 9-30 Basic circuits with the MOSFET. (*a*) Showing the reverse diode internal to all MOSFETs. (*b*) Three-phase inverter. (*c*) Single-phase inverter or four-quadrant d.c. motor drive. (*d*) Two-quadrant d.c. motor drive.

A more recent introduction is the MOSFET which is now used in low-power motor drives. Figure 9-30 shows some of the basic circuit layouts using MOSFETs. It must be remembered that internal to the MOSFET is a reverse conducting diode, hence eliminating the need to add feedback diodes in the inverter circuits. Figure 9-30*b* shows the usual three-phase inverter, the very fast switching action of the MOSFET making pulse-width modulation relatively easy to implement; hence the d.c. source voltage can be fixed in magnitude.

An application of the MOSFET, or other devices, can be found in the circuits of Fig. 9-30*c* and *d* for d.c. motor control. In Fig. 9-30*c*, using T_1 and T_2 only, the d.c. supply voltage can be chopped to give load voltage control, with any inductive current being maintained via the internal diode of one or both of the other MOSFETs. By appropriate control of the MOSFETs the load machine can motor or generate in either direction, giving four-quadrant operation.

The circuit of Fig. 9-30*d* is one in which by chopping T_1 on and off the mean motor voltage is controlled, the internal diode of T_2 acting as the commutating diode. The regenerative action takes place when T_1 is off and a reverse current in the motor is built up by turning T_2 on; then when T_2 is turned off the current is transferred to the reverse diode of T_1 back to the supply. The energy transfer is effectively in two stages. When T_2 is on the current rises, transferring energy from the mechanical system into the armature inductance; then the next stage is when T_2 is off and the energy is transferred from the inductance back into the d.c. source.

9-7 SLIP-RING INDUCTION MOTOR CONTROL

The slip-energy recovery system for speed control of a slip-ring motor is shown in Fig. 9-31, this form of control traditionally being known as the Kramer system. In Sec. 9-1-3 it was explained that, by taking power from the rotor, it was possible to vary the motor speed efficiently. The voltage at the slip-rings is at slip frequency, which is incompatible with the stator frequency, therefore the slip-ring voltage is rectified by means of a diode bridge into a d.c. link. The d.c. link power is transferred back into the main supply via a thyristor converter operating in the inverting mode.

The speed of the slip-energy recovery drive illustrated in Fig. 9-31a is set by the firing angle of the thyristors in the inverter, which thereby determines the d.c. link voltage. The d.c. link voltage is directly related to the slip-ring voltage, the slip of the motor adjusting itself to equal this voltage. Figure 9-31b gives typical torque-speed characteristics of a slip-ring induction motor with constant values of slip-ring voltage. As the load torque is increased, the slip will increase sufficiently to allow the current to build up to the level required to maintain the torque. As power can only be taken from the rotor, control is only possible below synchronous speed.

The power transfer via the rectifier-inverter is equal to the slip times the stator power, therefore if (say) control is only required down to 80% of synchronous speed, then the converter rating needs only to be 20% of the drive rating. However, to take advantage of this rating, at minimum speed, the firing advance on the converter must be at its minimum. To cater for the almost certain low voltage of the slip-ring circuit compared to the stator voltage, a transformer will be needed, interposed between the main supply and the converter, as shown in outline in Fig. 9-31a.

Only real power can be transferred through the d.c. link of Fig. 9-31a; hence the reactive power of the motor is entirely supplied from the stator, so the stator current must be at a lagging power factor. Also, the converter (when in the inverting mode) adds further to the reactive power demands on the supply. Hence, overall the current drawn from the main supply is at a worse lagging power factor than if resistors were used in the rotor circuit. The use of capacitors to improve this power factor is limited by the effects of the harmonics generated in the converter.

Where the converter is used for limited speed adjustment, say down to 70% synchronous speed, starting has to be by added rotor resistance, the slip-rings being automatically changed over to the converter when the minimum speed setting is reached. At speeds close to synchronous speed, the overlap period on the diode rectifier may be so long that four diodes are conducting together (overlap period well beyond 60°), in which case it is impossible to establish the d.c. link voltage, so no power is transferred. In practice, the top maximum speed at full-load torque with the converter in circuit is about 2% below that obtainable with short-circuited rings. The drive can, of course, run as a normal induction motor with short-circuited rings when no speed adjustment is required.

The closed-loop control system shown in conjunction with Fig. 9-31a operates in a similar manner to the earlier systems, a drop in speed automatically reducing

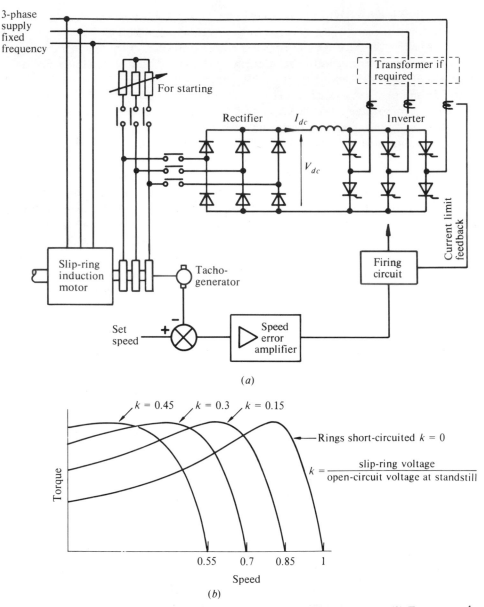

(a)

(b)

Figure 9-31 Slip-energy recovery scheme. (*a*) Circuit with control features. (*b*) Torque-speed characteristic at fixed slip-ring voltages.

the link voltage, hence allowing more current to circulate in the rotor. An inner loop limits the current in the converter to a safe level. The unidirectional power flow in the rotor circuit prohibits its use for regenerative braking.

Super-synchronous speeds are possible with the slip-ring induction motor by injecting power into the rotor circuit. For this to be possible, the arrangement of Fig. 9-32*a* can be used where the inverter can be self-commutated by the rotor voltage at speeds well away from synchronous speed. However, near to synchronous speed, the rotor voltage is low, and forced commutation must be employed in the inverter — which makes the scheme less attractive.

The slip-energy control scheme of Fig. 9-32*b* seems ideally suited as an application of the cycloconverter, being situated in the rotor circuit, provided the slip is less than (say) 0.33, giving the cycloconverter frequency ratio above 3 to 1. However, to utilize control to the full, below, above, and at synchronous speed, with motoring or regenerative braking, the control features are very complex, and the cost makes the use of this system prohibitive for most applications.

The power semiconductor circuits used in conjunction with the slip-ring induction motor are in effect attempting to achieve an electronic replacement for the mechanical commutator in the doubly-fed a.c. commutator motor. The commutator in the doubly-fed motor acts as a multi-phase frequency converter, enabling direct

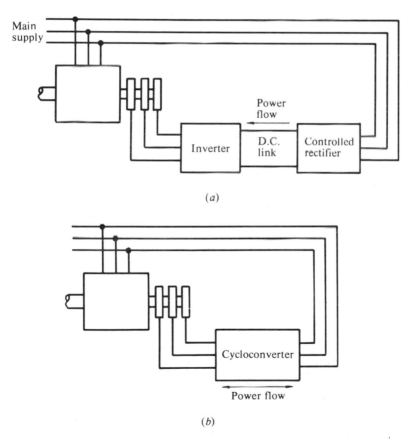

Figure **9-32** Alternative schemes for slip-ring motor. (*a*) Super-synchronous control. (*b*) With cycloconverter.

connection of the rotor to the main supply. By control of the in-phase and quadrature components of the rotor voltage at the commutator brushes, it is possible to achieve both speed and power factor control of the entire drive, both below and above synchronous speed. Theoretically, it is possible to replace the mechanical commutator by slip-rings and a cycloconverter, but in practice the cost and technical complexities prohibit or restrict the commercial applications of such schemes.

The application of the slip-ring motor is most attractive for pump and ventilation drives, which have a power requirement approximately the cube of the shaft speed; thus only small speed changes are necessary to give quite large load power variation.

9-8 BRUSHLESS SYNCHRONOUS MACHINES

The excited synchronous machine, either generator or motor, normally has its field on the rotor as described in Sec. 9-1-1. The field requires a supply of direct current, and this is obtained from an a.c. generator (exciter), fitted on the same shaft with the a.c. output from the exciter being converted to direct current by a rectifier mounted on the shaft or inside the synchronous machine rotor. The system is shown in Fig. 9-33, where it can be seen that, to avoid any sliding contacts, the exciter is inverted, that is, it has its field on the exciter stator. The description *brushless* arises from comparison with the pre-semiconductor era, when the exciter was a d.c. generator, which necessitated brushes on the d.c. generator-commutator and slip-rings on the synchronous machine.

The synchronous machine's field current is directly controlled by the current in the exciter field, and in this sense the exciter acts as a power amplifier. The diodes used in the rectifier are mounted to withstand the centrifugal forces and positioned to avoid creating any out-of-balance torques. If thyristors were used in the rectifiers, the control responses would be faster, but problems arise in trans-

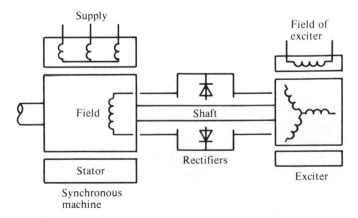

Figure 9-33 Brushless excitation system.

mitting the firing circuit control signals to the rotating system. The exciter is normally three-phase, with a frequency different to that of the main machine, so as to avoid possible undesirable interactions between the two elements.

Generator excitation does not pose any severe problems on the rectifier, as the field current to the exciter is not switched in until the generator is up to synchronous speed. However, it is possible that load or torque changes will induce high voltages into the generator field. These voltages, if allowed to appear in the nonconducting direction, could cause breakdown of a diode. To attenuate these voltages, a current path is provided by the resistor shown in Fig. 9-34a, damping any transient voltage. The ohmic value of the resistor is about ten times that of the field, being a compromise between damping the transient and avoiding excessive power loss.

Brushless excitation of motors involves more problems than those associated with the generator, as a magnetic field is set up by the stator currents during starting, so inducing a voltage at slip frequency in the field. During motor starting, the

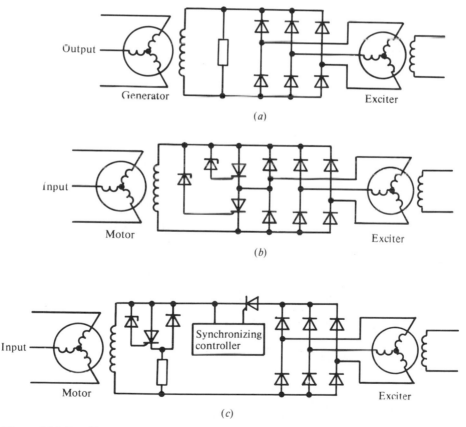

Figure 9-34 Rectifier arrangements for brushless machines. (*a*) Generator. (*b*) Motor with light starting torque. (*c*) Motor with heavy starting torque.

induced field voltage can be high, and destructive to any rectifier bridge. However, if the field is short-circuited during run-up, a current will flow, adding a torque to that of the induction motor torque produced in the cage winding.

The circuit of Fig. 9-34b can be used — here two thyristors are connected across the diode bridge. The gate firing current is supplied via Zener diodes which break down, that is, permit current flow into the gate when the winding voltage reaches a level in excess of the maximum direct voltage the field experiences when running normally in synchronism. The induced field voltage on open circuit will be several times its normal direct voltage, therefore the thyristors are fired early in the slip cycle. In the reverse half cycle of the slip voltage, the diodes will conduct. Hence, during starting, the field winding is effectively short-circuited. Once near to synchronous speed, the exciter is energized, the thyristors now being permanently in the off-state, as the Zener diodes are rated to withstand the excitation voltage. Two thyristors are used, so that each can be connected in inverse parallel with one set of the diodes in the diode bridge, so ensuring thyristor turn-off at the time of synchronization.

The starting and synchronizing torque can be increased if a resistor is placed in series with the field. Such an arrangement is shown in Fig. 9-34c, which also incorporates automatic synchronizing. The exciter field is on from the moment of starting, the thyristor in series with the diode bridge being off and so preventing any excitation current. In a like manner to that described for Fig. 9-34b, during run-up the field is short-circuited via the resistor. When the speed is near to synchronism, the synchronizing controller senses the low slip frequency and fires the series thyristor at the most favourable instant in the slip cycle, so exciting the motor field with direct current, pulling it into synchronism.

9-9 INVERTER-FED SYNCHRONOUS MOTOR DRIVES

Like the other inverter-fed a.c. motor drives the synchronous motor is fed via a rectifier, d.c. link, inverter, and on to the motor as shown in Fig. 9-35. However, as will be explained, the speed of the motor is proportional to the d.c. link voltage, the current is proportional to torque, and the excitation can be varied. The parameters controlling the motor are similar to those of the d.c. motor, so one will often see this drive described as a *brushless d.c. motor* drive. This is misleading, however, because the motor is of the same construction as the synchronous motor and in no way resembles the construction of a d.c. motor.

As explained in Sec. 9-1-1, the synchronous motor can be made to take its current at a leading power factor by having a high field excitation current. By having the motor stator current in advance of its voltage the inverter shown in Fig. 9-35 becomes self-commutating; the current in the thyristors is forced down to zero as a result of natural commutation. When a thyristor is fired to turn-on, the current is transferred to it from the previously conducting thyristor by the

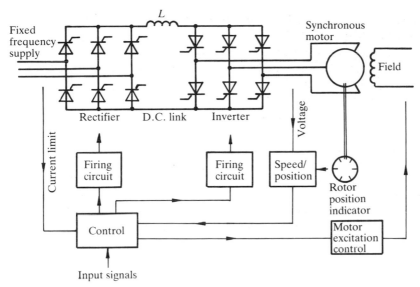

Figure 9-35 Circuit layout for the inverter-fed synchronous motor drive.

internally generated voltage of the motor; hence inexpensive conventional thyristors can be used, avoiding the need for any commutating circuits.

The inverter operation is like the inverting mode of a converter as described in Sec. 3-3, Figs. 3-7g and 3-26 illustrating typical voltage waveforms. The firing angle advance must be sufficient to allow adequate overlap time; this means that to commutate the thyristors the current must be leading the voltage, that is, the power is transferred from the d.c. side to the a.c. side operating at a leading power factor of typically 0.85, representing a firing angle advance of approximately 30°. The control circuits must ensure that the excitation at all speeds is sufficient to give the motor a power factor of about 0.85 leading. The I^2R losses of the motor will inevitably be higher than if it were possible to excite to unity power factor. If the motor were to be operated at unity or a lagging power factor then the inverter would have to be force-commutated in a similar manner to the cage induction motor drive with feedback diodes in the inverter.

The firing of the inverter thyristors is timed from the motor voltage and the rotor position sensor shown in Fig. 9-35; hence the inverter frequency is tied to that of the motor. An increase in link voltage causes a corresponding increase in speed, that is, frequency; hence the voltage/frequency ratio is constant, the same condition as for induction motors. The d.c. link voltage is determined by the firing angle in the supply rectifier, an increase in this voltage automatically accelerating the motor. Current limit is achieved in a like manner to the d.c. motor by feedback loops operating on the rectifier firing angles.

As far as the main a.c. input supply to the system is concerned, the characteristic is the same as for the d.c. motor, that is, a current magnitude matched to torque with a power factor which is lagging and proportional to speed.

The above description of operation is that which applies once the motor is running at above (say) 5 Hz when there is sufficient motor internal voltage to self-commutate the inverter thyristors. At starting and low speeds an alternate strategy has to be employed. When stationary, a constant direct current in the stator will produce a torque attempting to pull the rotor and its field into line with the stator field, as shown in Fig. 9-3. To ensure the correct inverter thyristors are fired, to give a torque, the physical position of the rotor is sensed by the rotor position indicator shown in Fig. 9-35. Since little commutating voltage (motor internal voltage) is available at low speed, the current has to be forced down to zero by delaying the firing in the rectifier, turning off the inverter thyristors. The thyristors are now fired to give the next sequence as determined by the rotor position. By pulsing the link current on and off in this manner the motor is accelerated to the speed where the motor's internal voltage will self-commutate the inverter thyristors.

The inverter-fed synchronous motor drive can be reversed by altering the firing sequence of the inverter thyristors. Referring to Fig. 9-35, the system can regenerate by reversing the roles of the converters with the d.c. link voltage reversed, passing power back from the motor (now generating) through the inverter (now rectifying), supply rectifier (now inverting), and back into the fixed frequency a.c. supply.

The exciter must produce field current when the drive is stationary, and to provide this the system shown in Fig. 9-36 can be used. The three-phase voltage to the three-phase stator winding of the exciter will set up a rotating flux, which by transformer action induces a three-phase voltage into the rotor winding, which is then rectified and fed into the motor field. Control of the thyristor voltage regulator will control the exciter flux and hence the field current. Once rotating, the induced voltage will change in magnitude, but provided that the exciter stator flux is arranged to rotate in the opposite direction to the shaft the exciter voltage will increase with speed. Further, if the pole-number of the exciter is less than that of the main motor the exciter voltage will only rise slightly with speed.

The inverter-fed synchronous motor drive is simple in its inverter requirement, is low cost, efficient, robust, and reliable, and can be used for systems up to the highest ratings. Its major disadvantage is that it presents the same problems as the

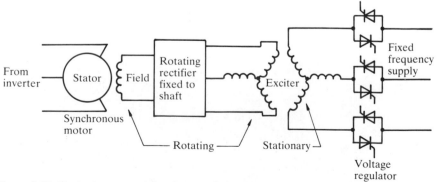

Figure 9-36 Excitation system of an inverter-fed synchronous motor.

rectifier-fed d.c. motor in respect to the a.c. supply of poor power factor at low speed.

The inverter-fed synchronous motor can be used for low power drives where a permanent magnet synchronous motor may be used.

9-10 RELUCTANCE AND STEPPER MOTOR DRIVES

Brief mention was made of the reluctance torque associated with the synchronous motor in Sec. 9-1-1, illustrated in Fig. 9-2. This principle of the development of torque in an asymmetrical rotor is exploited in two types of motor, the reluctance motor for higher powers and the stepper motor for application in very accurate position and control applications.

The basic construction of a four-phase reluctance motor is shown in Fig. 9-37, where it can be seen that there are eight poles, opposite pole windings being connected in series to form one phase of a four-phase arrangement labelled A, B, C, and D. The rotor has six projecting poles. The magnetic circuit is laminated but the rotor does not carry any winding; hence the construction is simple and robust, leading to an inexpensive design compared to other motors. The motor is often called a switched-reluctance motor because the windings are energized in turn from a switched d.c. source. When coils A are energized the rotor will align itself to the position shown in Fig. 9-37a to give minimum reluctance, that is, maximum flux for a given current. If, at this position, coils A are de-energized and coils B are energized, the flux pattern becomes that shown in Fig. 9-37b, with a torque being produced, causing clockwise movement until the poles align after 15° to that position shown in Fig. 9-37c when energizing coils C again produce torque. Hence by switching the coil currents continuously torque production results, with the speed being related synchronously to the switching rate.

The power control circuit for each set of coils is required to switch the d.c.

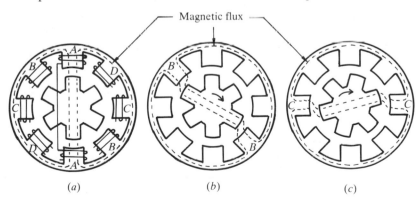

Figure 9-37 Construction and torque production in a four-phase switched-reluctance motor. (a) Coils A energized, rotor poles aligned to stator, zero torque. (b) Coils B energized, torque acting in the clockwise direction to align poles. (c) Rotor moved on by 15°, coils C energized, torque being produced.

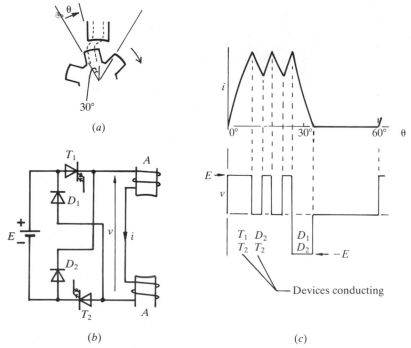

Figure 9-38 Chopper-type drive. (*a*) Positive torque developed only when teeth approaching alignment. (*b*) Circuit for one phase only. (*c*) Waveforms of the coil current and voltage.

supply on so as to establish coil current for the half period when the teeth are approaching alignment, as shown in Fig. 9-38*a*, and also during this time to control the mean level of the current by chopping the current on and off. Referring to Fig. 9-38*b* and *c*, the thyristors T_1 and T_2 are both turned on until the current reaches the predetermined current when T_1 is turned off, the current decaying via T_2 and diode D_2 until the control circuit senses the minimum predetermined level when T_1 is turned on again. Finally, when the coil current is to cease both T_1 and T_2 are turned off, with energy being returned to the voltage source via D_1 and D_2.

The number of thyristors and diodes can be halved to that shown in Fig. 9-38 by using the system shown in Fig. 9-39. The four sets of coils are connected to a common terminal which is maintained at a potential midway between the d.c. supply rails by two equal capacitors. Reference to the required on-periods of the coil currents (see Fig. 9-39*b*) shows that at all times current is flowing in either A or C, and in either B or D. The thyristors can chop the current to control its magnitude, the diodes conducting during the off-periods. The capacitors have the dual roles of maintaining the d.c. link voltage constant and acting as a current path during switching. It is necessary for there to be an even number of phases.

Gate turn-off thyristors are shown in Figs 9-38 and 9-39, but bipolar transistors can be used. If conventional thyristors are used then commutation networks must be added.

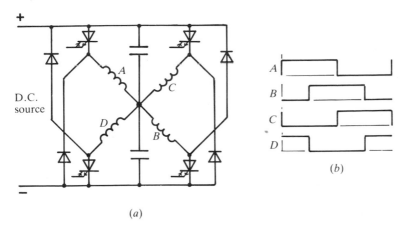

Figure 9-39 A centre-balanced split supply system. (*a*) Basic power circuit for a four-phase switched-reluctance motor. (*b*) Showing the on-periods for the current in each coil.

The current pulses to the stator coils of the switched-reluctance motor are timed by feedback signals from a rotor position sensor and positioned within the stepping cycle to maximize the torque. In a similar way to the inverter-fed synchronous motor, the voltage/frequency ratio is maintained constant, with only the voltage being controllable; hence an increase in the d.c. supply voltage will lead to a corresponding linear increase in speed. The torque level is proportional to the current; hence the overall characteristic is similar to that of the d.c. motor.

The switched-reluctance motor can be controlled to reverse or to give regenerative braking. Compared to the inverter-fed drives of the cage induction motor and the synchronous motor, the power electronics is much simpler in the switched-reluctance motor drive and the motor itself is less expensive. As one of the most recent additions to variable-speed drives, it has yet to be fully developed.

The stepper motor used for position or speed control can be of the reluctance type, explained earlier, or a hybrid type, as illustrated in Fig. 9-40 where the rotor is composed of two teethed rotor discs sandwiching a permanent magnet. In the hybrid type the torque is produced by both the permanent magnetic flux and the reluctance principle. In the motor illustrated in Fig. 9-40*a* the coil currents are as shown in Fig. 9-40*c*, and it can be reasoned that when the coil current is switched the rotor will step round by 18° With the small motors, to achieve a fast switching rate, the time constant of the coil circuit has to be reduced by the addition of a forcing resistor R, as illustrated in Fig. 9-40*d* and *e*. Figure 9-40*d* shows a simple arrangement with coil current flow in one direction only, whereas that of Fig. 9-40*e* can permit current flow in either direction. The addition of the forcing resistor greatly reduces the drive efficiency. The application of stepper motors is in those situations where ease and accuracy of control are paramount, rather than in situations requiring high torque and power for which they are not suited. The control elements can be either power transistors or MOSFETs.

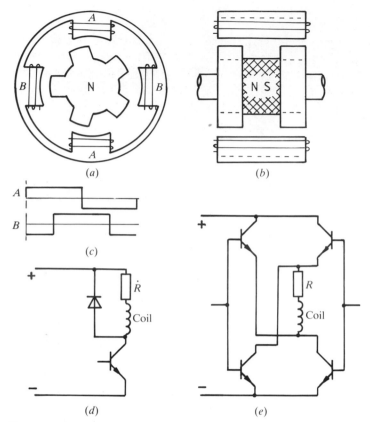

Figure 9-40 Hybrid-type stepper motor. (*a*) End view. (*b*) Side view showing the permanent magnet. (*c*) Coil current. (*d*) Simple uni-directional coil current control. (*e*) Bi-directional coil current control.

9-11 DRIVE CONSIDERATIONS

It has been seen in the previous sections that it is possible to use the induction motor in a drive and control it to provide any desired output, such as constant speed, constant torque, or braking. However, in general, the power semiconductor equipment and control electronics are more expensive and complex than those required with the d.c. motor, although the difference in cost between the cage induction motor and the d.c. motor will bring the overall costs closer. Applications, such as reeling drives requiring a constant power output, can be arranged by maintaining constant-power throughput in the d.c. link of the inverter.

Cost is certainly an important factor, but not the only criterion in choosing a particular drive for a given application. Environmental considerations may demand no rubbing contacts, with the consequent risk of sparking, so forcing the use of an a.c. motor. The supply authorities would look more favourably on a drive having a diode rectifier input, such as in some inverters, than the use of the phase-controlled thyristor rectifiers.

Inevitably the inverter waveforms to the motor contain harmonic components which increase the motor losses. In practice, the motor may have to be derated by 10% compared to normal sinewave excitation. The cooling of a motor often depends on air turbulence set up within the motor by a fan on the shaft. If a motor is run below its rated speed, then the cooling will be less efficient, and hence either a reduced torque rating is specified or the motor is force-cooled by external means.

A major consideration in drives is the running cost. If the overall efficiency can be raised as a result of reducing the drive losses then clearly this is a cost saving. An example of loss reduction is that which is possible with an underloaded motor. Taking, for example, a cage induction motor, its design will be such as to maximize its efficiency at or near to the rated load condition. The losses of such a motor fall broadly into two components, one the magnetizing loss which is a function of the voltage and the other the conductor loss which is a function of current (torque). When underloaded, a reduction in voltage to the motor will only slightly reduce the speed but will significantly reduce the magnetizing loss. A voltage reduction will increase the in-phase component of the current in order to supply the same mechanical power, but because the quadrature magnetizing component is reduced there may be little or no overall increase in current magnitude, so that overall the losses are reduced.

For a fixed-frequency supply the voltage to the motor can be reduced by phase angle delay using the circuit shown in Fig. 9-14, although this does introduce harmonic losses. If a gate turn-off thyristor were used in conjunction with diodes as shown in Fig. 6-2c then the current could be chopped on and off within the cycle to give voltage reduction without introducing the inherent poor power factor with conventional thyristors as shown in Fig. 6-9. Care must be taken in the control circuits to ensure full voltage is immediately restored if the load torque suddenly increases; otherwise the motor might stall.

With inverter-fed variable-speed drives, energy savings are possible by matching the voltage to the power demand at a given speed so as to maximize efficiency when underloaded. Examples where savings are possible are in pump and fan drives, where the torque is proportional to the square of the speed, giving considerable saving by the use of a variable-speed power electronics drive matched to give minimum motor loss at all speeds. To be an economic proposition the annual saving in cost of losses must exceed the annual capitalized cost of the additional power electronics equipment.

The advent of the microprocessor has expanded the versatility of performance and control of power electronics equipment. Many of the control functions previously performed by analogue electronics such as in selecting the switching instants in pulse-width modulation are now controlled by microprocessor systems. The microprocessor in association with computers can store data, handle complex sequencing, monitor for faults, and control to match the drive to the load, and may be programmed for the various demands of different operations using the same drive.

9-12 WORKED EXAMPLES

Example 9-1

A 4-pole, 3-phase induction motor has an equivalent (per phase) circuit at 50 Hz as shown in Fig. 9-41, neglecting magnetizing losses. Determine and plot the torque-speed characteristic at (i) 200 V/ph, 50 Hz; (ii) 100 V/ph, 25 Hz; (iii) 20 V/ph, 5 Hz; neglecting any harmonic content in the applied voltage.

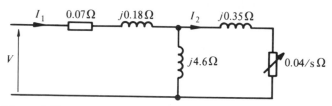

Figure 9-41

SOLUTION (i) The synchronous speed is 25 rev/s, the rotor speed being $25(1-s)$ rev/s. Analysing the circuit, the expression for I_2 in terms of the slip s gives

$$I_2 = \frac{200s}{0.075s + 0.0416 + j(0.544s - 0.0006)} \text{ A}$$

From Eq. (9-12), torque $T = \dfrac{I_2^2 R_2}{s\,\omega_{syn}}$ N m/phase, hence the total 3-phase torque is

$\dfrac{3I_2^2 \times 0.04}{50\pi s}$ N m.

(ii) The reduction of the frequency to 25 Hz reduces all the inductive reactance values in the equivalent circuit by the frequency ratio, that is, the values of the magnetizing reactance of $j4.6$ Ω at 50 Hz is now $j2.3$ Ω at 25 Hz. With the new voltage, the current is now

$$I_2 = \frac{100s}{0.075s + 0.0416 + j(0.272s - 0.0012)} \text{ A}$$

With a synchronous speed of 12.5 rev/s, the total torque is $T = \dfrac{3I_2^2 \times 0.04}{25\pi s}$ N m.

(iii) Here
$$I_2 = \frac{20s}{0.075s + 0.0416 + j(0.0544s - 0.006)} \text{ A}$$

and the total torque $T = \dfrac{3I_2^2 \times 0.04}{5\pi s}$ N m

The torque–speed characteristics for all three conditions are shown in Fig. 9-42. The effect of the stator impedance is to reduce the peak torque at the lower frequencies if a strictly constant voltage/frequency ratio is maintained. To achieve the

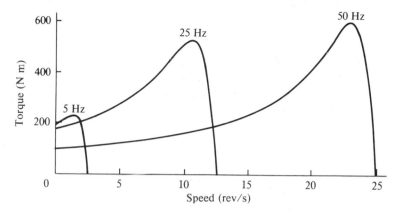

Figure 9-42

same torque at 5 Hz to that at 50 Hz, the voltage would have to be increased to $20 \times \sqrt{(595/227)} = 32.4$ V.

Example 9-2

If the input current to the equivalent circuit shown in Fig. 9-41, relating to Example 9-1, is held constant at 40 A, determine the torque-speed characteristic for (i) 50 Hz, (ii) 25 Hz, (iii) 5 Hz.

SOLUTION (i) With the input current I_1 constant at 40 A, the current

$$I_2 = \frac{j184s}{0.04 + j4.95s} \text{ A}$$

and the total torque $T = \dfrac{3I_2^2 \times 0.04}{50\pi s}$ N m

(ii) The change in frequency to 25 Hz makes

$$I_2 = \frac{j92s}{0.04 + j2.475s} \text{ A} \quad \text{and} \quad T = \frac{3I_2^2 \times 0.04}{25\pi s} \text{ N m}$$

(iii) The frequency, being 5 Hz, makes

$$I_2 = \frac{j18.4s}{0.04 + j0.495s} \text{ A} \quad \text{and} \quad T = \frac{3I_2^2 \times 0.04}{5\pi s} \text{ N m}$$

The rotor speed is the synchronous speed times $(1 - s)$.

The characteristics are shown in Fig. 9-43, where it can be seen that the peak torque occurs at low slip frequencies. As the slip nears zero, all the current is the magnetizing component, hence a high current setting will lead to saturation flux conditions.

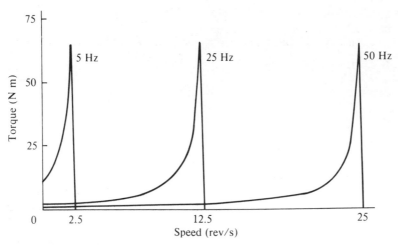

Figure 9-43

Example 9-3

An inverter supplies a 4-pole, cage induction motor rated at 220 V, 50 Hz. Determine the approximate output required of the inverter for motor speeds of (i) 900, (ii) 1200, (iii) 1500, (iv) 1800 rev/min.

SOLUTION Approximately, the slip may be neglected, with the output inverter frequency related to synchronous speed.
From Eq. (9-1),   frequency = (speed × pairs of poles)/60.
Also, at each condition, the voltage/frequency ratio = 220/50, hence the required inverter outputs are:

(i) frequency $= \dfrac{900 \times 2}{60} = 30$ Hz, voltage $= \dfrac{220}{50} \times 30 = 132$ V;

(ii) 40 Hz, 176 V; (iii) 50 Hz, 220 V; (iv) 60 Hz, 264 V.

Example 9-4

A 6-pole, star-connected induction motor has equivalent circuit parameters of: magnetizing inductance 132 mH, stator resistance 0.25 Ω, stator leakage inductance 4 mH, rotor resistance 0.20 Ω, and rotor inductance 3 mH, all referred to the stator. The motor is fed by a quasi-square-wave inverter at 40 Hz, such that the phase voltage is stepped as shown in Fig. 9-15a, having a top value of 280 V. Determine the input current waveform and find the torque at a slip of 0.04 by analysing its response to the harmonic components of the inverter output.

SOLUTION This form of analysis can only give an approximate guide to the motor's response, so it is adequate to use the approximate equivalent circuit of Fig. 9-12b:

$R_1 = 0.25\,\Omega$, $R_2 = 0.20\,\Omega$, $L_1 + L_2 = 4 + 3 = 7\,\text{mH}$, $L_m = 132\,\text{mH}$, and R_m is neglected.

The stepped voltage wave of Fig. 9-15a can be analysed in a similar manner to that used in Example 7-3 for Fig. 7-8b, giving $a_n = \dfrac{4}{n\pi}\left[\dfrac{280}{2}(1 + \cos\dfrac{n\pi}{3})\right]$, hence making the voltage expression for the inverter output as

$$v = 267.38\left[\sin\omega t + \frac{\sin 5\,\omega t}{5} + \frac{\sin 7\,\omega t}{7} + \frac{\sin 11\,\omega t}{11} + \frac{\sin 13\,\omega t}{13} + \dots\right]\text{V}$$

where $\omega = 2\pi f$, $f = 40$ Hz, the inverter output frequency.

The slip of 0.04 refers to the fundamental component.

Analysing the response to the fundamental component at 40 Hz, the current and voltage components in the equivalent circuit of Fig. 9-12b using peak values and taking the voltage V_1 as reference are: $V_1 = 267.38\underline{/0°}$ V, $s = 0.04$, $I_2 = 48.29\underline{/-18.53°}$ A, $I_0 = 8.06\underline{/-90°}$ A, $I_1 = 51.42\underline{/-27.07°}$ A. The synchronous speed is $(40/3) \times 2\pi = 26.6\pi$ rad/s. Using Eq. (9-12), the torque is

$$\left(\frac{48.29}{\sqrt{2}}\right)^2 \times \frac{0.2}{0.04} \times \frac{1}{26.6\pi} = 69.6\text{ N m/ph}$$

In finding the response to the fifth harmonic, the frequency f is $40 \times 5 = 200$ Hz. The synchronous speed is five times that of the fundamental but, as shown in relation to Eqs. (7-25) and (7-26), this component is of reverse sequence, hence the rotor rotates in the opposite direction to the harmonic flux. The slip then is

$$1 + \frac{(1 - 0.04)}{5} = 1.192$$

Using $f = 200$ and $s = 1.192$, analysis (peak values) of the equivalent circuit is: $V_1 = 267.38/5 = 53.48$ V, $s = 1.192$, $I_2 = 6.07\underline{/-87.28°}$ A, $I_0 = 0.32\underline{/-90°}$ A, $I_1 = 6.39\underline{/-87.42°}$ A, $\omega_{\text{syn}} = 26.6\pi \times 5$ rad/s; with a negative torque of 0.007 N m/ph.

The seventh harmonic produces a forward rotating flux, therefore here the slip is

$$1 - \frac{(1 - 0.04)}{7} = 0.863$$

Calculations for the seventh harmonic give the torque as only 0.002 N m/ph, hence at this condition the harmonic torques are negligible.

Continuing the input current analysis up to the 13th harmonic yields an expression of: $i = 51.42\sin(\omega t - 0.4725) + 6.39\sin 5(\omega t - 0.3052) + 3.26\sin 7(\omega t - 0.2191) + 1.32\sin 11(\omega t - 0.1409) + 0.95\sin 13(\omega t - 0.1194)$ A. The angle of lag of 0.3052 radian in, for example, the fifth harmonic is the angle of 87.42° divided by 5 to relate it to the fundamental reference $\omega = 2\pi40$.

The current is shown plotted in Fig. 9-44, using values from i above, the general shape being similar to that met in practice. It must be noted that the parameters in

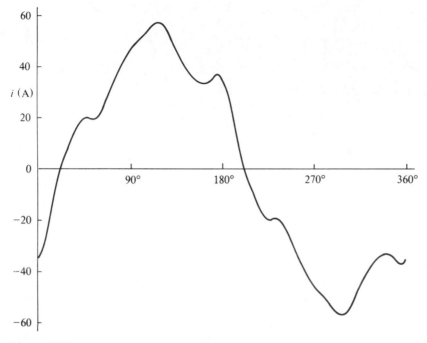

Figure 9.44

the equivalent circuit are not linear, their values being dependent on magnetic, load, and frequency conditions. However, this form of analysis gives a guide to motor response to a stepped voltage excitation.

Example 9-5

A 6-pole, 50 Hz, slip-ring induction motor is controlled by a slip-energy recovery scheme. Determine the angle of firing advance in the inverter at (i) 600, (ii) 800 rev/min, if the open-circuit standstill slip-ring voltage is 600 V, and the inverter is connected to a 415 V, 3-phase system. Neglect overlap and losses.

SOLUTION The system layout is that shown in Fig. 9-31a. From Eq. (9-1) the synchronous speed of the motor is $(50/3) \times 60 = 1000$ rev/min.

(i) From Eq. (9-6), at 600 rev/min the slip $= \dfrac{1000 - 600}{1000} = 0.4$. From Eq. (9-18), the rotor voltage at 600 rev/min is $600 \times 0.4 = 240$ V. Assuming 3-phase converter bridges and using Eq. (3-23), with $\alpha = 0$, $\gamma = 0$, and $p = 6$, the d.c. link voltage is

$$\frac{6 \times 240\sqrt{2}}{\pi} \sin \frac{\pi}{6} = 324 \text{ V. Using Eq. (3-24) with } \gamma = 0 \text{ and } p = 6, \text{ then } 324 =$$

$$\frac{6 \times 415\sqrt{2}}{\pi} \sin \frac{\pi}{6} \cos \beta, \text{ giving the angle of firing advance } \beta \text{ as } 54.7°.$$

(ii) At 800 rev/min, $s = 0.2$, rotor voltage $= 120$ V, $V_{dc} = 162$ V, and $\beta = 73.2°$.

Example 9-6

Recalculate the firing advance in Example 9-5(i) if there is an overlap of $20°$ in the rectifier and $5°$ in the inverter. Allow for diode and thyristor volt-drops of 1.5 V and 0.7 V respectively.

SOLUTION Using Eq. (3-23),

$$V_{\text{mean}} = \frac{6 \times 240\sqrt{2}}{2\pi} \sin \frac{\pi}{6} [1 + \cos 20°] - (2 \times 0.7) = 312.9 \text{ V}$$

Using Eq. (3-24),

$$312.9 - (2 \times 1.5) = \frac{6 \times 415\sqrt{2}}{2\pi} \sin \frac{\pi}{6} [\cos \beta + \cos (\beta - 5°)]$$

giving $\beta = 58.9°$.

Example 9-7

Using the data of Example 9-5 and taking the minimum required speed to be 600 rev/min, estimate the voltage ratio of a transformer to be interposed between the inverter and the supply. Also specify the power flow through the d.c. link as a ratio to the power input to the stator.

SOLUTION As both rectifier and inverter are the same 3-phase configurations, then the same relationship exists between the a.c. side and the d.c. link voltage when the firing angle advance is zero. Hence the voltage required at the transformer is 240 V, the slip-ring voltage at 600 rev/min. In practice, the firing advance will not be less than (say) $15°$ to allow for some overlap. The transformer ratio is approximately 415 V/240 V.

Using the power balance shown in Fig. 9-13c and taking the input stator power to be 100, the power out of the rotor is $100 \times 0.4 = 40$, and the shaft output power is $100 - 40 = 60$, neglecting losses within the motor.

Example 9-8

An inverter is feeding a 3-phase induction motor at a frequency of 52 Hz and at a voltage having a fundamental component of 208 V/phase. Determine the speed and fundamental component values of torque and current at a slip of 0.04. Use the equivalent circuit of Fig. 9-41 given with Example 9-1 for the motor data.

If the inverter output is suddenly changed to 48 Hz and 192 V/phase, determine the new torque and current assuming no speed change.

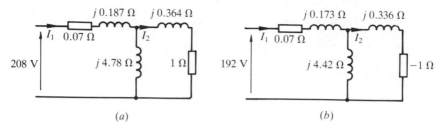

(a)   (b)

Figure 9-45

SOLUTION At 52 Hz, the speed of the 4-pole motor at a slip of 0.04 is $(52/2)$ $(1 - 0.04) = 24.96$ rev/s. The parameters in the equivalent circuit are as shown in Fig. 9-45a, giving a value of $183\underline{/-37°}$ A for I_1, with a torque calculated from Eq.

(9-12) of $T = \dfrac{167^2 \times 0.04}{0.04 \times 26 \times 2\pi} \times 3 = 514 \, \text{N m}.$

When the inverter frequency is reduced to 48 Hz the synchronous speed is 24 rev/s; hence with the motor speed at 24.96 rev/s the motor is running in a super-synchronous mode with a slip of $(24 - 24.96)/24 = -0.04$, that is, in a regenerative mode with negative slip. The equivalent circuit parameters are now as shown in Fig. 9-45b, noting that the equivalent resistor in the rotor circuit is negative, indicating it is a source of power. Using circuit theory I_1 is $208\underline{/-139.7°}$ A, which if reversed by $180°$ shows the generated current is 208 A at a power factor of 0.763 leading. The

torque is calculated in the same manner as before with $T = \dfrac{174^2 \times 0.04}{-0.04 \times 24 \times 2\pi} \times 3 =$

-602 N m, that is, the machine is in the braking mode and generating.

Example 9-9

A star-connected, 4-pole, 415 V, 50 Hz synchronous motor has a synchronous reactance of 0.5 Ω/phase and a rated stator current of 100 A. Determine the field excitation level required, compared to that required for 415 V open-circuit, for the motor to take current at 0.85 power factor leading at conditions of: (i) 50 Hz, 100 A; (ii) 30 Hz, 60 A. Neglect losses and also determine the load angle and torque.

SOLUTION The equivalent circuit to the synchronous motor is shown in Fig. 9-4a when losses are neglected. The excitation level is proportional to the internal voltage E.

(i) The given data is: frequency = 50 Hz, speed = 25 rev/s, $X = 0.5 \, \Omega$, $V = 415/$ $\sqrt{3} = 240$ V/phase, $\phi = \arccos 0.85 = 31.8°$, $I = 100$ A. Constructing the phasor diagram as in Fig. 9-4b gives $E = 270$ V/phase, with $\delta = 9.1°$, and from Eq. (9-5) a torque $T = 390$ N m. The excitation current is $270/240 = 1.124$ times that at open-circuit rated voltage.

(ii) At 30 Hz, speed = 15 rev/s, $V = 240 \times (30/50) = 144$ V/phase, $X = 0.5 \times$

$(30/50) = 0.3 \, \Omega$, $\phi = 31.8°$. Taking $I = 60 \, A$, the values calculated are $E = 154 \, V/$ phase, $\delta = 5.7°$, and $T = 234 \, N \, m$. The excitation current will be $154/144 = 1.071$ times that for open-circuit.

Example 9-10

In the elemental stepper motor shown in Fig. 9-46 determine the torque developed in terms of the current I and angle θ, given that the inductance as seen by the coil varies between a maximum of 6 mH to a minimum of 2 mH, with a sinusoidal variation between related to the angular position.

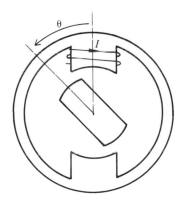

Figure 9-46

SOLUTION For such a singularly excited arrangement the reluctance torque developed is derived from the fundamental equation of torque $T = \dfrac{1}{2}I^2 \dfrac{dL}{d\theta}$.

From inspection of Fig. 9-46, the maximum value of inductance is at $\theta = 0°$ (poles in line) and the minimum value at $\theta = 90°$ (maximum air gap), giving an expresion for the inductance of $L = 4 + 2 \cos 2\theta$ mH.

The torque $T = \dfrac{1}{2}I^2 \dfrac{dL}{d\theta} = \dfrac{1}{2}I^2 \times 2 \times 2 \times (-\sin 2\theta)$, that is, a torque of $2I^2 \sin 2\theta$ mNm in a clockwise direction. Note that both at $\theta = 0°$ and $\theta = 90°$ the torque is zero, and that the torque is a maximum at $\theta = 45°$.

Example 9-11

Referring to Fig. 9-38, determine the waveform of the coil current for the system shown in respect to the reluctance motor. The inductance varies from a minimum value of 1.5 mH to a maximum value of 7.5 mH, and it may be assumed that the variation is linear with respect to the angular displacement. The d.c. supply is 160 V, and when established the current is to vary between a maximum of 100 A to a minimum of 60 A. The speed of the motor is 500 rev/min. Neglect all resistance and

device losses and assume that the thyristors are turned on at the instant when positive torque is available and that both thyristors are turned off to end the current when the rotor has travelled 90% into the positive torque sector.
Determine approximately the torque which will be developed by the 4-phase motor.

SOLUTION The time taken for the rotor to travel $30°$ is $(60/500) \times (30/360) = 0.01\,s = 10\,ms$. Taking $t = 0$ at the start of the current pulse when the inductance is a minimum, $L = 1.5 + \dfrac{(7.5 - 1.5)t}{0.01}\,mH = 0.0015 + 0.6t\,H$, the inductance reaching its maximum after $30°$ travel, that is, at $t = 0.01\,s = 10\,ms$.

The voltage developed across a varying inductor is given by the expression $v = \dfrac{d(Li)}{dt} = L\dfrac{di}{dt} + i\dfrac{dL}{dt}$, the component $i\dfrac{dL}{dt}$ being the back e.m.f.

We therefore have $160 = (0.0015 + 0.6t)\dfrac{di}{dt} + i(0.6)$,

from which $\dfrac{dt}{0.0015 + 0.6t} = \dfrac{di}{160 - 0.6i}$,

giving $\ln(0.0015 + 0.6t) = -\ln(160 - 0.6i) - 1.427$,
the arbitrary constant of -1.427 being calculated from the initial condition of $i = 0$ at $t = 0$.

When the current i reaches $100\,A$, the time $t = 1.5\,ms$ and the rotor has travelled $(1.5/10) \times 30 = 4.5°$, the inductance at this position being $2.4\,mH$.

For the next period the current freewheels down to $60\,A$, and taking $t = 0$ at the start of this period, $L = 0.0024 + 0.6t\,H$; using the equation $v = 0 = \dfrac{d(Li)}{dt} = L\dfrac{di}{dt} + i\dfrac{dL}{dt}$, we arrive at $\ln(0.0024 + 0.6t) = -\ln(0.6i) - 1.938$. When i has fallen to $60\,A$, $t = 2.66\,ms$, and the total rotor travel is now $4.5° + 8.0° = 12.5°$, with the inductance at this position being $4\,mH$.

For the next on-period we are back to the condition of $160 = \dfrac{d(Li)}{dt}$, with $L = 0.004 + 0.6t\,mH$ at the start $(t = 0)$, giving an on-time of $1.6\,ms$ to when the current again reaches $100\,A$, the rotor having now travelled to $17.3°$ with the inductance at $4.96\,mH$.

For the next off-period i falls to $60\,A$ after a further $5.5\,ms$, with rotor travel to $33.8°$, but this is beyond the $30°$ of positive torque. The thyristors are both turned off at $27°$ ($9\,ms$) which is 90% into the positive torque section, the current having then fallen to $71.9\,A$ in the $(27 - 17.3)°$ period of freewheeling. The current now decays via the diodes back through the voltage source, the equation now being $-160 = \dfrac{d(Li)}{dt}$, and $L = 0.0069 + 0.6t\,H$, the inductance being $6.9\,mH$ at the start of this period. Therefore $-160 = (0.0069 + 0.6t)\dfrac{di}{dt} + i(0.6)$; with an initial con-

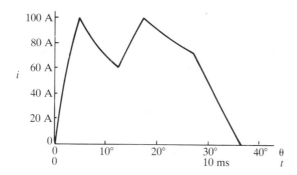

Figure 9-47

dition of 71.9 A, the current collapses to zero in 3.1 ms, that is, into the negative torque period for a time of 2.1 ms (6.3°).

The waveform of the current is shown in Fig. 9-47.

The torque can be determined by finding the mean power provided from the voltage source. From the current waveform the net energy taken from the d.c. source can be calculated from $\int vi\, dt$, energy being supplied during the periods between 0–1.5 ms and 4.2–5.8 ms and being returned between 9–12.1 ms. Approximately, the net energy per cycle is 17 J, giving over the cycle time of 20 ms a mean power of 850 W/phase. For four phases the total power is 3400 W at 500 rev/min yielding a torque of 65 N m (from power = $T\omega$).

The power semiconductor device requires protection against excessive voltage, current, and certain rates of change, so that the device is not damaged. Throughout the earlier chapters, much stress has been given to the current and voltage levels experienced by the devices in given applications, but, having chosen the appropriate device, no mention has been made of the manner in which the device is protected against damaging circumstances. An inspection of any cubicle containing power semiconductors will always show quite large inductors and capacitors whose sole purpose is protection, hence indicating the importance of protection.

Discussion of protection has been left to the end, as it is additional to the basic functioning of the many circuits. Although formulae specifying protection component values can be developed, choice of values used in practice are often empirical and based on a practical rather than theoretical basis. The basis of protection is outlined in this chapter, but without the preciseness of the earlier material.

10-1 CURRENT

A very onerous fault current to be protected against is that where a device provides a short-circuit path to the supply source. Figure 10-1a illustrates the possibility of short-circuits on an a.c. system which involves (say) a diode and, although the diode shown dotted would not be physically present in that position, it is quite possible in (say) a bridge circuit for alternative paths to exist for negative supply current flow. Figure 10-1b with a d.c. source is somewhat simpler, but the ultimate fault current is only limited by the resistance R.

The equation to the current flow in Fig. 10-1a depends on the angle at which the fault occurs in the alternating voltage cycle. The well-known equation for i is

$$ i = \frac{V_{\max}}{Z} \sin(\omega t - \phi) + I\, e^{-t/T} \tag{10-1} $$

where Z is the supply impedance, ϕ the steady-state current lag, I the transient component to satisfy initial conditions, and T the time constant. The derivation of this equation is illustrated in Example 2-1. If the resistance R is taken to be negligible and the fault occurs at the instant of peak voltage, then the peak current can after

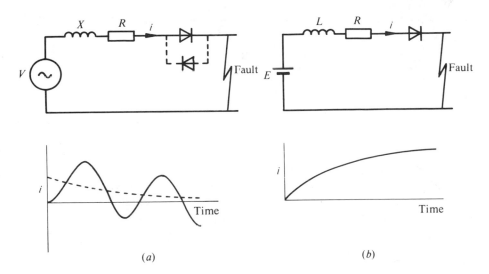

Figure 10-1 Short-circuit fault via supply. (*a*) A.C. source. (*b*) D.C. source.

one half cycle reach $(2V_{max})/X$. In practice, some resistance does exist and a factor of 1.8 may be used rather than 2, the transient component which gives the asymmetry in Fig. 10-1*a* dying to zero after a few cycles.

If the supply is strong, that is, has a low source impedance, then the fault current will be of such a magnitude that the fault must be cleared well before the first peak. As explained in Sec. 1-7 the thermal mass of a semiconductor device is low, hence an overcurrent will rapidly raise its temperature beyond the safe limit. Mechanical contactors frequently rely on clearing the fault at the first or (more likely) a subsequent current zero, but this is far too slow for semiconductor device protection. Breaking the fault current before the first zero means the breaking phenomenon is similar for both a.c. and d.c. circuit breaking.

The device used for current protection is the *fuse* and, before describing the fuse, it is necessary to consider carefully the requirements demanded of it, which are:

1. It must carry continuously the device rated current.
2. Its thermal storage capacity must be less than that of the device being protected, that is, the I^2t value let through by the fuse before the current is cleared must be less than the rated I^2t value of the device being protected.
3. The fuse voltage during arcing must be high enough to force the current down and dissipate the circuit energy.
4. After breaking the current, the fuse must be able to withstand any restriking voltage which appears across it.
   The I^2t (or $\int i^2\, dt$ value) mentioned in (2) was defined in Sec. 1-6.

The fuselink geometry is shown in Fig. 10-2, where it can be seen that the link

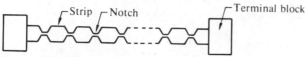

Figure 10-2 Fuselink geometry.

is a strip with several narrow notches. The fuse material is usually silver enclosed in a sand which absorbs the vaporization products of the arc. While carrying rated current, the heat generated in the notches is conducted to the wide sections and dissipated. However, during overcurrents, the notches melt and several arcs are struck in series.

The voltage drop across an arc is composed of the cathode drop region, the anode drop, and the resistive drop along the arc, amounting in total to (say) 50 V for silver. With several arcs in series, the current path is composed of many metal droplets forming many anode-cathode regions, and hence building up in total to a high voltage.

The stages in fusing are illustrated in Fig. 10-3. The onset of the fault current raises the fuse temperature until at time t_1 the notches in the fuse melt. Arcs are now formed, increasing the fuse voltage which can only be maintained by extracting the stored inductive energy in the faulted circuit through a decaying current. In time, the current decays to zero, when at t_2 the fault is cleared. Throughout the fusing period, the temperature of the device being protected is rising so the I^2t value of the fuse must include both the melting and arcing periods.

With less-severe faults, the fusing time will be longer, so it is essential to ensure that at all fault levels the fuse has an I^2t rating below that of the device being protected. With low-level overloads, the fuse may not be adequate due to (say) differing ambient temperatures, but here the slow-acting mechanical contactors would provide back-up protection.

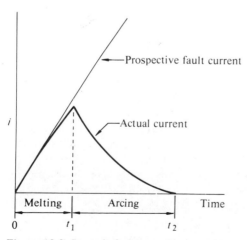

Figure 10-3 Stages in fusing.

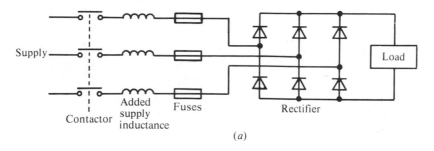

(a)

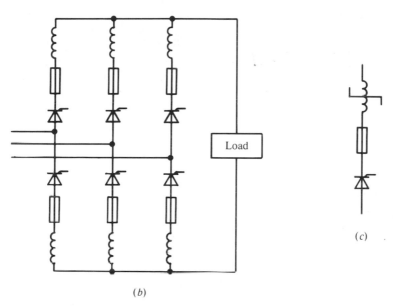

(b)

(c)

Figure 10-4 Positioning of fuses and added inductors. (*a*) Passive load. (*b*) Live (motor) load. (*c*) Possible position of a saturable reactor.

It may give adequate protection to place the fuses in the supply lines as shown in Fig. 10-4*a*, particularly with a passive load. However, with a live load, that is, a motor, it is possible for the load to provide the fault current, in which case each individual thyristor can be fused as shown in Fig. 10-4*b*. In general, the larger the equipment rating, the more tendency there is to fuse the devices individually.

Electromagnetic and explosive forces limit the maximum prospective current which a fuse can handle. It may be necessary to add inductance to limit the rate of rise of the fault current and hence avoid excessive stress on the fuse and device. The added inductance may be in the line, or in series with the individual device, and further, to avoid excessive power losses, a saturable reactor can be employed as shown in Fig. 10-4. The added inductance will, of course, increase the overlap so, as in many engineering designs, a compromise is required between circuit performance and protection.

Overcurrent protection of the power transistor presents a more difficult problem than for the thyristor. A rise in the collector current due to fault conditions may increase the collector-emitter voltage, particularly if the base current is insufficient to match the increased collector current demand. It is possible to have a high power loss within the transistor due to the rising collector-emitter voltage as the transistor goes out of saturation, without sufficient current to cause rapid fusing. It follows that a series fuse is inappropriate for the protection of the transistor. One solution is that illustrated in Fig. 9-29; here fault conditions are sensed by a rise in the collector-emitter voltage, in conjunction with high collector current, which by suitable circuitry fires the crowbar thyristor, which in turn causes blowing of the appropriate fuse-links, hence relieving the transistors of the overcurrent.

Less-severe overcurrents with transistors can be cleared by removing the base drive. A restraint to this protection is introduced by secondary breakdown, which can occur when the transistor is being turned off to block high voltages. Imperfections in the device structure cause uneven current distribution during turn-off, resulting in hot spots due to increased current density in those regions, with possible damage to the transistor.

10-2 VOLTAGE

The peak voltage rating of a device used in a particular application must obviously be greater than the peak voltage it experiences during the course of the circuit operation. In practice, a device experiencing (say) a peak off-state voltage of 500 V would be selected to have a rating between 750 and 1000 V, giving a safety factor of 1.5 to 2. However, transients within the circuit may momentarily raise the voltage to several thousand volts, and it is these voltages which must be prevented from appearing across the off-state device.

The origin of voltage transients having high dv/dt values may be from one of three sources:

1. Mains or supply source due to contactor switching, lightning, or other supply surges.
2. Load source, say the voltage arising from the commutator arcing on a d.c. motor.
3. From within the converter itself, due to switching of other devices, commutation oscillations, or the fuse arc voltage when it is clearing a fault.

To protect against all three sources of voltage transients, it is necessary to protect each device individually, as shown in Fig. 10-5. A capacitor C across the thyristor (or diode) means that any high dv/dt appearing at the thyristor terminals will set up an appropriate current ($= C\,dv/dt$) in the capacitor. The inductance in the circuit will severely limit the magnitude of the current to the capacitor and

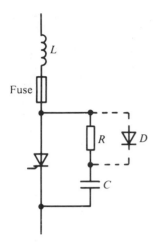

Figure 10-5 *R-C* protection of a single thyristor.

hence limit *dv/dt*. The *R-C* combination is often referred to as a *snubber network*, and also serves to limit the induced voltage spike produced during reverse recovery of the thyristor storage charge.

When the thyristor of Fig. 10-5 is fired, any charge on the capacitor will be discharged into the thyristor, possibly giving an excessively high *dI/dt*, but this can be suitably limited by the inclusion of the resistor *R*. A diode *D* may be included to by-pass *R* for improved *dv/dt* protection. Typically the values of the protection components will be: *C*, 0.01 to 1 μF; *R*, 10 to 1000 Ω; and *L*, 50 to 100 μH. The exact values chosen will depend on the circuit voltage and the stored energy capacity of the transient source, but often empirical formulae based on previous design experience are used.

For small power converters, it may be sufficient to use protection across the supply lines and/or the load lines as shown in Fig. 10-6. For large converters, suppression of external source transients (as shown in Fig. 10-6) will aid the individual device suppression circuit.

In Fig. 10-6*a*, it is possible for the circuit-breaker feeding the transformer to open and break (say) the magnetizing current before the current zero. If this happens, stored magnetic energy remains in the core, oscillating with the small stray capacitance, hence resulting in high voltage. By using added capacitors across the lines, the magnitude of the voltage is reduced to a safe level $(\frac{1}{2}LI^2 = \frac{1}{2}CV^2)$. The resistor *R* serves to damp out the oscillations. The *R-C* network across the load in Fig. 10-6*b* serves to absorb any transient energy originating in the load.

The selenium metal rectifier used extensively in the past for rectification may be used for protection. The metal-to-selenium junction has a low forward breakover voltage but, more importantly, a well-defined reverse breakdown voltage. The characteristic of the selenium diode is shown in Fig. 10-7, where the breakover or clamping voltage is of the order of 72 V for the type used in protection. If such a

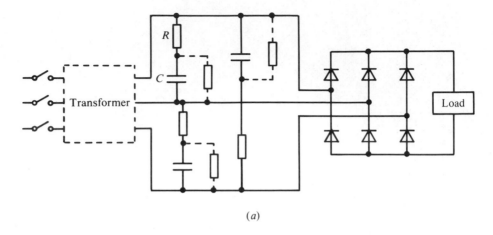

(a)

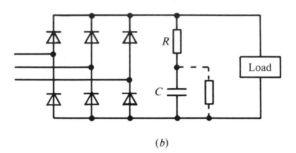

(b)

Figure 10-6 *R-C* suppression of external transients. (*a*) From mains source. (*b*) From load source.

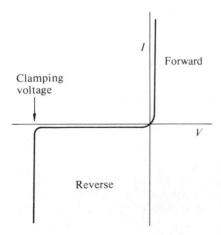

Figure 10-7 Selenium diode characteristic.

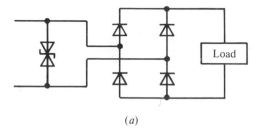

(*a*)

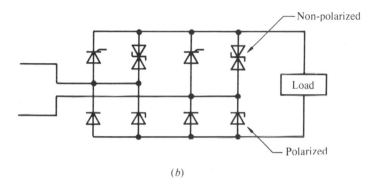

(*b*)

Figure 10-8 Positioning of selenium suppression devices. (*a*) A.C. supply transient suppression using non-polarized device. (*b*) Individual protection of thyristors and diodes.

device is placed in parallel with a power diode which has a higher peak reverse-voltage value, then any transient voltage will be clipped, so giving protection.

The selenium diode (or suppressor) must be capable of absorbing the surge energy without undue temperature rise. Each cell of a selenium diode is normally rated to 25 V a.c. (r.m.s.), so if (say) the a.c. supply in Fig. 10-8*a* is 150 V (r.m.s.), then the diode would require six cells in series. A six-cell diode would clip a voltage at $6 \times 72 = 432$ V, so a parallel diode rated at a peak reverse voltage of 450 V would be protected. For the application shown in Fig. 10-8*a*, or for thyristor protection as in Fig. 10-8*b*, the suppressor must be non-polarized, in which case cells are placed in series in opposite sense so that the clamping voltage can operate in either direction. With the diode as shown in Fig. 10-8*b*, the suppressor is a polarized diode, that is, cells in one direction only, as only reverse protection is required.

Compared to the *R*-*C* snubber circuit, the suppressor does not limit dv/dt to the same extent, due to their lower internal capacitance, but they do give well-defined voltage clipping. In protecting a device, the reliability of the suppressor is not so good as that of the *R*-*C* combination.

The avalanche diode is of different construction to the normal silicon diode, in as much that internal conditions differ during reverse voltage breakdown, so permitting the avalanche diode to absorb reverse current pulses without damage. The avalanche diode can be considered to be self-protecting.

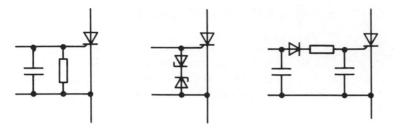

Figure 10-9 Gate protection.

In some applications, attention may have to be given to protecting the thyristor gate against transients, and three such circuits are shown in Fig. 10-9. Other variations of these circuits, using combinations of diodes, resistors, and capacitors, are possible.

To illustrate the complete protection of a converter, Fig. 10-10 shows elements giving complete protection to surge currents and voltages, dV/dt and dI/dt, and the contactors for the long-time low-level overload.

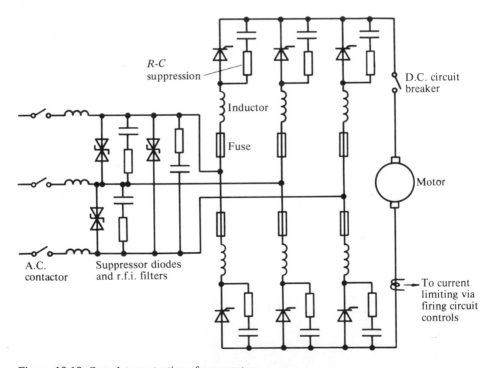

Figure 10-10 Complete protection of a converter.

10-3 GATE TURN-OFF THYRISTOR AND MOSFET

The gate turn-off thyristor is protected in a similar manner to the conventional thyristor for overcurrent, that is, by a very fast acting fuse. Additionally turn-off can be achieved by reverse gate current, but the time taken to detect the fault condition and initiate turn-off together with the inherent delay time of the device makes this method unreliable.

During normal turn-off of the gate turn-off thyristor it is essential to limit the rate at which the voltage rises; otherwise excessive power dissipation will occur in the narrow region into which the current is being squeezed.

The snubber network shown in Fig. 10-11a is generally used to limit the rate of voltage rise at turn-off. The size of the capacitor can be estimated from $I = C\, dv/dt$ where I is the anode current being switched and dv/dt the allowable rate of rise of voltage. The series resistor R is present to prevent the instant discharge of the capacitor-stored energy into the thyristor at turn-on, a rough guide to its size being estimated by taking the time constant RC as being about one-fifth of the turn-on time. The resistor R must be capable of dissipating a power of $\frac{1}{2}CV^2$ times the switching frequency. In practice the snubber network is a source of loss which needs to be minimized consistent with protection requirements. The diode is present to eliminate the delay the resistor introduces in protecting against rapid dv/dt. In bridge and other circuits it may be necessary to use the simple circuit of Fig. 10-11b, depending on the location of circuit and stray inductances. The snubber circuit must be located physically as near as possible to the device in order to minimize stray inductance effects.

In protecting the MOSFET against overcurrent a current-sensing control must be used which will turn off the MOSFET by reduction of the gate voltage. Like the power transistor a fuse is not appropriate because the rising current results in the appearance of a drain-source voltage which limits the current but results in destructive power generation within the device.

When switching off the current in the MOSFET even a low stray circuit inductance can in association with the very rapid switching time induce excessive drain-source voltages. Referring to Fig. 10-12a, a Zener diode acts as a voltage clamp. In

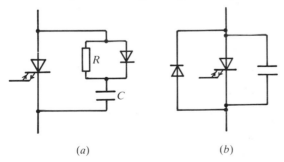

(a) (b)

Figure 10-11 Snubber networks for turn-off protection. (a) Polarised snubber for single device. (b) Slow rise parallel circuit.

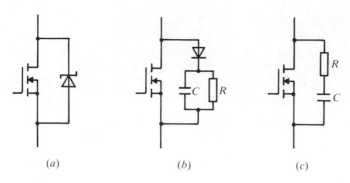

     (*a*)   (*b*)   (*c*)

Figure 10-12 Drain-source overvoltage protection for MOSFETs. (*a*) Zener diode. (*b*) Polarized capacitor circuit. (*c*) *RC* snubber network.

the circuit of Fig. 10-12*b* the capacitor limits the voltage rise with the resistor, dissipating the transient energy. In the circuit of Fig. 10-12*c* the *RC* snubber circuit limits the voltage rise but slows down the rate at which the MOSFET can be switched. Which of the alternative circuits shown in Fig. 10-12 is used is dependent on the circuit parameters, switching rates, and the losses incurred by the protection network.

TUTORIAL PROBLEMS

1. Discuss and explain the following information which is given in manufacturer's data relating to a thyristor: (i) on-state direct current, (ii) off-state direct current, (iii) latching current, (iv) continuous off-state direct forward voltage, (v) continuous direct reverse voltage, (vi) non-repetitive peak reverse voltage, (vii) dv/dt, (viii) di/dt, (ix) $i^2 t$, (x) turn-off time, (xi) gate firing voltage, (xii) gate firing current.

2. Describe the ideal form of gate current required to turn on a thyristor.

Draw and explain a firing circuit which may be used to fire a thyristor, making particular reference to the manner in which the firing delay angle is controlled.

3. A thyristor current rating is quoted as being 30 A mean for a sinusoidal current waveform extending from 90° to 180°. Determine the r.m.s. value of this current, and hence determine the level current the thyristor could be expected to carry for one-third of a cycle.

If the current (i amperes)–voltage (v volts) characteristic of the above thyristor can be quoted as $v = 1.0 + 0.007i$, determine the mean power loss for each of the above waveforms.

4. Describe the manner in which power semiconductor devices are cooled, and discuss the reasons why only short-time overloads are permitted.

A thyristor has a mean rated power loss of 25.2 W, with a juntion-to-base thermal resistance of 1.8 °C/W. If the ambient temperature is 30 °C, and the junction temperature is not to exceed 125 °C, select from the following the most suitable cooling fin.

Fin type A: thermal resistance 2.3 °C/W at 15 W, 2.0 °C/W at 30 W, 1.9 °C/W at 40 W.
Fin type B: 2.0 °C/W at 20 W, 1.8 °C/W at 35 W, 1.7 °C/W at 50 W.
Fin type C: 1.6 °C/W at 30 W, 1.45 °C/W at 45 W, 1.35 °C/W at 60 W.

5. Describe the characteristics of (i) the thyristor, (ii) the power transistor, and (iii) the gate turn-off thyristor. Compare and contrast these three devices.

6. A d.c. load requires a mean voltage of 200 V and takes a level current of 40 A. Design rectifier circuits to the following configurations, specifying the required a.c. supply voltage and the required ratings of the diodes. Neglect diode volt-drops and overlap.
   1. Single-phase, bi-phase half-wave
   2. Single-phase bridge
   3. Three-phase half-wave
   4. Three-phase bridge
   5. Double-star six-phase half-wave

7. Repeat the designs in Prob. 6, but assume a diode volt-drop of 0.7 V and an overlap angle of 15°.

8. Sketch the circuit of a single-phase bridge half-controlled rectifier which includes a commutating (flywheel) diode.

For (say) a firing delay angle of 100°, sketch the waveforms of the load voltage and current for (i) a pure resistance load, (ii) a light inductive load with continuous current, (iii) a small d.c. motor (constant field current) with discontinuous current.

Discuss with reference to the waveforms of the motor reasons why the frame size of the motor is larger when supplied from a thyristor converter than when supplied from a constant-level d.c. source.

9. An inductive d.c. load of 40 V at 50 A is to be supplied by a three-phase half-wave connection of diodes fed by a delta/inter-connected-star transformer from a 415 V three-phase supply.

Draw waveforms of (*a*) load voltage, (*b*) diode current, and (*c*) a.c. supply current.

Taking a diode volt-drop of 0.7 V and an overlap angle of 20°, estimate the following: (i) transformer secondary voltage, (ii) diode rating, (iii) a.c. supply r.m.s. current, (iv) transformer secondary rating, (v) transformer primary rating, (vi) transformer leakage reactance.

10. Draw the circuits of both the full- and half-controlled three-phase rectifier bridges. Describe the essential differences between the two circuits, and the function of the commutating diode in the half-controlled bridge.

For each of the above bridges, calculate the mean load voltage if the a.c. supply voltage is 220 V and the firing delay angle is 45°. Neglect device volt-drops and overlap. Sketch the load waveform shapes.

11. Derive formulae for the output voltage of half-controlled rectifier bridges for (a) single-phase, and (b) three-phase configuration. Sketch the load-voltage waveforms for firing delay angles of (i) 0°, (ii) 45°, (iii) 90°, and (iv) 135°.

Taking level load current, and neglecting overlap effects, calculate the percentage harmonic content of the a.c. supply current, and its power factor at each of the above firing delays.

12. Derive a general formula for the mean load voltage of a p-pulse fully-controlled rectifier in terms of the a.c. supply voltage, the firing delay angle α, and overlap angle γ.

Show that the rectifier may be considered as a source of voltage $V_0 \cos \alpha$, in series with an equivalent volt-drop of $(pIX)/2\pi$, where V_0 is the no-load voltage at zero firing delay, I the load current, and X the per-phase reactance of the supply.

13. Explain the cause and effect of overlap in a converter circuit.

Describe inversion and specify the conditions for it to take place.

Plot the thyristor voltage waveform in a three-phase fully-controlled bridge for the following conditions: (i) firing delay angle 30°, overlap angle 20°; (ii) firing advance angle 45°, overlap angle 15°.

14. Design a 12-pulse converter, specifying transformer and diode ratings, for supplying a d.c. load of 600 V, 3000 A. Take the supply to be 11 kV, three-phase, and allow for a drop in rectifier output voltage of 10% from no-load to full-load.

15. An uncontrolled 50-Hz, 6-pulse rectifier, with a peak voltage of 200 V in its output waveform, supplies a load of resistance 5 Ω, inductance 20 mH. The overlap at this condition is 20°. Determine the load-current waveform, and analyse the harmonic content of both the load voltage and current.

16. Discuss the relative merits of reducing the ripple content of a d.c. load voltage waveform by alternatively using filters or going to a higher-pulse rectifier.

A load of 25 Ω requires a mean voltage of 100 V. Design an LC filter which will reduce the ripple factor to 5% for (i) 2-pulse, (ii) 3-pulse rectifier, neglecting any overlap.

17. Discuss the relative merits of controlling a heating load by triacs operating in (i) phase angle delay, and (ii) integral cycle mode of control.

If a heating load (pure resistance) is set to one-third maximum power, determine the firing angle delay in phase angle control and the harmonic content of the current. Also determine the harmonic content of the current if integral cycle control is used with one on/off cycle occupying 24 cycles.

If the above heating load is 0.05 Ω, fed from a 240 V, 50 Hz supply of impedance (0.4 + $j0.25)\Omega$, estimate the harmonic voltage components appearing at the supply terminals for each form of control.

18. A three-phase star-connected heating load of 12 Ω per phase is fed via a fully-controlled inverse-parallel connected thyristor arrangement (Fig. 6-4a) from a 415 V, 50 Hz supply. Determine the firing delay angle required at 75% maximum power, and analyse the supply current for harmonic content and power factor for (i) neutral connected, and (ii) no neutral.

19. Discuss the use of inverters in industry for providing standby supplies. Describe the type of inverter which might be used for this purpose.

20. A d.c. chopper (Fig. 4-3) is used to control the power in a load of resistance $2\,\Omega$ fed from a 36 V battery. Allowing a turn-off time for each thyristor of $60\,\mu s$, estimate values for the capacitor and inductor, neglecting all losses. Plot waveforms to scale of the voltages and currents in the circuit.

21. The power in a $6\,\Omega$ load fed from a 48 V battery is controlled by a series resonance turn-off circuit (Fig. 4-8). Allowing a thyristor turn-off time of $50\,\mu s$, estimate suitable values for the inductor and capacitor. Determine the maximum rate at which the thyristor could be switched and the load power at this condition.

22. Repeat Prob. 21 using the parallel resonance turn-off circuit (Fig. 4-9).

23. Describe with the aid of circuit diagrams and waveforms the basic layout of the constant-voltage source three-phase inverter. Include discussion on the purpose of the feedback diodes and the need to maintain firing gate pulses over a long period.

Describe a suitable commutation circuit if thyristors are used in the above inverter.

24. An inverter fed at 300 V uses complementary impulse commutation for turn-off (Fig. 4-19). If the load is inductive, taking 30 A, estimate suitable values for the capacitor and inductor, allowing a turn-off time of $35\,\mu s$.

25. Repeat Prob. 24 using the impulse commutation circuit (Fig. 4-22).

26. A single-phase inverter (Fig. 5-20a) feeds a load of $5\,\Omega$, 0.02 H, from a 300 V d.c. source. If the output is to be quasi-square wave with an on-period of 0.8, determine the load current waveform and analyse its harmonic content when operating at 60 Hz. Neglect commutation effects and inverter losses. Determine the load power associated with each harmonic.

27. Describe with the aid of circuit diagrams and waveforms the basic layout of the constant-current source three-phase inverter.

Compare and contrast this inverter to the constant-voltage source inverter.

28. A six-pulse cycloconverter (Fig. 5-7b) feeds a three-phase load of 25 A at 240 V/phase. Determine the supply voltage and required ratings, taking an overlap angle of $15°$ at the instant of maximum load voltage.

29. Discuss with diagrams, where appropriate, solutions to the following problems relating to thyristors:

    (*a*) The connection of two or more thyristors in series.

    (*b*) The connection of two or more thyristors in parallel.

    (*c*) Protection against overloads and transients.

30. Discuss the application of power semiconductor devices to:

    (*a*) Induction heating.

    (*b*) Electrochemical applications.

    (*c*) H.V.D.C. transmission.

31. A d.c. motor with constant field excitation is fed via a resonant power controller from a battery source. The circuit description is: the battery is connected from its positive terminal into a series combination of a thyristor, an inductor, and a capacitor, back to the negative terminal; the motor in series with a diode is connected across the capacitor; a commutating (flywheel) diode is connected across the motor terminals.

Taking the motor inductance to be very much greater than that in the thyristor circuit, explain how the thyristor is naturally turned off and how the frequency of firing dictates the motor speed. Draw typical waveforms of the circuit voltages and currents.

32. Sketch the arrangement of using two parallel converters for a reversible d.c. motor drive. Show, in block diagram form, the closed loops required to maintain constant speed with varying torque, to limit the converter current, and to ensure the avoidance of short-circuits on the a.c. supply lines.

Describe the sequence of events, detailing the function of the various control loops, following a step input command to change the speed from maximum to reverse.

33. The armature of a small d.c. motor with constant field current is fed from a half-controlled single-phase bridge rectifier. Derive an expression for the armature current, and sketch its waveform, given the following data and conditions: a.c. supply 240 V, 50 Hz, thyristor fired at 90° delay, armature inductance 0.06 H, speed 140 rad/s, armature characteristic 0.9 V/rad/s. Neglect thyristor and diode volt-drops, and the armature resistance.

Indicate how a change in load torque will change the motor voltage waveform and speed, the firing delay angle remaining fixed.

34. By reference to a thyristor-controlled d.c. motor, explain how feedback loops can be added so as to limit the load current to a safe level and to maintain the speed constant during torque changes. Describe how the feedback loops eliminate the need for any special starting components.

A separately excited d.c. motor is fed from a 100 V d.c. chopper operating at a frequency of 1 kHz. The chopper is automatically turned off when the motor current is 14 A. If the speed is set to 120 rad/s, determine approximately the time taken for the motor to reach its steady-state speed from a stationary position. The motor/load details are: armature inductance 0.2 H, armature constant 0.5 V/rad/s (N m/A), inertia 0.04 kg m². The load torque is constant at 2.5 N m. Neglect all chopper and motor losses.

Determine the motor current waveform at 60 rad/s and at the steady running position.

35. The cage induction motor can be speed-controlled by:

(*a*) voltage level,
(*b*) inverter via a d.c. link,
(*c*) cycloconverter.

Assuming the supply to the above is the a.c. public supply system, describe and compare the three methods, detailing advantages, disadvantages, and range of control.

36. Describe the systems of drive control in

(*a*) a thyristor-controlled d.c. motor,
(*b*) a d.c. link inverter-fed cage induction motor.

Critically compare the above two forms of drive, both technically and economically.

37. Discuss the principal methods by which power semiconductor devices may be used to control and vary the speed of both cage and slip-ring induction motors.

38. Describe the harmonic pollution which power electronic equipment can impose on the a.c. public supply system. Discuss the effects of harmonics in the system and the possible means of reducing these harmonics.

ANSWERS

3. 67 A, 115 A, 61 W, 69.5 W.

4. Type B.

6. (1) 222 V/section, 28.3 A, 628 V, (2) 222 V, 28.3 A, 314 V, (3) 171 V/ph, 23.1 A, 419 V, (4) 148 V (line), 23.1 A, 209 V, (5) 171 V/ph, 11.5 A, 484 V.

7. (1) 226.8 V/section, 28.3 A, 641 V, (2) 227.6 V, 28.3 A, 322 V, (3) 174.6 V/ph, 23.1 A, 428 V, (4) 151.7 V/ph, 23.1 A, 215 V, (5) 174.6 V/ph, 11.5 A, 494 V.

9. (i) 35.9 V/ph, (ii) 28.9 A, 88 V, (iii) 3.53 A, (iv) 3.59 kVA, (v) 2.54 kVA, (vi) 0.053 Ω referred to secondary.

10. 105 V, 126.8 V.

11. See Fig. 7-10d, Example 7-7, Fig. 3-22, and Example 3-13.

14. Solution dependent on circuit chosen.

15. $i = 24.91 \sin(2\pi 50t + 0.498) + 24.56 \, e^{-250t}$ A, and $i = 21.57 \sin(2\pi 50t + 0.672) + 23.84 \, e^{-250t}$ A, respectively for the two waveform periods. 300 Hz: 11.3 V (r.m.s.), 0.3 A (r.m.s.). 600 Hz: 2.7 V (r.m.s.), 0.035 A (r.m.s.).

16. Several solutions are possible.

17. See Fig. 7-11, see Example 7-9 with $M = 24$ and $N = 8$.

18. (i) See Fig. 7-11, (ii) See Fig. 6-28; 0.893; 50 Hz, 16.37 A; 250 Hz, 2.61 A.

20. 43 μF, 34 μH.

21, 22, 24, 25. Several solutions are possible.

26. $i = 60 - 88.5 e^{-250t}$ A, and $i = 43.3 e^{-250t}$ A. 50 Hz: 257 V, 28.4 A, 4030 W. 150 Hz: 53 V, 2.3 A, 26 W. 250 Hz: 0 V, 0 A, 0 W. 350 Hz: 22.7 V, 0.4 A, 1 W.

28. 255.7 V, 362 V, 20 A.

33. $i = 18 \sin 2\pi 50t - 2100t$ A, to $t = 5$ ms; then $i = 7.5 - 2100t$ A, to $t = 3.57$ ms.

34. 1.093 s.

Advance, firing angle. The angle at which a thyristor starts conduction in advance of, and relative to, the instant when the thyristor forward voltage falls to zero in a converter. Used in relation to the inverting mode of operation.

Ambient. The temperature of the general mass of air or cooling medium into which heat is finally transferred from a hot body.

Anti-parallel. The parallel connection of two unidirectional devices or converters, so making the overall characteristic bidirectional.

Asymmetrical. Not uniform in characteristic or construction.

Auxiliary thyristor. A second or further thyristor, probably linked to the commutation capacitor, used to initiate and execute the turn-off of a main-load current-carrying thyristor.

Avalanche. A chain reaction which occurs when minority carriers are accelerated by a high electric field, so liberating further carriers leading to a sharp increase in reverse current and breakdown.

Avalanche diode. May temporarily be exposed to voltages of the order of the breakdown voltage.

Back e.m.f. An expression commonly used for the voltage internally generated in a machine as a consequence of its rotation.

Back-to-back. See Anti-parallel.

Base drive. The source of the base current to fully turn-on (saturate) a transistor.

Bidirectional. Capable of current conduction in both directions, for example, a triac.

Braking. Reversing the power flow in a motor, so reversing the shaft torque and causing it to generate.

Breakdown voltage. The reverse voltage level at which the reverse current increases rapidly in a diode.

Breakover. The change from the non-conducting state to the conducting state in a forward-biased thyristor.

Bridge. Full-wave converter.

Brushless d.c. motor. A term often used for an inverter-fed synchronous motor.

Burst firing. See Integral cycle.

By-pass diode. See Commutating diode.

Chopping. A technique of rapidly switching on and off a source of voltage.

Clamping voltage. Introduction of a voltage reference level to a pulsed or transient waveform to limit the peak value.

Closed-loop. A system in which information regarding the state of the controlled quantity is fed back to the controlling element.

Commutating diode. A diode placed across a d.c. load to permit transfer of load current away from the source, and so allow the thyristors in the d.c. source to turn off. In addition to permitting commutation of the source thyristors, the diode will prevent reversal of the load voltage.

Commutating reactance. The value of the reactance delaying the commutation of current from one device to another in a converter.

Commutation angle. See Overlap angle.

Complementary thyristor. The other series-connected thyristor in an arm of a bridge circuit.

Conducting angle. The angle over which a device conducts.

Constant-current inverter. An inverter fed from a d.c. source with a large series inductor, so that over each inverter cycle the source current remains almost constant.

Constant-voltage inverter. An inverter fed from a d.c. source with a large parallel capacitor, so that over each inverter cycle the source voltage remains almost constant.

Converter. A circuit which converts a.c. to d.c., or vice versa.

Critical damping. Damping which gives most rapid transient response without overshoot or oscillation.

Crowbar. A device or action that places a high overload on the supply, so actuating protective devices. An expression that arises from placing a crowbar across power lines to short-circuit them.

Cycle syncopation. See Integral cycle.

Cycloconverter. Frequency converter which has no intermediate d.c. state.

Cycloinverter. A cycloconverter which is capable of converting power to a higher frequency.

Damping. Extraction of energy from an oscillating system.

Darlington. A power transistor incorporating a driver transistor on the same chip, or two separate transistors mounted in a single housing, giving high current amplification.

D.C. link. Intermediate d.c. stage between two systems of differing frequency.

Delay, firing angle. The retarded angle at which a thyristor starts conduction relative to that instant when the thyristor forward voltage becomes positive, that is, the angle by which thyristor conduction is delayed to if it were a diode.

di/dt. Rate of rise of current. A device has a critical di/dt rating which, if exceeded, can cause overheating during turn-on.

Displacement angle. The angle by which a symmetrical current is delayed relative to the voltage in an a.c. line.

Displacement current. Displacement of electrons in a dielectric, which sets up a magnetic field as if a current were flowing.

Displacement factor. The cosine of the displacement angle. With a sinusoidal voltage and harmonics present in the current, it is the cosine of the delay angle of the fundamental component.

Distortion factor. Ratio of fundamental r.m.s. current to the total r.m.s. current.

Doping. The addition of impurities to a semiconductor to achieve a desired characteristic such as to produce *N*- or *P*-type materials.

Double-way. See Full-wave.

dv/dt. Rate of rise of voltage. A thyristor has a rated forward limit which, if exceeded, may breakover the junction without the injection of gate current.

Electron. The elementary negative particle.

End-stop pulse. Delivered automatically by the firing circuit at an angle of firing advance which is just sufficient to ensure commutation.

Error signal. Difference between the controlled (output) quantity and desired (input) quantity in a closed-loop system.

Extinction angle. See Recovery angle.

Extrinsic semiconductor. A doped material of high conductivity.

Feedback. Information on the state of the controlled quantity conveyed back to the controlling element.

Feedback diode. In an inverter to permit reverse power flow.

Filter. Attenuates currents of selected frequencies.

Firing. Injection of gate current to turn on a forward-biased thyristor or triac.

Firing angle. See Delay, and Advance, firing angle.

Flywheel diode. See Commutating diode.

Forward volt-drop. Value at rated current.

Four-quadrant. Drive which can motor or generate in either direction of rotation.

Fourier. Analysis of a cyclic wave into a harmonic series.

Freewheeling diode. See Commutating diode.

Full-wave. A converter in which the a.c. input current is bidirectional.

Fully-controlled. Converter in which power flow can be in either direction, that is, a bidirectional converter.

GTO. Gate turn-off thyristor.

Half-controlled. Unidirectional converter, reverse power flow not possible.

Half-wave. A converter in which the a.c. input current (transformer secondary) is unidirectional.

Harmonic. A component at an exact multiple of the fundamental.

Heatsink. A mass of metal that is added to a device for the purpose of absorbing and dissipating heat.

Holding current. The value below which thyristor current will cease when the reduction rate is slow.

Hole. A mobile vacancy in the electronic valence structure of a semiconductor. A hole exists when an atom has less than its normal number of electrons. A hole is equivalent to a positive charge.

Infinite load-inductance. A load in which the cyclic variation in current is negligibly small.

Inrush current. The initial heavy current that flows when, for example, a transformer, motor, or incandescent lamp is switched on.

$i^2 t$. $\int i^2 t \, dt$, a measure of maximum permissible surge energy a device can safety pass, or at which a fuse will clear.

Integral cycle. A.C. load control in which regulation is by alternating a continuous whole number of half cycles on and off.

Interphase transformer. Centre-tapped reactor placed between parallel groups of a converter to permit independent commutation of the devices in each group.

Intrinsic semiconductor. A chemically pure material of poor conductivity.

Inverse-parallel. See Anti-parallel.

Inversion. Reverse power flow in a fully-controlled converter. Occurs when the firing delay angle exceeds $90°$.

Inverter. A circuit which converts d.c. power to a.c. power by sequentially switching devices within the circuit.

Jitter. Small rapid variations in a waveform due to fluctuations in firing angle, supply voltage, or other causes.

Latching current. The minimum forward current to maintain conduction in a thyristor after the removal of gate current.

Leakage current. Undesirable flow of current through a device. That component of alternating current that passes through a rectifier without being rectified.

Losses. Unwanted power dissipation.

Majority carriers. The type of carriers that constitute more than half the total number of carriers in a semiconductor device. Electrons in N-type, holes in P-type.

Maximum reverse current. Rated value of the small current that flows in a reverse-biased diode or thyristor.

Mean current rating. Device forward-current rating based on the mean value of a half sinewave of (say) 40 to 400 Hz, which the device can carry without overheating.

Minority carriers. The type of carriers that constitute less than half the total number of carriers in a semiconductor device. Electrons in P-type, holes in N-type.

N-type. An extrinsic semiconductor in which the conduction electron density exceeds the hole density.

Notched. Pulses of equal length.

Open-loop. Control system in which no information regarding the controlled quantity is known to the controlling element.

Overdamped. Damping greater than that required for critical damping.

Overlap angle. Commutation period in converters during which time two or more devices are conducting simultaneously.

P-type. An extrinsic semiconductor in which the conduction hole density exceeds the electron density.

Parallel-capacitor. Connection of a reverse-charged capacitor in parallel with a thyristor to facilitate turn-off.

Peak inverse voltage. See Peak reverse voltage.

Peak repetitive forward blocking voltage (abbr. P.F.V.). The maximum instantaneous cyclic voltage which a forward-biased thyristor can withstand.

(*Continues*)

Normally a thyristor is chosen to have a value twice that of the peak forward voltage experienced by the thyristor in the working circuit.

Peak repetitive reverse blocking voltage. (abbr. P.R.V.). See Peak repetitive forward blocking voltage, but applies to a reverse-biased diode or thyristor.

Peak transient voltage. Non-repetitive, occasional peak instantaneous voltage the device can withstand without breakdown.

Phase angle control. A.C. load control in which regulation is by delaying conduction in each half cycle.

Point of common coupling. For a consumer, it is the point in the public supply network to which other consumers' loads are connected.

Polarized. Having non-symmetrical characteristic.

Power factor. The ratio of mean power to the product of the r.m.s. values of voltage and current.

Power transistor. A junction transistor designed to handle high currents and power. Used chiefly in low (audio)-frequency and switching applications.

Pulse-number. Repetition cyclic rate in the output voltage waveform of a converter as a multiple of the supply frequency.

Pulse-width modulation. A square wave sinusoidally modulated to give varying on and off periods within the cycle, so as to eliminate the low-order harmonics.

Q-factor. A figure of merit for an energy-storing device. Equal to the reactance divided by the resistance. Shows the rate of decay of stored energy.

Quasi-square wave. Similar to a square wave, but with two equal zero periods in each cycle.

Radio-frequency interference. Undesired electromagnetic radiation of energy in the communication-frequency bands.

Ramp. Quantity rising linearly with time.

Recovery angle. Angle after turn-off in which a thyristor experiences a reverse voltage in a converter.

Rectifier. A circuit which converts a.c. to d.c.

Regulation. Control or adjustment. Characteristic of regulated load.

Restriking voltage. The voltage that seeks to re-establish conduction.

Reverse conducting thyristor. An integrated device incorporating a thyristor and an anti-parallel diode on the same silicon wafer.

Reverse recovery charge. The charge carrier quantity stored in the junction, giving rise to a reverse current before the junction recovers its blocking state, when current reduction is fast at turn-off.

Ringing. An oscillatory transient.

Ripple factor. Ratio of r.m.s. value of the alternating component to the mean value of a direct voltage waveform.

R.M.S. current rating. Maximum permissible forward r.m.s. current.

Safe operating area. The area within the collector current versus collector-emitter voltage characteristic in which the power transistor can be operated out of saturation during switching.

Saturable reactor. An iron-cored reactor in which the flux reaches saturation below the rated current level.

Saturation voltage. Collector-emitter (or base-emitter) voltage when transistor is saturated at defined levels of collector and base currents.

Silicon controlled rectifier (symbol SCR). An expression used for the thyristor before the name *thyristor* was adopted.

Single-way. See Half-wave.

Smoothing. Attenuation of the ripple in a direct-voltage waveform.

SMPS. Switched mode power supply.

Snubber. RC series network placed in parallel with a device to protect against overvoltage transients.

Soft-start. The gradual application of power when switching in a load.

Sub-harmonic. A harmonic at a fractional ratio of the exciting frequency. Occurs in the current associated with integral cycle control.

Subsynchronous. Below synchronous speed.

Supersynchronous. Above synchronous speed.

Suppressor, surge. A device that responds to the rate of change of current or voltage to prevent a rise above a predetermined level.

Surge current rating. The peak value of a single half sinewave current of specified length (say 10 ms) which a device can tolerate with the junction temperature at rated value at the start of the surge.

Sustaining voltage. Maximum rated collector-emitter voltage with the base open-circuited.

Switching loss. The energy dissipated within a device during switching from off- to on-state or vice versa.

Synchronous. In step with the exciting frequency.

Synthesis. Putting together component parts to form the whole.

Temperature, maximum junction. Maximum operating junction temperature.

Thermal resistance. Ratio of temperature difference to the heat power flow between two interfaces.

Triggering. (See also Firing). To initiate a sudden change.

Turn-on time. The time from the moment turn-on is initiated to when the device is freely conducting.

Turn-off time. The time between the reduction of current to zero and the reapplication of forward voltage. A thyristor or triac has a minimum rated value.

Uncontrolled. Rectifier with diodes only.

Undamped. No damping.

Underdamped. Damping less than that required for critical damping. Transient response is oscillatory.

Unidirectional. Flow in one direction only.

Zener diode. A special type of diode in which the reverse avalanche breakdown voltage is almost constant, irrespective of current. Used for regulation and voltage limiting. May be considered an avalanche diode.

REFERENCES

The following books will provide further reading to supplement that given in the text.

1. Blicher, A., *Thyristor Physics,* Springer-Verlag, New York, 1976.
2. Ghandhi, S. K., *Semiconductor Power Devices,* Wiley, New York, 1977.
3. Rissik, H., *Mercury-Arc Current Converters,* Pitman, London, 1941.
4. Pelly, B. R., *Thyristor Phase-Controlled Converters and Cycloconverters,* Wiley, New York 1971.
5. McMurray, W., *The Theory and Design of Cycloconverters,* M.I.T. Press, 1972.
6. Bedford, B. D. and R. G. Hoft, *Principles of Inverter Circuits,* Wiley, New York, 1964.
7. Stevenson, W. D., *Elements of Power System Analysis,* McGraw-Hill, New York, 1975.
8. Shepherd, W. and P. Zand, *Energy Flow and Power Factor in Nonsinusoidal Circuits,* Cambridge University Press, 1979.
9. Morton, A. H., *Advanced Electrical Engineering,* Pitman, London, 1966, p. 221.
10. Fitzgerald, A. E., *Electrical Machinery,* McGraw-Hill, New York, 1971.
11. Emanuel, P. and E. Leff, *Introduction to Feedback Control Systems,* McGraw-Hill, New York, 1979.
12. *Silicon Controlled Rectifier Manual,* General Electric Company, New York.
13. Mullard Technical Handbook, *Semiconductor Devices,* Mullard Ltd., London.

BIBLIOGRAPHY

DEVICES

Cornick, J. A. F. and M. J. Ramsbottom, 'Behaviour of thyristors when turned on by gate current', *Proc. IEE,* **123**, 12, 1365–1367, 1976.

Morris, R., 'Power transistors as viable alternatives to thyristors', *IEETE Electrotechnology,* 12–14, Jan. 1979.

Beatty, B. A., S. Krishna and M. S. Adler, 'Second breakdown in power transistors due to avalanche injection', *IEEE Trans.,* IECI–24, 4, 306–312, 1977.

Yair, A. and M. Steinkoler, 'Improved pulse delay circuit for phase-controlled rectifiers and a.c. voltage controller', *IEEE Trans,* IECI–24, 2, 200–202, 1977.

Murugesan, S. and C. Kameswara Rao, 'Simple adaptive analog and digital trigger circuit for thyristors working under wide range of supply frequencies', *IEEE Trans.,* IECI–24, 1, 46–49, 1977.

Ilango, B., et al., 'Firing circuit for three-phase thyristor-bridge rectifier', *IEEE Trans.,* IECI–25, 1, 45–49, 1978.

Olivier, G., V. R. Stefanovic and M. A. Jamil, 'Digitally controlled thyristor current source', *IEEE Trans.,* IECI–26, 3, 185–191, 1979.

Newell, W. E., 'Dissipation in solid-state devices', *IEEE Trans.,* IA–12, 4, 386–396, 1976.

Newell, W. E., 'Transient thermal analysis of solid-state power devices', *IEEE Trans.,* IA–12, 4, 405–420, 1976.

Tani, T., et al., 'Measuring system for dynamic characteristics of semiconductor switching elements and switching loss of thyristors', *IEEE Trans.,* IA–11, 6, 720–727, 1975.

Cornick, J. A. F. and M. J. Ramsbottom, 'Instantaneous temperature rise in thyristors under invertor and chopper operating conditions', *Proc. IEE,* **119**, 8, 1141–1148, 1972.

McLaughlin, M. H. and E. E. Vonzastrow, 'Power semiconductor equipment cooling methods and application criteria', *IEEE Trans.*, IA–11, 5, 546–555, 1975.

Van der Broeck, H. W., J. D. van Wyk and J. J. Schoeman, 'On the steady-state and dynamic characteristics of bipolar transistor power switches in low-loss technology', *Proc. IEE*, **132B**, 5, 251–259, 1985.

Hashimoto, O. *et al.*, 'Turn-on and turn-off characteristics of a 4.5 kV 3000 A gate turn-off thyristor', *IEEE Trans.*, IA–22, 3, 478–482, 1986.

Ho, E. Y. Y. and P. C. Sen, 'Effect of gate-drive circuits on GTO thyristor characteristic', *IEEE Trans*, IE–33, 3, 325–331, 1986.

Pong, M. H. and R. D. Jackson, 'Computer-aided design of power electronic circuits', *Proc. IEE*, **132B**, 6, 301–306, 1985.

CONVERTER CIRCUITS

Wells, R., 'Interphase transformers in power rectifier circuits', *Electrical Review, London*, **200**, 7, 28–30, 1977.

Yair, A., W. Alpert and J. Ben Uri, 'Bridge rectifiers with double and multiple supply', *Proc. IEE*, **116**, 5, 811–821, 1969.

Freris, L. L., 'Multigroup converters with series a.c. connection', *Proc. IEE*, **118**, 9, 1971.

Farrer, W. and D. F. Andrew, 'Fully controlled regenerative bridges with half-controlled characteristics', *Proc. IEE*, **125**, 2, 109–112, 1978.

Wells, R., 'Effect of transformer reactance on rectifier output characteristics', *Electrical Review, London*, 285–287, 1975.

Jones, V. H. and W. J. Bonwick, 'Three-phase bridge rectifiers with complex source impedance', *Proc. IEE*, **122**, 6, 630–636, 1975.

Stefanovic, V. R., 'Power factor improvement with a modified phase-controlled converter', *IEEE Trans.*, IA–15, 2, 193–201, 1979.

Kataoka, T., K. Mizumachi and S. Miyairi, 'A pulsewidth controlled a.c. to d.c. converter to improve power factor of a.c. line current', *IEEE Trans.*, IA–15, 6, 670–675, 1979.

Shepherd, W. and P. Zakikhani, 'Power-factor compensation of thyristor-controlled single-phase load', *Proc. IEE*, **120**, 2, 245–246, 1973.

Mirbod, A. and A. El-Amawy, 'A general-purpose microprocessor-based control circuit for a three-phase controlled rectifier bridge', *IEEE Trans.*, IE–33, 3, 310–317, 1986.

COMMUTATION

Davis, R. M. and J. R. Melling, 'Quantitative comparison of commutation circuits for bridge inverters', *Proc. IEE*, **124**, 3, 237–246, 1977.

McMurray, W. 'Thyristor commutation in d.c. choppers, a comparative study', *IEEE Trans.*, IA–4, 6, 547–558, 1978.

Robinson, W. M., 'Standards for commutating capacitors', *IEEE Trans.*, IA–12, 1, 17–27, 1976.

CYCLOCONVERTERS

Bland, R. J., 'Factors affecting the operation of a phase-controlled cycloconverter', *Proc. IEE*, **114**, 12, 1908–1916, 1967.

Hirane, Y. and W. Shepherd, 'Theoretical assessment of a variable-frequency envelope cycloconverter', *IEEE Trans.*, IECI–25, 3, 238–246, 1978.

Slonim, M. A. and P. P. Biringer, 'Steady-state processes in cycloconverters; Part I, High frequency; Part II, Low frequency', *IEEE Trans.*, IECI–28, 2, 126–136, 1981.

INVERTERS

Epstein, E., A. Yair and A. Alexandrovitz, 'Analysis of a reactive current source used to improve current drawn by static inverters', *IEEE Trans.*, IECI–26, 3, 172–177, 1979.

Bansal, S. C. and U. M. Rao, 'Evaluation of p.w.m. inverter schemes', *Proc. IEE*, **125**, 4, 328–334, 1978.

Pollack, J. J., 'Advanced pulsewidth modulated inverter techniques', *IEEE Trans.*, IA–8, 2, 145–154, 1972.

Daniels, A. R. and D. T. Slattery, 'New power convertor technique employing power transistors', *Proc. IEE*, **125**, 2, 146–150, 1978.

Daniels, A. R. and D. T. Slattery, 'Application of power transistors to polyphase regenerative power convertors', *Proc. IEE*, **125**, 7, 643–647, 1978.

Calkin, E. T. and B. H. Hamilton, 'A conceptually new approach for regulated d.c. to d.c. converters employing transistor switches and pulsewidth control', *IEEE Trans.*, IA–12, 4, 369–377, 1976.

Evans, P. D. and R. J. Hill-Collingham, 'Some aspects of power transistor inverter design', *IEE Elec. Power App.*, **2**, 3, 73–80, 1979.

Wilson, J. W. A., 'The forced-commutated inverter as a regenerative rectifier', *IEEE Trans.*, IA–14, 4, 335–340, 1978.

Yoshida, Y., K. Mohri and K. Yoshino, 'PWM inverter using high-gain pulse-triggered power transistors and a new PWM control method', *IEEE Trans.*, IE–33, 2, 132–137, 1986.

Bowes, S.R. and T. Davies, 'Microprocessor-based development system for PWM variable-speed drives', *Proc. IEE*, **132B**, 1, 18–45, 1985.

Bowes, S.R. and A. Midoun, 'New PWM switching strategy for microprocessor controlled inverter drives', *Proc. IEE*, **133B**, 4, 237–254, 1986.

NON-MOTOR APPLICATIONS

Khalifa, M., et al., 'Solid-state a.c. circuit breaker', *Proc. IEE*, **126**, 1, 75–76, 1979.

Okeke, S. N., 'Application of thyristor inverters in induction heating and melting', *IEE Electronics & Power*, 217–221, Mar. 1978.

Roda, M. R. and G. N. Revankar, 'Voltage-fed discontinuous current mode high-frequency inverter for induction heating', *IEEE Trans.*, IECI–25, 3, 226–232, 1978.

Revankar, G. N. and D. S. Trasi, 'Symmetrically pulse width modulated a.c. chopper', *IEEE Trans.*, IECI–24, 1, 39–45, 1977.

Chauprade, R., 'Inverters for uninterruptible power supplies', *IEEE Trans.*, IA–13, 4, 281–297, 1977.

Thomas, B. G., 'High-power, high-current, diode rectifiers', *IEE Electronics & Power*, 307–311, Apr. 1977.

Wells, R., 'Current balance in rectifier circuits', *Electrical Review, London*, 569–571, 1975.

Zakarevicius, R. A., 'Calculation of the switching-on overvoltages in a series-connected thyristor string', *IEEE Trans.*, IA–13, 5, 407–417, 1977.

Corbyn, D. B., 'D.C. power control for aluminium and electrolytic loads', *Proc. IEE*, **115**, 11, 1693–1704, 1968.

Bayliss, C. R., 'Modern techniques in electrolytic refining of copper', *IEE Electronics & Power*, 773–776, Dec. 1976.

Moore, A. H., 'Parallel operation of electrochemical rectifiers', *IEEE Trans.*, IA–15, 6, 656–663, 1979.

Gardner, G. E. and D. Fairmaner, 'Alternative convertor for h.v.d.c. transmission', *Proc. IEE*, **115**, 9, 1289–1296, 1968.

'Skagerrak h.v.d.c. link provides benefits to Norway and Denmark', *Electrical Review, London*, **201**, 16, 23–25, 1977.

'Testing GEC thyristor valves for the 2000 MW h.v.d.c. cross-channel link', *Electrical Review, London*, **203**, 7, 35–37, 1978.

Zhao, K. B., P. C. Sen and G. Premchandran, 'A thyristor inverter for medium-frequency induction heating', *IEEE Trans.*, IE–31, 1, 34–36, 1984.

Cuk, S. and R. D. Middlebrook, 'Advances in switched-mode power conversion. Parts I and II', *IEEE Trans.,* IE–30, 1, 10–29, 1983.

Griffith, D. C. and B. P. Wallace, 'Development trends in medium and large UPSs., *IEE Electronics & Power,* 455–458, June 1986.

HARMONICS

Corbyn, D. B., 'This business of harmonics', *IEE Electronics & Power,* 219–223, June 1972.

'Limits for harmonics in the United Kingdom electricity supply system', Engineering recommendation G5/3, *The Electricity Council, London,* 1976.

Dobinson, L. G., 'Closer accord on harmonics', *IEE Electronics & Power,* 567–572, May 1975.

Yacamini, R. and J. C. de Oliveira, 'Harmonics produced by direct current in convertor transformers', *Proc. IEE,* **125,** 9, 873–878, 1978.

Krishnamurthy, K. A., G. K. Dubey and G. N. Revankar, 'Convertor control with selective reduction of line harmonics', *Proc. IEE,* **125,** 2, 141–145, 1978.

Steeper, D. E. and R. P. Stratford, 'Reactive compensation and harmonic suppression for industrial power system using thyristor converters', *IEEE Trans.,* IA–12, 3, 232–254, 1976.

Bogle, A. G., 'Rectifier circuit performance: some new approximate formulas', *Proc. IEE,* **124,** 12, 1127–1134, 1977.

Konttinen, K., M. Ampuja and S. Purves, 'Ironing out distribution disturbances with harmonic filters', *Electrical Review, London,* **202,** 2, 28–30, 1978.

Chowdhuri, P. and D. F. Williamson, 'Electrical interference from thyristor-controlled d.c. propulsion system of a transit car', *IEEE Trans.,* IA–13, 6, 539–550, 1977.

Hall, J. K. and D. S. Palmer, 'Electrical noise generated by thyristor control', *Proc. IEE,* **123,** 8, 781–786, 1976.

Uceda, J., F. Aldana and P. Martinez, 'Active filters for static power converters', *Proc. IEE,* **130B,** 5, 347–354, 1983.

Schroeder, E. R. and R. V. DeVore, 'Radio interference on DC lines from HVDC converter stations', *Proc. IEE,* **129C,** 5, 221–227, 1982.

Yacamini, R. and J. C. de Oliveira, 'Comprehensive calculation of convertor harmonics with system impedances and control representation', *Proc. IEE,* **133B,** 2, 95–102, 1986.

D.C. MACHINES

Smith, G. A., 'Thyristors in the control of d.c. machines', *IEETE Journal, London,* 28–38, May 1968.

Mehta, P. and K. Mukhopadhyay, 'Improvement in d.c. motor performance by asymmetrical triggering', *IEEE Trans.,* IA–11, 2, 172–181, 1975.

Krishnan, T. and B. Ramaswami, 'Speed control of d.c. motor using thyristor dual converter', *IEEE Trans.,* IECI–23, 4, 391–399, 1976.

Sen, P. C. and M. L. MacDonald, 'Thyristorized d.c. drives with regenerative braking and speed reversal', *IEEE Trans.,* IECI–25, 4, 347–354, 1978.

Bates, J. J. and J. Stanway, 'Development of a 300 kW 3000 rev/min d.c. machine using thyristor-assisted commutation', *Proc. IEE,* **123,** 1, 76–80, 1976.

Bennell, F. T., 'Rectifiers for railway-traction substations', *IEE Elec. Power App.,* **2,** 1, 22–26, 1979.

Bezold, K-H., J. Forster and H. Zander, 'Thyristor converters for traction d.c. motor drives', *IEEE Trans.,* IA–9, 5, 612–617, 1973.

Farrer, W., 'D.C.-to-d.c. thyristor chopper for traction application', *Proc. IEE,* **123,** 3, 239–244, 1976.

Bailey, R. B., D. F. Williamson and T. D. Stitt, 'A modern chopper propulsion system for rapid transit application with high regeneration capability.' *IEEE Trans.,* IA–14, 6, 573–580, 1978.

Whiting, J. M. W., 'A regenerative braking system for d.c. railway traction', *IEE Electronics & Power,* 710–714, Oct. 1979.

Bose, B. K. and R. L. Steigerwald, 'A d.c. motor control system for electric vehicle drive', *IEEE Trans.,* IA–4, 6, 565– 572, 1978.

Barak, M., 'Fuel economy through battery power', *Electrical Review, London,* 535–539, 1975.

Chalmers, B. J., K. Pacey and J. P. Gibson, 'Brushless d.c. traction drive', *Proc. IEE,* **122,** 7, 733–738, 1975.

Bowler, P., 'Power transistors in variable speed drives', *IEE Electronics & Power,* 730–736, Oct. 1978.

Gokhale, K. P. and G. N. Revankar, 'Microprocessor-controlled separately excited DC-motor drive system', *Proc. IEE,* **129B,** 6, 344–352, 1982.

Hill, R. J. and P. Cork, 'Chopper control of DC disc-armature motor using power MOSFETs'. *Proc. IEE,* **132B,** 2, 93–100, 1985.

Inaba, H. *et al.,* 'A new speed control system for DC motors using GTO converter and its application to elevators', *IEEE Trans.,* IA–21, 2, 391–397, 1985.

A.C. MACHINES

Rahman, B. E. and W. Shepherd, 'Thyristor and diode controlled variable voltage drives for 3-phase induction motors', *Proc. IEE,* **124,** 9, 784–790, 1977.

Bradley, D. A., et al., 'Adjustable-frequency invertors and their application to variable-speed drives', *Proc. IEE,* **111,** 11, 1833–1846, 1964.

Phillips, K. P., 'Current-source converter for a.c. motor drives', *IEEE Trans,* IA–8, 6, 679–683, 1972.

Revankar, G. N. and A. Bashir, 'Effect of circuit and induction motor parameters on current source inverter operation', *IEEE Trans.,* IECI–24, 1, 126–132, 1977.

Nabae, A., 'Pulse-amplitude modulated current-source inverters for a.c. drives', *IEEE Trans.,* IA–15, 4, 404–411, 1979.

Cornell, P. and T. A. Lipo, 'Modelling and design of controlled current induction motor drive systems', *IEEE Trans.,* IA–13, 4, 321–330, 1977.

Bose, B. K. and T. A. Lipo, 'Control and simulation of a current-field linear induction machine', *IEEE Trans.,* IA–15, 6, 591–600, 1979.

Shimer, D. W. and L. J. Jacovides, 'An improved triggering method for a high-power cyclo-converter-induction motor drive', *IEEE Trans.,* IA–15, 5, 472–481, 1979.

Holmes, P. G. and R. G. Stephens, 'Wide speed-range operation of a chopper-controlled induction motor using asymmetrical chopping', *IEE Elec. Power App.,* **1,** 4, 121–126, 1978.

Wani, N. S. and M. Ramamoorty, 'Chopper controlled slipring induction motor', *IEEE Trans.,* IECI–24, 2, 153–161, 1977.

Shepherd, W. and A. Q. Khalil, 'Capacitive compensation of thyristor-controlled slip-energy-recovery system', *Proc. IEE,* **177,** 5, 948–956, 1970.

Smith, G. A., 'Static Scherbius system of induction-motor speed control', *Proc. IEE,* **124,** 6, 557–560, 1977.

Bonwick, W. J., 'Voltage waveform distortion in synchronous generators with rectifier loading', *IEE Proc.,* **127B,** 1, 13–19, 1980.

Williamson, A. C., N. A. H. Issa and A. R. A. M. Makky, 'Variable-speed inverter-fed synchronous motor employing natural commutation', *Proc. IEE,* **125,** 2, 113–120, 1978.

Issa, N. A. H. and A. C. Williamson, 'Control of a naturally commutated inverter-fed variable-speed synchronous motor., *IEE Elec. Power App.* **2,** 6, 199–204, 1979.

Allen, G. F. H., 'Brushless excitation systems for synchronous machines', *IEE Electronics & Power,* 866–869, Sep. 1975.

Spencer, P. T., 'Microprocessors applied to variable-speed-drive systems', *IEE Electronics & Power,* 140–143, Feb. 1983.

Williamson, A. C. and K. M. S. Al-Khalidi, 'Starting of convertor-fed synchronous machine drives', *Proc. IEE,* **132B,** 4, 209–214, 1985.

Lawrenson, P. J. *et al.,* 'Variable-speed switched reluctance motors', *Proc. IEE,* **127B,** 4, 253–265, 1980.

Davis, R. M., W. F. Ray and R. J. Blake, 'Inverter drive for switched reluctance motor: circuits and component ratings', *Proc. IEE,* **128B,** 2, 126–136, 1981.

Peak, S. C. and J. L. Oldenkamp, 'A study of losses in a transistorized inverter-induction motor drives system', *IEEE Trans.,* IA–21, 1, 248–258, 1985.

Park, M. H. and S. K. Sul, 'Microprocessor-based optimal-efficiency drive of an induction motor', *IEEE Trans.,* IE–31, 1, 69–73, 1984.

PROTECTION

Milward, V. E. and P. G. Rushall, 'Design and protection of a.c./d.c. thyristor convertors', *IEE Electronics & Power,* 728–732, June 1975.

Reimers, E., 'A low-cost thyristor fuse', *IEEE Trans.,* IA–12, 2, 172–179, 1976.

De Bruyne, P. and P. Wetzel, 'Improved overvoltage protection in power electronics using active protection devices', *IEE Elec. Power App.,* **2,** 1, 29–36, 1979.

Finney, D. 'Modular motor control design simplifies maintenance', *Electrical Review, London,* **203,** 3, 26–29, 1978.

Howe, A. F. and P. G. Newbery, 'Semiconductor fuses and their applications', *Proc. IEE,* **127B,** 3, 155–168, 1980.

Rajashekara, K. S., J. Vithyathil and V. Rajagopalan, 'Protection and switching-aid networks for transistor bridge inverters', *IEEE Trans.,* IE–33, 2, 185–192, 1986.

Conference Records

A valuable source of reference is to be found in the following records of conferences which are held annually, biennially or less frequently.

'Power Electronics – Power Semiconductors and their Applications', Conference Records, *IEE, London.*

'Electrical Variable-Speed Drives', Conference Records, *IEE, London.*

'Power Electronic Specialists' Conference Records, *IEEE, U.S.A.*

'International Semiconductor Power Converter Conference Records', *IEEE/IAS, U.S.A.*

'Industry Application Society Conference Records, *IEEE, U.S.A.*

INDEX

Acceleration, 303, 343
A.C. machines, 330
Ambient, 21
Angle:
  firing advance, 101
  firing delay, 40, 54
  overlap, 93
Anti-parallel converters, 304–306
Applications:
  electrochemical, 239–242
  heating, 221–230
  machine, 313, 370
  non-motor, 220
Arc voltage, 384
Asymmetrical thyristor, 22, 23, 197
Asymmetrical voltage, 270
Auxiliary thyristor, 126, 143
Avalanche diode, 389

Back e.m.f., 293
Barrier potential, 1, 2
Base current, 7, 8
Bibliography, 404
Bidirectional, 398
Bi-phase half-wave, 42–44, 74
Blocking, 4
Braking, 295, 338
  d.c. injection, 338, 339
  d.c. motor, 295, 299, 358
  dynamic, 295, 299, 311
  induction motor, 338, 348, 353
  plugging, 295
  regenerative, 295, 299, 318, 348, 358
  synchronous machine, 334
Breakover voltage, 3
Bridge, 37
Brushless d.c. motor, 364
Brushless machines, 362–364
  generator, 363
  motor, 363–364
Burst firing, 222
By-pass diode, 38

Cage induction motor, 334–340
  (*see also* Induction motor)
Chopper:
  d.c., 126, 147–152
  motor control, 308–312, 326–328
  two-phase, 310
Chopping, 244, 358
Clamping voltage, 387

Closed-loop control, 301, 316, 347, 359
  (*see also* Feedback)
Collector, 6–8
Collector-emitter voltage, 6–8
Commutation, 92, 125
  (*see also* Turn-off)
Commutation angle (*see* Overlap)
Commutation failure, 123
Commutating diode, 38
Commutating reactance, 95
Complex wave, 259
  (*see also* Harmonics)
  power flow, 261, 286
  r.m.s. value, 261
Complementary impulse commutation, 138–145
Complementary thyristor, 183
Conducting angle, 399
Constant-current inverter, 195–197, 215–218, 352–356
Constant-voltage inverter, 343–352
Contactor, 220, 221
  mechanical, 220, 384
  semiconductor, 220, 221, 246
Converters, 92
  (*see also* Rectifiers)
  choice, 68
  equivalent circuit, 104, 107
  fully-controlled, 37
  half-controlled, 37, 38
  inverting mode, 92, 99–102
  uncontrolled, 37, 38
Cooling, 19–22, 236
Coupled pulse, 135–138, 157–159
Critically damped, 147
Crowbar protection, 357, 386
Current feedback, 302
Current limiting, 302–304, 327
Current transformer:
  a.c., 302
  d.c., 240
Cycle syncopation, 222
Cycloconverter, 164–180, 200–203
  blocked group, 167–174
  circulating current, 174–177, 203
  control, 177, 178
  envelope, 178–180, 203
  firing angles, 177, 178
  motor control, 356, 361
  negative group, 164, 165
  positive group, 164, 165

410

rating, 199
  voltage, 171, 199, 201
Cycloinverter, 230

Darlington, 10
D.C. chopper (see Chopper)
D.C. circuit breaking, 383
D.C. current transformer, 240
D.C. line commutation, 125
D.C. link, 344
D.C. machines, 293
  construction, 294
  equations, 293–295
D.C. motor:
  armature voltage, 296–298
  braking, 295, 299
  harmonics, 297, 298, 322
  regeneration, 295, 318, 358
  speed adjustment, 294, 295, 358
  speed-torque, 316, 317, 319–323
  starting, 297, 304, 327
  waveforms, 296–298, 315–325
Diac, 223
di/dt, 17, 18, 387
Diesel-electric locomotives, 308
Diode, 1, 2
Displacement angle, 99, 331
Displacement current, 18
Displacement factor, 99
Distortion factor, 99, 116
Doping, 1
Double-star, 55–59, 81, 82
Double-way, 37
Doubly-fed motor, 361
Drives:
  comparisons, 313, 370
  considerations, 313, 370
  constant power, 305, 313, 370
  efficiency, 371
dv/dt, 18, 386

Electrified railways, 306–308
Electrochemical applications, 239–242
Electrochemical forming (machine), 241, 242
Electrolysis, 239
Electroplating, 239
Emitter, 7
End-stop pulse, 102
Environmental considerations, 370
Error signal, 301
Excitation, 333, 363
Exciter, 362, 363, 366
Exponential formulae, 145, 146
Extinction angle, 101

Feedback, 240, 301–305, 347, 348
  control, 301–305, 316, 347, 354, 359
  error signal, 301
  speed, 301–305, 316, 360

Feedback diodes, 182
Field effect transistor, 25
Field weakening, 295
Filters, 266–270, 287, 288
  a.c. line, 269
  high-pass, 269
  inverter output, 268, 269, 291, 292
  low-pass, 268, 291
  resonant-arm, 269, 292
  tuned, 269
Firing advance angle, 101
Firing circuits:
  control features, 16, 17
  master-slave, 239
  requirements, 13, 306
  starting pulses, 62, 223
  typical, 14–16
Firing delay angle, 40, 54
Flicker, lamp, 266
Fluctuations, voltage, 266
Flywheel diode, 38
Fork-lift truck, 312
Forward voltage, 4
Forward volt-drop, 19
Four-quadrant drives, 313, 353, 358
Fourier analysis, 260, 270–282
Fourier series, 260, 270, 273
Freewheeling diode, 38
Frequency conversion, 164
Full-wave, 37
Fully-controlled, 37
Fuse, 383–385

Gate:
  current, 4, 13
  characteristic, 10–12
  load line, 13, 14
  protection, 390
  pulse, 13, 14
  rating, 11
Gate turn-off thyristor, 5, 22 25, 197, 391
Gear changing, 350
Glossary of terms, 398
Graphical analysis, 260, 261, 275

Half-controlled, 37, 38
Half-wave, 37
Harmonics, 259
  a.c. motors, 346, 350, 374, 375
  a.c. supply, 263–266, 274–285
  analysis, 259–261, 270–282
  d.c. motors, 297, 298, 322
  distortion factor, 261
  graphical analysis, 260, 261, 275
  high-order, 189, 282
  inverters, 189, 289–292
  line traps, 269
  load voltage, 262, 272, 276–278
  low-order, 189, 272

Harmonics (*continued*)
  power, 261, 286
  resonance, 264
  sub-harmonic, 266, 350
  torque, 297, 343, 346, 375
  triplen, 265
Heat sink, 21
Heat transfer, 19
Heating, 221, 229, 247, 253
High-voltage bridge, 90
Holding current, 3, 4
Holes, 1
H.V.D.C. transmission, 242, 243, 258

Illumination, 223
Impulse commutation, 138–145, 160–163
Incandescent lamps, 223
Induction heaters, 229, 230, 253–255
Induction motors:
  braking, 338, 348, 353
  cage, 334–340
  characteristic, 336–341, 372–374
  construction, 334
  cycloconverter control, 356
  efficiency, 371
  equations, 335–341
  equivalent circuit, 336, 339, 340, 372
  harmonics, 346, 350, 374, 375
  inverter drives, 343–356
  linear, 308
  regeneration, 338, 348, 353
  slip-ring, 341, 342, 359–362, 376, 377
  soft-starting, 343
  speed control, 337–342
  speed-torque, 337–341, 372–374
  starting, 343, 348
  voltage/frequency, 336, 374
  voltage regulation, 342
  waveforms, 344, 351, 354, 374–376
Infinite load inductance, 263
Inhibition, 301
Inrush current, 400
Instability, 346, 350, 354
i^2t, 18, 383
Integral cycle control, 222, 280, 281
Interconnected-star, 52
Interphase transformer, 56–59, 82
Inverse-parallel, 5, 221, 249, 250
Inversion, 99–102, 119–124
Inverters:
  cage motor, 343–356
  centre-tapped, 180–183, 204–210
  constant-current, 195, 196, 215–217, 352–356
  constant-voltage, 343–352
  control, 185, 346
  d.c. link, 344
  drives, 343–356
  filtering, 268, 269, 291, 292

firings instants, 186, 213
harmonics, 187, 289–292
losses, 189, 346, 351, 357
notched, 186, 212, 346
overloads, 234
phase-shifted, 182, 186, 191, 346
pulse-width modulated, 186, 193, 194, 212, 346, 350, 351, 356
quasi-square wave, 185, 190, 214, 346
reverse power flow, 199, 218
single-phase bridge, 183–189, 211
standby, 233, 234
three-phase bridge, 189–194, 213, 214
transistorized, 197–199, 357
turn-off, 138–145
Inverting mode, 92
Isolating switch, 220

Jitter, 262, 285
Junction breakdown, 18
Junction, *P-N*, 1, 2
Junction temperature, 18, 20

Kramer system, 359

Latching current, 2, 3, 22
Leakage current, 2
Leakage reactance, 92
Lighting, 223
  emergency, 234
Linear induction motor, 308
Load angle, 332, 333
Load current:
  continuous, 44
  discontinuous, 44
  level, 50
Load inductance, 40
Load voltage waveform, 97
Locomotive, 306–308, 328
Losses:
  commutation, 189
  conduction, 19, 30
  device, 19, 24, 26, 27, 89
  inverter, 189, 346, 351, 357
  motor, 371
  switching, 19, 29, 31–34
  transistor, 8–10, 26, 27

McMurray-Bedford, 138
Majority carriers, 1
Master-slave firing, 239
Mean current, 18, 297
Mean voltage, 37, 260
Microprocessor, 352, 371
Minority carriers, 1, 2
MOSFET, 25–27, 198, 358, 391
Multipliers, voltage, 232, 233, 256

N-type, 1
Negative sequence, 265

Noise, 197, 352
Notched inverter, 186, 212, 346

Oil cooling, 236
Open-loop control, 301, 349
Optical link, 238
Optimum flux, 336
Oscillation formulae, 146
Overdamped response, 147
Overhauling load, 348
Overlap, 95–98, 108–110
Overloads, 20, 302

P-pulse waveforms, 83, 105, 106, 262
P-type, 1
Parallel-capacitance turn-off, 125–130,
        147–152, 180–182
Parallel diodes, 234–236, 257
Parallel thyristors, 236
Peak forward voltage, 18
Peak inverse voltage, 401
Peak reverse voltage, 18
Peak transient voltage, 402
Periodic current reversal, 241
Phase angle control, 221, 247, 248, 279, 280
Plugging, 295
Point of common coupling, 266
Polarized, 389
Potential barrier, 1, 2
Power, 99, 261
Power factor, 99, 116–119
Power MOSFET, 25–27, 198, 358, 391
Power transistor, 6–10
Protection, 382
    back-up, 384
    crowbar, 357, 386
    d.c., 383
    gate, 390
    overcurrent, 382–386
    overvoltage, 386–389
    peak voltage, 386
    power transistor, 386
    R-C snubber, 387, 391
    secondary breakdown, 386
Pulse-number, 38
Pulse-width modulation, 186, 193, 194, 212,
        346, 350, 351, 356
    firing instants, 186, 213
    harmonics, 189, 289–291

Q-factor, 402
Quasi-square wave, 185, 190, 214, 346

Radio-frequency interference, 266, 270, 282
Railways, electrification, 306–308
Ramp, 304, 349
Rating:
    device, 17–19
    motor, 298, 317, 371

    transformer, 66
RC and RL response formulae, 145, 146
RCL response formulae, 146, 147
Reactor:
    (see also Interphase transformer)
    cycloconverter, 174, 203
Recovery angle, 101
Rectifiers, 37
    bi-phase-phase half-wave, 42–44, 74
    bridge, 37
    choice, 68
    double-star, 55–59, 81, 82
    double-way, 37
    equivalent circuit, 104, 107
    fork, 55, 56
    fully-controlled, 37
    half-controlled, 37, 38
    half-wave, 37
    high-voltage bridge, 90
    losses, 89, 114
    p-pulse equations, 83, 105, 106
    pulse-number, 38
    single-phase bridge, 44–49
        fully-controlled, 46, 47, 75
        half-controlled, 48, 49, 75, 77
        uncontrolled, 44–46, 75
    single-phase half-wave, 38–42
    single-way, 37
    six-phase half-wave, 54–59, 80
    three-phase bridge, 60–65
        fully-controlled, 62, 63, 84, 85
        half-controlled, 64, 65, 86, 87
        uncontrolled, 60–62, 84
    three-phase half-wave, 49–54, 79
    twelve-pulse circuits, 65–67, 89, 90
    uncontrolled, 37, 38
Rectifying mode, 92
Refining, 239
Regenerative braking, 295, 299, 318, 348, 358
Regulation, 102–105, 107, 230
Reluctance motor, 331, 367
Resonant turn-off, 130–135, 152–156
Restriking voltage, 220, 383
Reverse conducting thyristor, 5
Reverse recovery, 17, 33, 34
Reversing drives, 299–301, 348
R.M.S. current, 18, 84
Ripple factor, 268, 287
Ripple voltage, 88
Road vehicles, 308
Rotating field, 330, 331

Safe operating area, 9, 198
Saturable reactor, 133, 154
Saturation, transistor, 8
Saturation voltage, 8, 386
Secondary breakdown, 386
Selenium metal rectifier, 387–389
Self commutation, 364

Series connection, 236–239, 257
Series motor, 295
Short-circuit, 382
Silicon, 1
Silicon controlled rectifier, 403
Single-way, 37
Single-phase bridge:
  rectifier, 44–49
  inverter, 183–189, 211
Six-phase half-wave, 54–59, 80
Slip, 335
Slip compensation, 348
Slip-energy recovery, 359, 376, 377
Slip-ring induction motor, 341, 342, 359–362,
  376, 377
Small d.c. motor, 297, 313–315
Smoothing, 267, 287, 288
Snubber circuit, 387, 391
Soft-start, 304
Speed feedback, 301
Standby inverters, 233, 234
Starting firing pulses, 62, 223
Stepper motor, 367–370
Storage charge, 4, 17, 387
Stray inductance, 391
Sub-harmonic, 165, 266
Subsynchronous speed, 342, 359–362
Supersynchronous speed, 355, 360
Supply impedance, 264, 282
Supply reactance, 92, 95
Suppression, 220, 389
Suppressor, 389
Surge current, 403
Sustaining voltage, 403
Switched mode power supplies, 244–246
Switched-reluctance motor, 367
Switching losses, 19, 29, 31–34
Switching time, 10, 198
Symmetrical voltage, 270
Synchronous machine, 330–334
  braking, 334
  brushless, 362–364
  construction, 330–332
  drives, 349, 364–367
  equations, 333
  generator, 363
  inverter-fed, 364–367
  load angle, 332, 333
  motor, 332
  open-loop control, 349
  reactance, 333
  rotor angle feedback, 365
  speed, 330, 331
  synchronization, 364
Synchronous speed, 330
Syncopation, cycle, 222
Synthesis, 164

Tacho-generator, 301

Temperature:
  junction, 20
  overload, 20–22
  rise, 19–22
Thermal:
  agitation, 2
  resistance, 19, 34, 35
  storage, 20, 21
  time constant, 21, 222
  transient time constant, 22, 35
Three-phase bridge:
  rectifier, 60–65
  inverter, 187–194, 213, 214
Three-phase half-wave, 49
Thyristor, 3–5
  gating, 10
  gate turn-off, 5, 22–25, 197
  reverse conducting, 5
Torque:
  harmonic, 297, 343, 346, 375
  control, 305, 347
Traction drives, 306–312
Transformers:
  ampere-turn balance, 182
  interconnected-star, 52
  interphase, 56–59, 82
  leakage reactance, 92
  pulse, 46
  rating, 66
  tapped, 231, 252, 253
Transistor, 6–10
  comparison to thyristor, 10
  power, 6–10
  unijunction, 15
Transient:
  asymmetry, 383
  ratings, 19, 20
  voltage, 19, 386
Transport, 306, 312
Triac, 5, 6, 221
Trickle charge, 233
Turn-off, 4, 17, 25, 125
  a.c. line, 38
  coupled-pulse, 135–138, 157–159
  d.c. line, 125
  gate, 24
  resonance, 130–135, 152–156
Turn-on, 18, 23
Twelve-pulse circuits, 65–67, 89, 90

Uncontrolled, 37, 38
Undamped response, 146
Unidirectional, 38
Unijunction transistor, 15
Uninterruptible supply, 233

Virtual junction, 20
Voltage multiples, 232, 233, 256
Voltage regulation, 230–232, 252, 342

fully-controlled, 223–227, 248–251
  half-controlled, 223, 227, 228, 248
Voltage regulator, 230–232, 248, 296
Volt-drop, 68, 104, 116
Volt-seconds, 95

Ward-Leonard drive, 313
Welding, spot, 220, 223

Zener diode, 14, 364
Zig-zag, 52